IRREVERSIBLE THERMODYNAMICS

Y. L. Yao

Formerly at
Canada Centre for Mineral and Energy Technology,
Department of Energy, Mines and Resources,
Ottawa, Canada

SCIENCE PRESS

Beijing, China, 1981

Distributed by

VAN NOSTRAND REINHOLD COMPANY

New York Cincinnati Toronto London Melbourne

Published by Science Press, Beijing
Distributed by Van Nostrand Reinhold Company
New York, Cincinnati, Toronto, London, Melbourne
Printed by C & C Joint Printing Co., (H.K.) Ltd.

First published 1981
ISBN 0-442-20074-9
Science Press Book No. 1968-8

PREFACE

There are some ten books published bearing the title of irreversible thermodynamics or its equivalent; however, I feel that there is room for another book on the subject which can be used as a quick reference work for researchers to look for any of the usual formulae encountered in irreversible thermodynamics.

The first part of the book discusses the general principles of irreversible thermodynamics, and the second part is directed to specific applications. The selection of applications emphasizes those for which treatment by linear thermodynamics is indispensable or is superior to the old method, often found to be incorrect. To limit the size of the work, classical and statistical thermodynamics are not treated; there are already many good books on classical thermodynamics and it is felt that there is a wide gap between statistical and irreversible thermodynamics. On the other hand, nonlinear thermodynamics is still in the process of growth and is briefly treated by incorporating bond graphs in the theoretical part of the book.

The reader will find the results given in this book not much different from other books of the same nature. The difference lies in the method of presentation, such as the inner product for rate of entropy-production, the stochastic matrix for detailed balancing, the differentiated orthogonal matrix for the modification of the Onsager law, the general proof of the Curie law and the restraints on the isotropic tensor.

I am greatly indebted to Canada Centre for Mineral and Energy Technology, Department of Energy, Mines and Resources,

Ottawa, Canada, as this book was written while I was employed there and was one of their research projects, MRP/PMRL–77–2(J). I want to express my thanks to Dr. W. N. Roberts, who spent considerable time in editing the manuscript. My special thanks are due to Mrs. Marilyn Fraser and Mrs. Elaine Atkinson for the skillful typing.

General References

[1] I. Prigogine: Étude Thermodynamique des Phénomenes Irréversibles, Desoer, Paris (1947).

[2] I. Prigogine: Introduction to Thermodynamics of Irreversible Processses, Interscience, New York (1961).
The more recent and shorter book gives highlights of his classic research and touches the frontier of nonlinear thermodynamics. Because of brevity, the reader has to consult the earlier book and other references for details.

[3] K. G. Denbigh: The Thermodynamics of the Steady State, Methuen, London (1951).
If one wishes to know what irreversible thermodynamics is about, this pocket-book is highly recommended. Mathematics is kept at a minimum and the presentation is lucid.

[4] S. R. de Groot: Thermodynamics of Irreversible Processes, North-Holland, Amsterdam (1952).
This book was written when irreversible thermodynamics was still in its formative stages. In the last chapter the author attacks the irreversible thermodynamics of virtual entropy transfer, which was pioneered by W. Thomson (Lord Kelvin) and followed by H. von Helmholtz, E. D. Easter, G. Wagner and others.

[5] S. R. de Groot and P. Mazur: Non-Equilibrium Thermodynamics, Interscience, New York (1952).
This book is the most comprehensive treatise on linear thermodynamics available. Principles and practice are separated. In the theoretical part thermodynamics is so intermingled with probability theory, kinetic theory and stochastic processes that reading is not easy. In applications, the presentation is more sophisticated than that in the earlier book written by S. R. de Groot. The elegant formalism in diffusion is especially recommended.

[6] R. O. Davis: The Macroscopic Theory of Irreversibility, *Reports on Progress in Physics*, **19**(1956).
This review paper shows that little headway is made in relaxation theory if the hypothesis of linearity and the Onsager law are abandoned.

[7] J. Meixner and H. G. Reik: Thermodynamik der irreversible Prozesse, Handbuch der Physik, Band **III/2**, Edited by S. Fluegg, Springer, Berlin (1957).
In this review the author foresees the connection between thermodynamics and electrical network theory, and there is a short discussion on statistical and relativistic thermodynamics.

[8] D. G. Miller: Thermodynamics of Irreversible Processes, *Chem. Revs.*, **60**, 15 (1960).

[9] D. G. Miller: The Onsager Relations-Experimental Evidence, Symposium

on the Foundations of Continuum Thermodynamics (1973).
These two review papers summarize the experimental support of the Onsager law, whatever may be the misgivings of its proof microscopically.

[10] D. D. Fitts: Nonequilibrium Thermodynamics, McGraw-Hill, New York, (1962).
This book represents the thoughts of the American school led by J. G. Kirkwood and leans toward applications in chemistry. Phenomena occuring exclusively in the solid state are not treated. The topics are nicely arranged.

[11] R. Hasse: Thermodynamik der irreversiblen Prozesse, D. Steinkopff, Darmstadt (1963).

[12] R. Hasse: Thermodynamics of Irreversible Processes, Addison-Wesley, Reading, Mass. (1969).
The twin books, the latter being the translation of the former with more references added, devote about one fifth of their presentation to classic thermodynamics. There is more emphasis on physical chemistry than that in the other books.

[13] A. Katchalsky and P. F. Curran: Nonequilibrium Thermodynamics in Biophysics, Harvard Univ. Press, Cambridge, Mass. (1965).
After a quick survey of classical and irreversible thermodynamics, the authors turn their attention to some biological topics.

[14] L. Tiza: Generalized Thermodynamics, M. I. T. Press, Cambridge, Mass. (1966).
Part B of this book is devoted to irreversible thermodynamics treated statistically.

[15] R. J. Tykodi: Thermodynamics of Steady States, MacMillian, New York (1967).
Without mentioning bond graphs, the author uses block diagrams to solve thermodynamical problems.

[16] C. A. Truesdell: Rational Thermodynamics, A Course of Lectures on Selected Topics, McGraw-Hill, New York (1969).
In four lectures, the author ridicules the significance of the Curie law, questions the inadequacy of the choice of fluxes and forces in linear thermodynamics and deplores the lack of applications of linear thermodynamics in engineering. His unfounded opinion on the first two topics should not deter readers from his sound presentation of the third. He demonstrates that the time is ripe for the application of nonlinear thermodynamics to practical problems.

[17] S. Wiśniewski, B. Staniszewski and R. Szymanik: Thermodynamics of Nonequilibrium Processes, D. Reidel, Boston (1976).
The authors present irreversible thermodynamics from the point of view of an engineer.

LIST OF SYMBOLS

Symbols pertaining to one section and not causing confusion with those in the other sections, are not listed. The bracket after the explanation has two numbers, the first number referring to the section and the second to the equation introduced.

α	Coefficient of thermal expansion	(3, 6)
γ	Activity coefficient	(38, 54)
	Adiabatic exponent	(30, 24)
$\gamma_\pm$	Mean activity coefficient	(41, 39)
ε	Degree of advancement	(25, 3)
	Permittivity	(11, 13)
	+1 for the force being the even function of velocity and −1 for the odd function	(17, 16)
ζ	Scalar potential	(7, 41)
	Attenuation	(30, 5)
η	Shear viscosity coefficient	(5, 6)
η_1	First viscosity coefficient	(5, 13)
η_2	Second viscosity coefficient	(5, 13)
η_r	Rotational viscosity coefficient	(5, 7)
η_V	Volume or bulk viscosity coefficient	(5, 5)
$\boldsymbol{\theta}$	Angle of rotation	(7, 22)
θ	Thermal conductivity	(46, 1)
θ'	Thermal conductivity when concentration gradient is not set up	(46, 5)
θ''	Thermal conductivity when diffusion ceases	(46, 10)
$\dot{\boldsymbol{\theta}} \equiv \boldsymbol{\omega}$	Angular velocity	(7, 23)
$\ddot{\theta}$	Angular acceleration	(7, 39)
κ	Electrical susceptibility	(11, 15)
	Compressibility coefficient	(3, 5)

Λ	Reduced elastic coefficient tensor	(34, 18)
	Mass stoichiometric coefficient	(25, 7)
$\bar{\Lambda}$	Atomic stoichiometric coefficient	(25, 2)
$\bar{\Lambda}_{\pm}$	Mean atomic stoichiometric coefficient	(41, 38)
λ	Lamé's constant	(5, 12)
	Characteristic roots of a matrix	(20, 4)
μ	Specific chemical potential	(2, 1)
	Permeability	(11, 14)
$\bar{\mu}$	Atomic chemical potential	(38, 30)
Π	Pressure tensor	(5, 3)
	Peltier heat	(49, 29)
π	Probability	(16, 25)
	Osmotic pressure	(23, 57)
	Number of apparently independent components for a permutation sequence in a partition of an isotropic tensor	(13, 1)
	Thermoelectric power	(49, 27)
ρ	Total density	(38, 13)
	Stored energy	(21, 5)
ρ_i	Partial density of the component i	(38, 13)
Σ	Thomson heat	(49, 31)
σ	Rate of entropy-production per unit volume per unit time	(1, 6)
Γ	Relaxation time	(29, 9)
$\boldsymbol{\Phi}$	Magnetic flux linkage	(7, 11)
Φ	Valency	(39, 2)
χ	Magnetic susceptibility	(11, 16)
	Differential coefficient of pressure to volume	(31, 9)
ψ	Potential energy per unit mass	(6, 6)
	Planck potential	(51, 12)
Ω	The ohm	(22, Table 1)
$\boldsymbol{\Omega}$	Directed area	(4, 4)
ω	Angular velocity	(30, 4)
	Permeability	(23, 57)
$\boldsymbol{A}$	Vector potential	(11, 5)

A	Matrix to reduce the dimensions from n to $(n-1)$	(38, 11)
	Affinity	(25, 10)
	The ampere	(22, Table 1)
$\bar{A}$	Atomic affinity	(25, 4)
$\boldsymbol{a}$	Acceleration	(7, 38)
a	Activity	(41, 35)
	Absorption per unit wave length	(30, 7)
a_0	Dimensionless absorption	(30, 8)
a'	Dimensionless absorption	(30, 9)
$\bar{a}$	Absorption	(30, 10)
$a_\pm$	Mean activity	(41, 36)
$\boldsymbol{B}$	Magnetic induction	(11, 3)
B	$\equiv L/\rho T$	(29, 8)
	Matrix to convert the mass flux from one reference velocity to another	(38, 7)
C	Capacitance	(22, 18)
	Speed of sound	(30, 16)
C	The coulomb	(22, Table 1)
C_p	Specific heat at constant pressure	(3, 3)
C_V	Specific heat at constant volume	(3, 4)
c	Speed of light	(11, 10)
	Mass fraction	(4, 29)
$\mathscr{C}$	Elastic coefficients	(34, 3)
D	$\equiv D^{0\rho}$ Standard diffusion coefficient	(38, 51)
	$\equiv d/dt'$ Differential operator with respect to world time t'	(12, 50)
$\boldsymbol{D}$	Electrical displacement	(11, 13)
	Propagation velocity of the front of a shock wave	(37, 15)
d	Dispersion	(30, 6)
$\mathscr{D}$	Modified standard diffusion coefficient	(38, 65)
E	Total energy per unit mass	(6, 8)
	Young's modulus	(34, 9)
E'	Total energy per unit mass in electrical systems	(11, 44)
$\boldsymbol{E}$	Electrical field	(7, 41)
e	Strain	(34, 1)

F	Helmholtz free energy per unit mass	(3, 18)
F	The farad	(22, Table 1)
$\boldsymbol{F}$	Force per unit mass	(5, 1)
f	Frequency	(30, 3)
$\mathscr{F}$	Rayleigh's dissipation function	(14, 2)
	Electrical charge carried by one gram equivalent	(39, 2)
G	Gibbs free energy per unit mass	(3, 19)
	Lamé's constant or modulus of rigidity	(5, 12)
	$\equiv A\mu$	(38, 35)
$\bar{G}$	Gibbs free energy per mole	(25, 4)
g	Gibbs free energy for an arbitrary quantity of mass	(4, 2)
$\boldsymbol{g}$	Group velocity	(36, 19)
H	Enthalpy per unit mass	(3, 17)
H	The henry	(22, Table 1)
H'	Enthalpy per unit mass in electrical systems	(11, 47)
H^*	Enthalpy of transfer	(22, 21)
$\boldsymbol{H}$	Magnetic field	(11, 2)
$\mathscr{H}$	Hamiltonian function	(7, 2)
$\boldsymbol{I}$	Current density (charge flux)	(11, 2)
	World mass flux vector	(12, 27)
$\boldsymbol{I}^q$	World heat flux vector	(12, 45)
i	Current	(22, 14)
$\boldsymbol{\mathscr{I}}$	Magnetic polarization	(11, 14)
J	The joule	(11, Table 1)
$\boldsymbol{J}$	Flux	(1, 6)
$\boldsymbol{J}^{abs}$	Flux in the absolute sense	(47, 12)
$\boldsymbol{J}_{el}$	Charge flux	(11, 23)
$\boldsymbol{J}_k \equiv \boldsymbol{J}_k^c$	Flux of the component k based on the barycentric velocity	(4, 37)
$\boldsymbol{J}_k^{\bar{c}}$	$\boldsymbol{J}_k^c$ with the partial atomic density	(38, 25)
$\boldsymbol{J}_k^n$	Flux of the component k based on the molar velocity	(4, 38)
$\boldsymbol{J}_k^{\bar{n}}$	$\boldsymbol{J}_k^n$ with the partial atomic density	(38, 26)
$\boldsymbol{J}_k^r$	Flux of the component k based on the velocity	

	of the nth component	(4, 40)
$\boldsymbol{J}_k^{\bar{r}}$	$\boldsymbol{J}_k^r$ with the partial atomic density	(38, 28)
$\boldsymbol{J}_k^0$	Flux of the component k based on the volume velocity	(4, 39)
$\boldsymbol{J}_k^{\bar{0}}$	$\boldsymbol{J}_k^0$ with the partial atomic density	(38, 27)
$\boldsymbol{J}_q$	Heat flux	(6, 10)
$\boldsymbol{J}_q^{abs}$	Heat flux in the absolute sense	(47, 13)
$\boldsymbol{J}_q^r$	Reduced heat flux	(6, 10)
$\boldsymbol{J}_S$	Entropy flux	(8, 7)
$\boldsymbol{J}_S^{abs}$	Entropy flux in the absolute sense	(47, 14)
$\boldsymbol{J}_{S,\,tot}$	Total entropy flux	(10, 11)
$\boldsymbol{j}$	Absolute mass flux per unit volume per unit time	(53, 2)
$\boldsymbol{j}_q$	Absolute heat flux per unit volume per unit time	(53, 2)
j	Rate of chemical reaction per unit volume per unit time	(4, 13)
K	Equilibrium constant	(26, 10)
$\boldsymbol{K}$	Complex propagation vector	(30, 1)
	World force vector	(12, 7)
K_V	Bulk modulus	(34, 8)
k	Boltzmann's constant	(16, 6)
	Electrical conductivity	(39, 10)
	Phase constant	(30, 3)
	Reaction constant	(26, 1)
	Spring constant	(7, 1)
$\boldsymbol{k}$	Real propagation vector	(30, 2)
L	Inductance	(22, 19)
	Phenomenological coefficients	(14, 1)
l	Wave length	(30, 3)
$\mathscr{L}$	Lagrangian function	(22, 2)
M	Molecular weight	(38, 14)
$\boldsymbol{M}$	Angular momentum	(7, 28)
m	Mass	(4, 24)
	Number of entries in the partition of an isotro-	

	pic tensor	(13, 2)
	The meter	(22, Table 1)
$\mathscr{M}$	Molality	(41, 37)
$\mathscr{M}_{\pm}$	Mean molality	(41, 38)
N	Number of moles per unit volume	(38, 13)
	The newton	(22, Table 1)
n	Molar fraction	(4, 30)
	Number of components in a system	(4, 14)
	Number of dimensions of a tensor	(13, 2)
	Number of turns in a coil	(7, 11)
$\mathbf{n}$	Wave normal	(34, 17)
P	Total entropy produced per unit time	(1, 12)
	Viscous pressure tensor	(5, 3)
p	Pressure	(5, 3)
	Poisson's ratio	(34, 10)
	Rank of a tensor	(13, 1)
$\mathbf{p}$	Generalized momentum	(7, 2)
$\mathscr{P}$	Electrical polarization	(11, 13)
Q	An orthogonal matrix	(1, 11)
	Dissipation coefficient	(36, 23)
	Proper quotient	(26, 6)
	Heat	(1, 2)
$\mathbf{Q}$	Volume flow	(22, Table 1)
Q^*	Heat of transfer	(22, 20)
q	Electrical charge	(7, 41)
	Radiation coefficient	(32, 8)
$\mathbf{q}$	Generalized coordinate	(7, 2)
R	Gas constant	(41, 40)
	Resistance	(22, 17)
	Transition probability	(16, 26)
$\mathbf{R}$	Position vector in polar coordinates	(21, 7)
r	Resistivity	(35, 9)
$\mathbf{r}$	Position vector	(7, 21)
$\mathscr{R}$	Routhian function	(22, 6)
	Rankine-Hugoniot's function	(37, 18)

S	Entropy per unit mass	(2, 1)
	Number of apparently independent components in an isotropic tensor	(13, 2)
	Stress	(34, 3)
S^*	Entropy of transfer	(22, 22)
ΔS_{sys}	Change of entropy of the system	(1, 1)
ΔS_{sur}	Change of entropy of the surroundings	(1, 3)
ΔS_e	Change of external entropy	(1, 4)
ΔS_i	Change of internal entropy	(1, 4)
$\mathscr{S}$	Stokes' number	(32, 13)
	Soret coefficient	(48, 10)
T	Kinetic energy	(12, 9)
	Modified Maxwell's stree tensor	(11, 32)
	Temperature	(2, 1)
t	Time	(4, 1)
	Number of independent components in an isotropic tensor	(13, 13)
	Transference number	(39, 9)
t'	World time	(12, 2)
$\mathscr{T}$	Torque	(7, 29)
U	Internal energy per unit mass	(2, 1)
	Electrical mobility	(39, 15)
$\boldsymbol{u}$	Velocity behind the front of a shock wave	(37, 15)
	Displacement vector	(5, 12)
V	The volt	(22, Table 1)
	Volume	(4, 21)
$\overline{V}$	Atomic volume	(38, 30)
V_k	Partial volume of the component k	(4, 22)
$\boldsymbol{V}$	Velocity vector in polar coordinates	(21, 7)
$\boldsymbol{v}$	Velocity	(7, 36)
W	Mobility	(39, 12)
	World momentum-energy tensor	(12, 31)
X	$\equiv \omega(\eta_V + 4/3\eta)/\rho C_0^2$	(32)
$\boldsymbol{X}$	Thermodynamic force	(1, 6)
Y	$\equiv \theta/(\eta_V + 4/3\eta)C_p$	(32)

Z	Electrical charge per unit mass	(39, 1)
$\bar{z}$	Electric charge per unit mole	(39, 1)

CONTENTS

Part I Principles

Part II Applications

Chapter VII Scalar Processes

Chapter VIII Vectorial Processes

Chapter IX Tensorial Processes

Chapter X Heterogeneous Processes

PART I

PRINCIPLES

CHAPTER I
INTRODUCTION

§1 *TRANSITION FROM CLASSICAL TO IRREVERSIBLE THERMODYNAMICS*

Some authors claim that thermodynamics has no field of study of its own. This statement is neither entirely accurate nor entirely false. The three laws of thermodynamics (including the zeroth law) have been used implicitly in all other disciplines of science, yet only in thermodynamics has the meaning of entropy been made precise. Moreover, it is thermodynamics, having little field of its own, which could correlate all branches of science. The concept of entropy, properly strengthened, provides a macroscopic framework of describing the rate and the driving force of all natural processes. Incidentally there are no other branches except in hydrodynamics and electrodynamics, having such a richness in the number of formulae derived.

Pedagogically, classical and irreversible thermodynamics (short for thermodynamics of irreversible processes) are treated as separate subjects. This is probably because few textbooks cover both extensively. However, this should not mean that they are not related. On the contrary, irreversible thermodynamics is a generalization of classical thermodynamics. Once the formulation of equations in irreversible thermodynamics is complete, it is easy to revert to reversible processes. The concept of entropy flow vector and inner entropy production was conceived by Bertrand[1] in 1887 before classical thermodynamics was fully developed, yet the birth of irreversible thermodynamics had to wait for the classical work of Onsager[2] in 1931. In entropy flow, we treat entropy as a fluid, just as heat (energy) was once so treated. Energy and entropy are put on the same footing. Either both have reality or both have not. On the other hand, in contrast to energy, entropy can be created but not destroyed. Onsager's contribution is to deduce the rate of entropy-production and then to find the linear relation. Furthermore, Onsager proposed a reciprocal relationship, unlike other kinds, of macroscopic symmetry. This gives new information which cannot be explained in any other way.

While linear thermodynamics is fruitful in correlating many unrelated irreversible processes, it is seldom more than a good approximation. Historically the equations of mechanical actions were developed long before any attention was given to electrical systems. In the early days it was natural to deal with electrical problems in terms of mechanical concepts. The situation changed after vacuum tubes, transistors, tunnel diodes and other electronic controlling devices were invented. That such revolutionary devices had similar circuit representations indicated that the network theory was developed to a higher state than the corresponding theory of mechanical systems. Paynter[3] devised bond graphs for the flow of energy or power in electrical and non-electrical systems. The bond graph is especially adapted for nonlinear constitutive relations and the computing algorithm. The full impact of the bond graph on thermodynamics remains to be seen.

We distinguish among isolated systems, which can exchange neither energy nor matter with the surroundings; closed systems, which can exchange energy only; and open systems, which can exchange both. In classical thermodynamics, the entropy of isolated systems for irreversible processes can only increase. Thus for isolated systems,

$$\Delta S_{sys} \geqq 0, \tag{1}$$

where the equality sign is retained for reversible processes. The change of entropy of the closed systems is always greater than that of the surroundings if the processes are irreversible or entropy can be created. Thus for the closed systems,

$$\Delta S_{sys} \geqq \Delta Q / T_{sur}, \tag{2}$$

where ΔQ is heat absorbed by the system from the surroundings and T_{sur} is the absolute temperature of the surroundings. Equations (1) and (2) can be combined:

$$\Delta S_{sys} + \Delta S_{sur} \geqq 0. \tag{3}$$

In open systems, eq. (2) is not valid because another term is needed to represent entropy associating with matter transferred into or out of the system.

Experimentally an isolated system is unattainable. We know that cosmic rays or high-energy particles constantly penetrate into the earth. Also open systems cannot be ignored. For biological specimens the exchange of matter with the environment is vital for their existence. In irreversible thermodynamics we split ΔS_{sys} into change of external entropy due to flow of matter and energy from or into the system, and change of internal entropy due to production inside the system, or

$$\Delta S_{sys} = \Delta S_e + \Delta S_i . \tag{4}$$

Then formally ΔS_i is no longer tangled with ΔS_{sur} and the inequality is replaced by an equality.

Our definition of irreversible processes is that, while nothing can be said about ΔS_{sys} and ΔS_e in eq. (4), being positive, negative or zero, ΔS_i is positive definite, or

$$\Delta S_i > 0 \text{ in general.} \tag{5}$$

Next we introduce σ, the rate of increase of internal entropy per unit time per unit volume, or entropy-production for short. We also generalize $T\sigma$ as a negative dissipation function (in mechanics the dissipated energy carries a negative sign). de Donder's dissipation of power is the special case for chemical reactions and Rayleigh's dissipation function the special case for viscous fluids.

In searching for a positive definite function to represent σ, we choose an inner product. It is an ordinary product of two scalar quantities, a scalar product of two vectorial quantities, or a double contracted product of two second-rank tensorial quantities. Alternatively, it is a product of two numbers or matrices, or the integration of a product of two functions. In symbols,

$$\sigma = \frac{\partial \Delta S_{\mathrm{i}}}{\partial t} = \frac{1}{T} [\boldsymbol{J} | - \boldsymbol{X}], \tag{6}$$

where $\boldsymbol{J}$ is the flux, i. e., heat, matter and so forth, passing through a surface in the normal direction per unit time per unit area and $\boldsymbol{X}$ is the thermodynamical or driving force for an irreversible process. Throughout this book the word "force" unmodified means thermodynamical force.

The defining equation (6) is useful for several reasons:

(A) Because the inner product is positive or zero, eq. (6) includes eq. (5).

(B) The symmetry of inner product yields

$$[\boldsymbol{J} | \boldsymbol{X}] = [\boldsymbol{X} | \boldsymbol{J}]. \tag{7}$$

Thus $\boldsymbol{X}$ and $\boldsymbol{J}$ are put on the same footing. For some irreversible processes, e. g., flow of viscous fluids, the assignment of $\boldsymbol{X}$ and $\boldsymbol{J}$ is arbitrary.

(C) There are infinite ways of choosing $\boldsymbol{J}$ and $\boldsymbol{X}$. To preserve the inner product, two sets of $\boldsymbol{J}$ and $\boldsymbol{X}$ are transformed contragrediently. Actually this may be considered as another definition of the inner product. In symbols, if

$$\boldsymbol{J}' = \boldsymbol{C}\boldsymbol{J}, \tag{8}$$

$\boldsymbol{C}$ being any non-singular matrix, then

$$X' = (C^{\mathrm{T}})^{-1}X, \tag{9}$$

where $(C^{\mathrm{T}})^{-1}$ is the reciprocal of the transposed matrix C. This is checked by

$$[J'|X'] = [CJ|(C^{\mathrm{T}})^{-1}X] = [J|C^{\mathrm{T}}(C^{\mathrm{T}})^{-1}X] = [J|X], \tag{10}$$

or the inner product is invariant.

If C is an orthogonal matrix Q, then

$$(Q^{\mathrm{T}})^{-1} = Q. \tag{11}$$

The distinction between contra-gredience and co-gredience breaks down.

(D) Any 3×3 nonzero matrix can be split into a symmetric and a skew-symmetric matrix. The skew-symmetric matrix is necessarily without a trace, i. e., it has a zero trace. The symmetric matrix can be further split into a scalar matrix and a symmetric matrix without a trace. For example,

$$\begin{bmatrix} 1 & 2 & 3 \\ 4 & 5 & 6 \\ 7 & 8 & 9 \end{bmatrix} = \begin{bmatrix} 5 & 0 & 0 \\ 0 & 5 & 0 \\ 0 & 0 & 5 \end{bmatrix} + \begin{bmatrix} -4 & 3 & 5 \\ 3 & 0 & 7 \\ 5 & 7 & 4 \end{bmatrix} + \begin{bmatrix} 0 & -1 & -2 \\ 1 & 0 & -1 \\ 2 & 1 & 0 \end{bmatrix}.$$

If the pair of the inner product are matrices, their symmetry must coincide. For example, if we have an inner product

$$[\mathrm{Grad}^{\mathrm{sy},0}\, v|\mathrm{Grad}\, v],$$

it is simplified to

$$[\mathrm{Grad}^{\mathrm{sy},0}\, v|\mathrm{Grad}^{\mathrm{sy},0}\, v],$$

because the inner product of the matrix of symmetric velocity gradient without a trace and the scalar matrix or the skew-symmetric matrix is zero. The extension of σ to P, entropy-production per unit time, is immediate. P is just the integral version of the rate over some region R with volume V,

$$P = \int_{R} \sigma dV. \tag{12}$$

Thus σ is for a particle, but P is for a continuous medium. The task for irreversible thermodynamics is to implement the defining equation (6).

References

[1] J. L. F. Bertrand: Thermodynamique, Gauthier, Paris Chap. 12, (1887).
[2] L. Onsager: *Phys. Rev.*, **31**, 405, (1931); *ibid.*, **31**, 2265 (1931).
[3] H. M. Paynter: Analysis and Design of Engineering Systems, M. I. T. Press, Cambridge, Mass. (1961).

CHAPTER II
PRESUMPTIONS

§2 *POSTULATE OF LOCAL EQUILIBRIUM*

It is assumed that a system undergoing irreversible processes can be divided into subsystems. The subsystems must not be so large that "surface" interactions from subsystem to subsystem cause inhomogeneities. The subsystems must not be so small that they cannot be treated macroscopically. The postulate of local equilibrium stipulates equilibrium prevailing in every subsystem.

The postulate permits us to apply all the results of classical thermodynamics, with suitable modifications, to a given subsystem. In particular, the Gibbs equation

$$dS = \frac{1}{T}\,dU + \frac{1}{T}\,pdV - \frac{1}{T}\sum_k \mu_k dc_k \qquad (1)$$

is of special importance, where S is entropy, U internal energy, all referred to unit mass; μ is chemical potential, p is pressure and c_k is mass concentration of the component k. In most applications, the differentials can also be read as differences. As an example, the best established mechanism for the isothermal oxidation of metals is that of scale thickening controlled by migration due to electrochemical reaction. According to the postulate, local equilibrium prevails at every thin slice of the oxide. Equilibrium is also established at the metal-oxide and oxide-oxygen interfaces. But the whole oxide is not at equilibrium because of nonuniformity of electrochemical potential. By further using the principle of electroneutrality, a parabolic law of oxidation can be established (§42).

It is possible, sometimes by ingenious means[1], to find a reversible path for any irreversible process. Thus we know how much entropy is produced from the initial to the final state. We are able to specify entropy produced at any particular state during the infinite sequence from the initial to the final state, or the entropy-production assigned to a subsystem is meaningful. If the total production of entropy is low, there is not much difference between the local rate of production in a subsystem and the average rate in the whole system. For example, if an ideal gas is expanded isothermally or is forced into a porous plug (Joule-Thomson effect), the instantaneous rate can be taken as the average rate. But if the gas is expanded into vacuum (Gay-Lussac effect), the rate of expansion is so fast initially that it is difficult to specify an instantaneous rate. Besides turbulent flow, the flame front of burning gases and shock-wave phenomena will invalidate the postulate.

Intuitively, the postulate holds if the average rate of entropy-production remains constant, i.e., at steady equilibrium. By steady equilibrium is meant the macroscopic parameters have time-independent values at every subsystem, despite the non-uniformity of the parameters. The steady equilibrium can only be maintained by constraints, i.e., boundary conditions. When the constraints are removed, the steady equilibrium spirals into static equilibrium. For example, a constant flow of heat passes if the temperatures of two boundaries are maintained at two different values. The temperature of the system will eventually approach the time-independent values (steady equilibrium). If the boundary conditions are removed, the temperature gradient is eliminated to produce one nearly uniform temperature (thermo-static equilibrium).

Besides experimental verification, an alternate approach is to calculate entropy by kinetic methods. Prigogine[2] calculated the kinetic entropy during a steady flow of gas and Callen[3] calculated for the steady flow of electrons in metals. Both showed that the local equilibrium gives only a first approximation to the kinetic theory. Because the kinetic theory is also arbitrary, we consider local equilibrium as a postulate.

References

[1] J. D. Frost: Entropy, Philips Technical Library, Eindhoven, Holland (1962).
[2] I. Prigogine: *Physica*, **15**, 272 (1949).
[3] H. B. Callen: Thesis, M. I. T., Cambridge, Mass. (1948).

§3 *HYPOTHESIS OF INCOMPLETE REACTION*

Kinematically one or more parameters, in addition to those necessary for equilibrium, may be required to describe how far the instantaneous state of the whole system deviates from equilibrium. Such a system is called "incomplete". The hypothesis of incomplete reaction is useful for those irreversible processes for which the causes of dissipation cannot be precisely established.

First we review some stability conditions for equilibrium. Gibbs has shown that the second-order virtual changes for the specific Gibbs free energy G or Helmholtz free energy F is negative definite,

$$\delta^2 G \leqq 0, \tag{1}$$

$$\delta^2 F \leqq 0. \tag{2}$$

The symbol δ is used to denote virtual but not actual change. For simplicity we confine ourselves to a one-component system. Let C_p and C_V be the specific heats at constant pressure and volume respectively, α_p be the coefficient of thermal expansion at constant pressure, and κ_S and κ_T be the coefficients of compressibility at constant entropy and temperature respectively. The following inequalities can be established:

$$\frac{C_p}{T} = \left(\frac{\partial S}{\partial T}\right)_p = -\left(\frac{\partial^2 G}{\partial T^2}\right)_p \geqq 0, \tag{3}$$

$$\frac{C_V}{T} = \left(\frac{\partial S}{\partial T}\right)_V = -\left(\frac{\partial^2 F}{\partial T^2}\right)_V \geqq 0, \tag{4}$$

$$\kappa_T = -\frac{1}{V}\left(\frac{\partial V}{\partial p}\right)_T = -\frac{1}{V}\left(\frac{\partial^2 G}{\partial p^2}\right)_T \geqq 0, \tag{5}$$

$$\alpha_p = \frac{1}{V}\left(\frac{\partial V}{\partial T}\right)_p \geqq 0 \text{ or } \leqq 0, \tag{6}$$

$$C_p - C_V = \frac{TV\alpha_p^2}{\kappa_T} \geqq 0. \tag{7}$$

Also

$$\frac{\kappa_S}{\kappa_T} = \frac{\left(\frac{\partial V}{\partial p}\right)_S}{\left(\frac{\partial V}{\partial p}\right)_T} = \frac{\partial(V, S)\partial(p, T)}{\partial(p, S)\partial(V, T)}. \tag{8}$$

In eq. (8) we have written partial differentials as a determinant of the Jacobian, namely,

$$\frac{\partial(V, S)}{\partial(p, S)} = \begin{vmatrix} \left(\frac{\partial V}{\partial p}\right)_S & \left(\frac{\partial V}{\partial S}\right)_p \\ \left(\frac{\partial S}{\partial p}\right)_S & \left(\frac{\partial S}{\partial S}\right)_p \end{vmatrix} = \left(\frac{\partial V}{\partial p}\right)_S. \tag{9}$$

But

$$\frac{C_V}{C_p} = \frac{\left(\frac{\partial S}{\partial T}\right)_V}{\left(\frac{\partial S}{\partial T}\right)_p} = \frac{\partial(S, V)\partial(T, p)}{\partial(T, V)\partial(S, p)}, \tag{10}$$

we have

$$\frac{\kappa_T}{\kappa_S} = \frac{C_p}{C_V} \tag{11}$$

from eqs. (8) and (10). From eqs. (7) and (11),

$$\kappa_T \geqq \kappa_S. \tag{12}$$

Let us choose T and p as independent variables of a one-component system,

$$\begin{bmatrix} -dS \\ dV \end{bmatrix} = \begin{bmatrix} -\left(\frac{\partial S}{\partial T}\right)_p & -\left(\frac{\partial S}{\partial p}\right)_T \\ \left(\frac{\partial V}{\partial T}\right)_p & \left(\frac{\partial V}{\partial p}\right)_T \end{bmatrix} \begin{bmatrix} dT \\ dp \end{bmatrix}$$

$$= \begin{bmatrix} -\frac{C_p}{T} & \alpha_p V \\ \alpha_p V & -\kappa_T V \end{bmatrix} \begin{bmatrix} dT \\ dp \end{bmatrix}. \tag{13}$$

We have written the matrix of the partial differentials as a Jacobian. We know that the necessary and sufficient condition for the Jacobian to be singular is that a functional relation exists between S and V. As written in Eq. (13), the matrix is not singular.

For chemical reactions we choose the rate of change of degree of advancement ε, varying between $+1$ and -1 and measured from equilibrium state, as flux; and affinity A, the difference between Gibbs free energy of products and that of reactants, as force (§25). In the language of chemistry, we distinguish between thermodynamical coefficients at constant degree of advancement, such as $(\partial S/\partial T)_{p\varepsilon}$, and those at constant affinity, such as $(\partial S/\partial T)_{pA}$. The former coefficients are "frozen" and the latter are equilibrium coefficients. We also define the differential ratio of affinity to degree of advancement as an ordering coefficient, such as $(\partial A/\partial \varepsilon)_{pT}$. It will be shown later (§29) that the ordering coefficient is negative definite. Thus

$$\Delta C_p = C_{pA} - C_{p\varepsilon} = T\left(\frac{\partial S}{\partial T}\right)_{pA} - T\left(\frac{\partial S}{\partial T}\right)_{p\varepsilon}$$

$$= T\left(\frac{\partial S}{\partial \varepsilon}\right)_{pT}\left(\frac{\partial \varepsilon}{\partial T}\right)_{pA}$$

$$= -T\left(\frac{\partial S}{\partial \varepsilon}\right)_{pT}\left(\frac{\partial A}{\partial T}\right)_{p\varepsilon}\left(\frac{\partial \varepsilon}{\partial A}\right)_{pT}$$

$$= -T\left(\frac{\partial S}{\partial \varepsilon}\right)_{pT}^2\left(\frac{\partial \varepsilon}{\partial A}\right)_{pT} > 0. \tag{14}$$

We have used one of Maxwell's relations,

$$\left(\frac{\partial S}{\partial \varepsilon}\right)_{pT} = \left(\frac{\partial A}{\partial T}\right)_{p\varepsilon}, \tag{15}$$

which are derived from the following equations:

$$dU = TdS - pdV - Ad\varepsilon \tag{16}$$

$$dH = TdS + Vdp - Ad\varepsilon \tag{17}$$

$$dF = -SdT - pdV - Ad\varepsilon \tag{18}$$

$$dG = -SdT + Vdp - Ad\varepsilon \tag{19}$$

$$d(U + A\varepsilon) = TdS - pdV + \varepsilon dA \tag{20}$$

$$d(H + A\varepsilon) = TdS + Vdp + \varepsilon dA \tag{21}$$

$$d(F + A\varepsilon) = -SdT - pdV + \varepsilon dA \tag{22}$$

$$d(G + A\varepsilon) = -SdT + Vdp + \varepsilon dA, \tag{23}$$

where H is specific enthalpy. Similarly,

$$\Delta\kappa_T = \kappa_{TA} - \kappa_{T\varepsilon} = -\left(\frac{1}{V}\right)\left(\frac{\partial V}{\partial \varepsilon}\right)_{Tp}^2\left(\frac{\partial \varepsilon}{\partial A}\right)_{Tp} > 0. \tag{24}$$

$$\Delta\alpha_p = \alpha_{pA} - \alpha_{p\varepsilon} = -\left(\frac{1}{V}\right)\left(\frac{\partial V}{\partial \varepsilon}\right)_{pT}\left(\frac{\partial S}{\partial \varepsilon}\right)_{pT}\left(\frac{\partial \varepsilon}{\partial A}\right)_{pT}$$

$$\geqq 0 \text{ or } \leqq 0. \tag{25}$$

From eqs. (14), (24), and (25),

$$\Delta C_p = TV(\Delta\alpha_p)^2/\Delta\kappa_T. \tag{26}$$

Equation (26) means that, measured from the "complete" or frozen condition, C, κ_T, and α_p cannot proceed independently in any stage of development. They must keep in step. Equation (26) is reminescent of eq. (7) for the static condition and superficially similar to the condition of singularity of the matrix in eq. (13).

A typical example of incomplete reaction is in glasses[1]. If any intensive property such as C_p, α_p, or κ_T is plotted against temperature, there is a temperature range in which there is discontinuity in going from a supercooled liquid to a solid. If the measurements are made more slowly (depending on temperature), the discontinuity is increasingly inhibited. We may visualize the discontinuity as the change of the ordering coefficient, which is a function of temperature. Because the rate of entropy-production is $-(\rho A/T)\partial\varepsilon/\partial t$, where ρ is the density, a variation of ordering coefficient means a change of rate of entropy-production. For the changes taking place at constant temperature or pressure, this could be given a simple graphical interpretation. When the glass moves along the (V, p) plane with constant temperature, the rate of entropy-production is equal to the area swept out divided by temperature. When the glass moves along the (H, T) plane with constant pressure, the rate is equal to the area divided by temperature squared. When the glass moves along the (V, T) plane with constant pressure, the rate is equal to the area multiplied by $\Delta\alpha_p/T\Delta\kappa_T$, being equal to $(1/T)\ (\partial p/\partial T)_V$. This type of construction has been used to estimate the possible errors in the measured zero-point entropy due to irreversible effects. The causes of the irreversible effects cannot be definitely specified.

It is assumed that we could extend incomplete chemical reactions to all incomplete irreversible processes. For example, in the language of metallurgical engineering we know the development of creep (relaxation) if the creep (relaxation) function and the ordering coefficient are known. Unlike chemical reactions, the choice of the ordering coefficient is not unique, because we cannot write an equation of a physical process with stoichiometric coefficients. Furthermore, while one scalar ordering coefficient is sufficient for one chemical reaction, one or more coefficients may be required for a scalar physical process [This may be the reason why eq. (26) is not experimentally verified[1]. The situation is even worse for a tensorial physical process.

Reference

[1] R. O. Davies and G. O. Jones: *Advances in Phys.*, **2**, 370 (1953).

CHAPTER III
CONSERVATION LAWS

§4 *CONSERVATION OF MASS*

The equations derived below are hydrodynamical equations. There are two kinds of equations in hydrodynamics: one is Eulerian and the other Lagrangian. In the Eulerian equations, partial derivatives of time, $\partial/\partial t$, are used, which measure the total change observed by a stationary observer at a point fixed with respect to the external coordinate system. In the Lagrangian equations, substantial derivatives of time, d/dt, are used, which measure the local change observed by an observer moving with the non-stationary frame of reference, having the center of mass as the origin. We have

$$\frac{d}{dt} = \frac{\partial}{\partial t} + \boldsymbol{v} \cdot \text{grad}, \tag{1}$$

where $\boldsymbol{v}$ is the barycentric velocity.

Let g be any extensive property referred to an arbitrary quantity of mass,

$$\frac{dg}{dt} = \frac{d}{dt}\int \rho G dV, \tag{2}$$

where G is g divided by the total mass, or a specific quantity; ρ is the density and V is an arbitrary volume. If we assume that the order of differention and integration can be interchanged,

$$\frac{dg}{dt} = \int \frac{\partial(\rho G)}{\partial t}\, dV. \tag{3}$$

Because the change in g is due to the flow of g from outside and the production of g from within,

$$\frac{dg}{dt} = -\int \boldsymbol{\phi} \cdot d\boldsymbol{\Omega} + \int \sigma dV, \tag{4}$$

where $\boldsymbol{\phi}$ is a vector inclined to the surface $\boldsymbol{\Omega}$ enclosing the volume V, negative for the inward flow, and σ is the production per unit volume (Alternatively we may write the directed area $d\boldsymbol{\Omega}$ as $\boldsymbol{n}\, dA$, where dA is the magnitude of $d\boldsymbol{\Omega}$ and $\boldsymbol{n}$ is the unit vector perpendicular to dA). We have assumed that a streaming flow exists on a minute area. By applying the Gauss theorem, from eqs. (3) and (4),

$$\int \frac{\partial(\rho G)}{\partial t} dV = -\int \operatorname{div} \boldsymbol{\phi} dV + \int \sigma dV. \tag{5}$$

Because V is arbitrary, we have

$$\frac{\partial(\rho G)}{\partial t} = -\operatorname{div} \boldsymbol{\phi} + \sigma. \tag{6}$$

Next we consider dg/dt for a moving frame of reference. In contrast to eq. (3), we have

$$\frac{dg}{dt} = \int \rho \frac{dG}{dt} dV', \tag{7}$$

where V' is an arbitrary but fixed volume. Also, in contrast to eq. (4), the actual flow of fluid within the moving frame is not observed by a moving observer and must be deducted,

$$\int \rho \frac{dG}{dt} dV' = -\int (\boldsymbol{\phi} - \rho G \boldsymbol{v}) \cdot d\boldsymbol{\Omega}' + \int \sigma dV'. \tag{8}$$

By applying the Gauss theorem,

$$\int \rho \frac{dG}{dt} dV' = -\int \operatorname{div} (\boldsymbol{\phi} - \rho G \boldsymbol{v}) dV' + \int \sigma dV'. \tag{9}$$

Because V' is arbitrary,

$$\rho \frac{dG}{dt} = -\operatorname{div}(\boldsymbol{\phi} - \rho G \boldsymbol{v}) + \sigma. \tag{10}$$

From eqs. (6) and (10),

$$\rho \frac{dG}{dt} = \frac{\partial(\rho G)}{\partial t} + \operatorname{div} \rho G \boldsymbol{v}. \tag{11}$$

By setting $G = 1$ in eq. (11),

$$\frac{\partial \rho}{\partial t} = -\operatorname{div} \rho \boldsymbol{v}. \tag{12}$$

The physical meaning of eq. (12) is that the change of density is due to convection based on barycentric velocity. There is neither a sink nor a source for the total mass.

By setting $G = c_k$ in eq. (6), where c_k is the mass fraction of the component k,

$$\frac{\partial \rho_k}{\partial t} = -\operatorname{div} \rho_k \boldsymbol{v} - \operatorname{div} \boldsymbol{J}_k + \sum_r \Lambda_{kr} j_r, \tag{13}$$

where ρ_k is the partial density for the component k, $\boldsymbol{J}_k$ is the diffusion flux, Λ_{kr} is the mass stoichiometric coefficient of the component k participating in the reaction r and j_r is the rate of the chemical reaction r.

To understand eq. (13), we digress for a while to show the relations between the total and the partial quantities. Because, at constant temperature and pressure, the extensive quantities are homogeneous functions of the first degree with respect to m_k, the mass of the component k; from the Euler theorem,

$$g = \sum_k^n \left(\frac{\partial g}{\partial m_k}\right)_{Tpm_j, j \neq k} m_k = \sum_k^n G_k m_k. \tag{14}$$

Dividing by the total mass m,

$$G = \sum_k^n G_k c_k. \tag{15}$$

Multiplying eq. (15) by ρ,

$$\rho G = \sum_k^n \rho_k G_k, \tag{16}$$

the relation

$$\rho_k = \rho c_k \tag{17}$$

having been used. Also eq. (15) can be differentiated,

$$dG = \sum_k^n (G_k dc_k + c_k dG_k) = \sum_k^n G_k dc_k, \tag{18}$$

the Gibbs-Duhem relation having been used. Because

$$\sum_k^n c_k = 1, \tag{19}$$

the dependence can be removed by writing

$$dG = \sum_k^{n-1} (G_k - G_n) dc_k. \tag{20}$$

Equations (15), (16) and (18) can be applied to any extensive quantity, such as entropy and free energy. In the special example of specific volume V,

$$V = \sum_k^n V_k c_k, \tag{21}$$

$$\sum_k^n \rho_k V_k = 1, \tag{22}$$

$$\sum_k^n V_k dc_k = 0. \tag{23}$$

In the special example of m, we have only

$$m = \sum_k^n m_k. \tag{24}$$

We digress again to discuss the reference velocities in diffusion. In eq. (13) $\boldsymbol{J}_k$ is referred to the barycentric velocity $\boldsymbol{v}$, because hydrodynamical equations are reduced to simple forms. However, in the sense of the Galilean transformation, the barycentric velocity is not the only reference velocity. In general,

$$\boldsymbol{J}_k^a = \rho_k(\boldsymbol{v}_k - \boldsymbol{v}^a), \tag{25}$$

where $\boldsymbol{v}_k$ is the partial velocity of the component k and $\boldsymbol{J}_k^a$ is the flux based on the reference velocity $\boldsymbol{v}^a$. The reference velocity is so chosen that

$$\boldsymbol{v}^a = \sum_k^n a_k \boldsymbol{v}_k, \tag{26}$$

and

$$\sum_k^n a_k = 1. \tag{27}$$

The restraint on $\boldsymbol{J}_k^a$ is

$$\sum_k^n \frac{a_k}{c_k} \boldsymbol{J}_k^a = \rho \sum_k^n a_k(\boldsymbol{v}_k - \boldsymbol{v}^a) = 0. \tag{28}$$

Specifically for eq. (27),

$$\sum_k^n c_k = 1, \tag{29}$$

$$\sum_k^n n_k = 1, \tag{30}$$

$$\sum_k^n \rho_k V_k = 1, \tag{31}$$

$$\sum_k^n \delta_{kn} = \delta_{nn} = 1, \tag{32}$$

where n is the molar concentration and δ is the Kronecker delta (The confusion about n as the molar concentration and as a component is temporary). Specifically for eq. (26),

$$\sum_{k}^{n} c_k \boldsymbol{v}_k = \boldsymbol{v}, \tag{33}$$

$$\sum_{k}^{n} n_k \boldsymbol{v}_k = \boldsymbol{v}^n, \tag{34}$$

$$\sum_{k}^{n} \rho_k V_k \boldsymbol{v}_k = \boldsymbol{v}^0, \tag{35}$$

$$\sum_{k}^{n} \delta_{kn} \boldsymbol{v}_k = \boldsymbol{v}_n, \tag{36}$$

where $\boldsymbol{v}$ is the barycentric velocity, $\boldsymbol{v}^n$ is the molar velocity, $\boldsymbol{v}^0$ is the volume velocity and $\boldsymbol{v}_n$ is the velocity of the nth component.

Specifically for eq. (25),

$$\boldsymbol{J}_k = \rho_k(\boldsymbol{v}_k - \boldsymbol{v}), \tag{37}$$

$$\boldsymbol{J}_k^n = \rho_k(\boldsymbol{v}_k - \boldsymbol{v}^n), \tag{38}$$

$$\boldsymbol{J}_k^0 = \rho_k(\boldsymbol{v}_k - \boldsymbol{v}^0), \tag{39}$$

$$\boldsymbol{J}_k^r = \rho_k(\boldsymbol{v}_k - \boldsymbol{v}_n), \tag{40}$$

where $\boldsymbol{J}_k$ is the flux based on a mass frame of reference, useful in simplifying hydrodynamical equations; $\boldsymbol{J}_k^n$ is the flux based on a molar frame of reference, useful in kinetic theory; $\boldsymbol{J}_k^0$ is the flux based on a volume frame of reference, useful in dealing with diffusion in which the mean velocity is practically equal to the convection velocity (the flow of volume relative to a fixed point of the apparatus); and $\boldsymbol{J}_k^n$ is the flux referred to the relatively less active component, useful in eliminating the electrically neutral or the slowly moving component. Specifically for eq. (28),

$$\sum_{k}^{n} \boldsymbol{J}_k = 0, \tag{41}$$

$$\sum_{k}^{n} \boldsymbol{J}_k^n / M_k = 0, \tag{42}$$

$$\sum_{k}^{n} V_k \boldsymbol{J}_k^0 = 0, \tag{43}$$

$$\boldsymbol{J}_n^r = 0, \tag{44}$$

where M_k is molecular weight of the component k.

Returning to eq. (13), we interpret from it that the flow is due to convection and the production is due to diffusion and chemical reactions. Because the splitting into flow and production is arbitrary, we could also interpret from it that the flow is due to convection and diffusion (selective convection) and the production is due to chemical reactions. From eq. (37), eq. (13) can also be written as

$$\frac{\partial \rho_k}{\partial t} = -\operatorname{div} \rho_k \boldsymbol{v}_k + \sum_r \Lambda_{kr} j_r. \tag{45}$$

We refer the flow to the absolute velocity.

From eqs. (1) and (12),

$$\frac{d\rho}{dt} = -\rho \operatorname{div} \boldsymbol{v}. \tag{46}$$

We have used the identity

$$\operatorname{div} a\boldsymbol{b} = a \operatorname{div} \boldsymbol{b} + \boldsymbol{b} \cdot \operatorname{grad} a. \tag{47}$$

By using eqs. (1) and (47), eq. (13) becomes

$$\frac{d\rho_k}{dt} = -\rho_k \operatorname{div} \boldsymbol{v} - \operatorname{div} \boldsymbol{J}_k + \sum_r \Lambda_{kr} j_r. \tag{48}$$

By summing k of eq. (13) or eq. (45), we recover eq. (12) By summing k of eq. (48), we recover eq. (46). We have used the relations

$$\sum_k^n \rho_k = \rho, \tag{49}$$

$$\sum_k^n \rho_k \boldsymbol{v}_k = \rho \boldsymbol{v}, \tag{50}$$

$$\sum_{k,r}^n \Lambda_{kr} j_r = 0. \tag{51}$$

Equation (51) means that there is no change in the total mass by considering all reactions and all components.

By using eqs. (11) and (13),

$$\rho \frac{dc_k}{dt} = -\text{div}\, \boldsymbol{J}_k + \sum_r \Lambda_{kr} j_r. \tag{52}$$

Finally, by using eq. (46),

$$\rho \frac{dV}{dt} = -\frac{1}{\rho}\frac{d\rho}{dt} = \text{div}\, \boldsymbol{v}. \tag{53}$$

The physical meaning of eq. (53) is as follows: If $\text{div}\, \boldsymbol{v} > 0$ or < 0 at some point, the point is called a source or a sink of $\boldsymbol{v}$ respectively, and $\text{div}\, \boldsymbol{v} = (1/V)dV/dt$ is called the dilation rate. If $\text{div}\, \boldsymbol{v} = 0$, $\boldsymbol{v}$ is called solenoidal ($\boldsymbol{v}$ is called irrotational if $\text{curl}\, \boldsymbol{v} = 0$.), and the flow of a fluid of constant density is incompressible.

We collect now the continuity equations derived before:

$$\frac{\partial \rho}{\partial t} = -\text{div}\, \rho \boldsymbol{v}, \tag{12}$$

$$\frac{d\rho}{dt} = -\rho\, \text{div}\, \boldsymbol{v}\,, \tag{46}$$

$$\frac{\partial \rho_k}{\partial t} = -\text{div}\, \rho_k \boldsymbol{v} - \text{div}\, \boldsymbol{J}_k + \sum_r \Lambda_{kr} j_r \tag{13}$$

$$= -\,\text{div}\, \rho_k \boldsymbol{v}_k + \sum_r \Lambda_{kr} j_r, \tag{45}$$

$$\frac{d\rho_k}{dt} = -\rho_k \operatorname{div} \boldsymbol{v} - \operatorname{div} \boldsymbol{J}_k + \sum_r \Lambda_{kr} j_r, \tag{48}$$

$$\rho \frac{dc_k}{dt} = -\operatorname{div} \boldsymbol{J}_k + \sum_r \Lambda_{kr} j_r, \tag{52}$$

$$\rho \frac{dV}{dt} = \operatorname{div} \boldsymbol{v}. \tag{53}$$

§5 BALANCE OF MOMENTUM

The balance equation for the mechanical forces in fluids is

$$\rho \frac{d\boldsymbol{v}}{dt} = -\operatorname{Div} \Pi + \sum_k \rho_k \boldsymbol{F}_k, \tag{1}$$

where Π is the pressure tensor and $\boldsymbol{F}$ is the mechanical force. In matrix form,

$$\Pi = \begin{bmatrix} P_{11} + p & P_{12} & P_{13} \\ P_{21} & P_{22} + p & P_{23} \\ P_{31} & P_{32} & P_{33} + p \end{bmatrix} \tag{2}$$

or

$$\Pi - p\overleftrightarrow{1} = P, \tag{3}$$

where P is the viscous pressure tensor. Unless P vanishes, the mean pressure, $\frac{1}{3}\operatorname{tr} \Pi$, is not the hydrostatic pressure; however, it is the pressure in the thermodynamic sense, i.e., a function of state variables. Equation (1) was proposed by Euler. As written, the momentum is not invariant. The physical meaning of eq. (1) is that, if there is no outside mechanical force, the surface force of $-\operatorname{Div} \Pi$ due to mechanical action of contiguous material on a volume element and the body force acting on the same volume element is balanced.

Even if there is no irreversibility in fluids, grad$p$ is not necessarily zero because the hydrostatic pressure may be non-uniform, nor is div$\boldsymbol{v}$ because the fluid may be compressible; however, there are no viscous pressure and velocity gradients. This suggests that there is some subtle relation between the two. The quotient of the viscous pressure tensor and the velocity gradient is called the viscosity coefficient, which in general is a fourth-rank tensor. According to §13, an isotropic fourth-rank tensor has three independent components. Formally we split Π into a scalar matrix, a symmetric matrix without a trace and a skew symmetric matrix

$$\Pi = \left(p + \frac{1}{3} \operatorname{tr} P\right) 1 + P^{0,\mathrm{sy}} + P^{\mathrm{sk}}, \tag{4}$$

and the three relations are

$$\frac{1}{3} \operatorname{tr} P = -\eta_V \operatorname{tr} \operatorname{Grad} \boldsymbol{v}, \tag{5}$$

$$P^{0,\mathrm{sy}} = -2\eta \operatorname{Grad}^{0,\mathrm{sy}} \boldsymbol{v}, \tag{6}$$

$$P^{\mathrm{sk}} = -2\eta_r \operatorname{Grad}^{\mathrm{sk}} \boldsymbol{v}, \tag{7}$$

where η_V is the volume or bulk, η the shear, and η_r the rotational viscosity coefficients. The kinematic viscosity coefficients are viscosity coefficients divided by densities. At the moment, the relations (5—7) are general and we have not made any assumptions about the coefficients.

Equations (5—7) for viscous fluids resemble those for elastic solids. At low temperatures, very viscous fluids, say glasses, are almost indistinguishable from other solids in their behaviour. In general, fluids are even more elastic than solids with respect to compression. In solids, tension is considered positive but in fluids it is a negative pressure. As to the differences, in solids, except by special application, there is no hydrostatic pressure. Also the fluids cannot withstand static shear stresses. However, a fluid in streaming motion exerts shear stresses on its boundaries, which are proportional to the rates of strain (Newton's law). Thus in solids derivatives of the displacement $\boldsymbol{u}$ rather than the velocity $\boldsymbol{v}$ appear. Furthermore, the stresses S and the strains e are symmetric (See §34), so there is no equation for solids corresponding to eq. (7).

Corresponding to eqs. (4—6),

$$S = S^0 + \frac{1}{3}(\text{tr } S)\overleftrightarrow{1}, \tag{8}$$

$$\text{tr } S = \beta \, \text{tr}(\text{Grad } \boldsymbol{u}) = \beta \, \text{div } \boldsymbol{u}, \tag{9}$$

$$S^0 = \alpha \, \text{Grad}^0 \boldsymbol{u}, \tag{10}$$

where α is the deviatoric coefficient and β is the dilational coefficient. Combining eqs. (8—10),

$$S = \alpha \left[\text{Grad } \boldsymbol{u} - \frac{1}{3}(\text{div } \boldsymbol{u})\overleftrightarrow{1} \right] + \frac{1}{3}\beta(\text{div } \boldsymbol{u})\overleftrightarrow{1}, \tag{11}$$

as compared with stresses of an isotropic solid expressed in Lamé's constants λ and G,

$$S = 2G \, \text{Grad } \boldsymbol{u} + \lambda \, (\text{div } \boldsymbol{u})\overleftrightarrow{1}. \tag{12}$$

Corresponding to eq. (12), many authors define

$$-P^{sy} = 2\eta_1 \, \text{Grad } \boldsymbol{v} + \eta_2 \, (\text{div } \boldsymbol{v})\overleftrightarrow{1}, \tag{13}$$

where η_1 and η_2 are the first and second viscosity coefficients respectively. In hydrodynamics, the second coefficient is usually ignored, either because the fluid is incompressible or because the Stokes relation is assumed to be valid. The Stokes relation is $\eta_2 + \frac{2}{3}\eta_1 = 0$, although Stokes himself did not use it. In 1942 Tisza[1] pointed out the significance of the second coefficient in sound absorption in fluids. Because $K_V = \lambda + \frac{2}{3}G$ (See §34), where K_V is the bulk modulus, from eqs. (11—13), the equivalence between the viscosity coefficients in fluids and the elastic coefficients in solids can be shown in Table 1.

Table 1

Equivalence between the Viscosity Coefficients in Fluids and the Elastic Coefficients in Solids

Viscosity Coefficients	η_V	η	$\eta_V - \frac{2}{3}\eta$
	$\eta_2 + \frac{2}{3}\eta_1$	η_1	η_2
Elastic Coefficients	$\frac{1}{3}\beta$	$\frac{1}{2}\alpha$	$\frac{1}{3}(\beta - \alpha)$
	K_V	G	λ

The Table 1 is useful in that the solution of certain sinusoidal viscoelastic problems can be obtained from the known solution of the corresponding elastic problems by a "correspondence principle"[2].

Because

$$\mathrm{Div}\,(A\overleftrightarrow{1}) = \sum_{ij}\left(\frac{\partial A_i}{\partial x_i}\right)\delta_{ij} = \sum_i \frac{\partial A_i}{\partial x_j} = \mathrm{grad}\,A, \tag{14}$$

$$\mathrm{Grad}^{0,\mathrm{sy}}\,\boldsymbol{v} = \sum_{ij}\left[\frac{1}{2}\left(\frac{\partial v_i}{\partial x_j} + \frac{\partial v_j}{\partial x_i}\right) - \frac{1}{3}\frac{\partial v_i}{\partial x_i}\right], \tag{15}$$

$$\mathrm{Grad}^{\mathrm{sk}}\,\boldsymbol{v} = \sum_{ij}\left[\frac{1}{2}\left(\frac{\partial v_i}{\partial x_j} - \frac{\partial v_j}{\partial x_i}\right)\right] = \frac{1}{2}(\mathrm{curl}\,\boldsymbol{v})^{\mathrm{ps}}, \tag{16}$$

where the superscript ps denotes a pseudo-tensor (skew-symmetric) of the second rank, which is defined as follows:

$$\boldsymbol{u} \times \boldsymbol{v} = \boldsymbol{u} \cdot \boldsymbol{v}^{\mathrm{ps}} = \boldsymbol{u}^{\mathrm{ps}} \cdot \boldsymbol{v}$$

$$= (u_1, u_2, u_3)\begin{bmatrix} 0 & -v_3 & v_2 \\ v_3 & 0 & -v_1 \\ -v_2 & v_1 & 0 \end{bmatrix} = \begin{bmatrix} 0 & -u_3 & u_2 \\ u_3 & 0 & -u_1 \\ -u_2 & u_1 & 0 \end{bmatrix}\begin{bmatrix} v_1 \\ v_2 \\ v_3 \end{bmatrix}, \tag{17}$$

from eq. (14),

$$-\mathrm{Div}\,p\overleftrightarrow{1} = -\mathrm{grad}\,p, \tag{18}$$

from eqs. (5) and (14),

$$-\operatorname{Div}\left(\frac{1}{3}\operatorname{tr}P\right)\overleftrightarrow{\mathrm{I}} = \operatorname{grad}(\eta_V \operatorname{div}\boldsymbol{v})$$

$$= \sum_{ij} \frac{\partial}{\partial x_i}\left(\frac{\eta_V\,\partial v_j}{\partial x_j}\right)$$

$$= \sum_{ij}\left(\frac{\partial \eta_V}{\partial x_i}\right)\left(\frac{\partial v_j}{\partial x_j}\right) + \eta_V \sum_{ij}\left(\frac{\partial}{\partial x_i}\right)\left(\frac{\partial v_j}{\partial x_j}\right)$$

$$= (\operatorname{div}\boldsymbol{v})\operatorname{grad}\eta_V + \eta_V \operatorname{grad}\operatorname{div}\boldsymbol{v}, \tag{19}$$

from eqs. (6) and (15),

$$-\operatorname{Div}P^{0,\mathrm{sy}} = \operatorname{Div}(2\eta\,\operatorname{Grad}^{0,\mathrm{sy}}\boldsymbol{v})$$

$$= \sum_{ij}\frac{\partial}{\partial x_i}2\eta\left[\frac{1}{2}\left(\frac{\partial v_i}{\partial x_j} + \frac{\partial v_j}{\partial x_i}\right) - \frac{1}{3}\frac{\partial v_i}{\partial x_i}\right]$$

$$= \operatorname{grad}2\eta\cdot\operatorname{Grad}^{0,\mathrm{sy}}\boldsymbol{v} + \frac{\eta}{3}\operatorname{grad}\operatorname{div}\boldsymbol{v}$$

$$+ \eta\operatorname{div}\operatorname{Grad}\boldsymbol{v}, \tag{20}$$

and from eqs. (7) and (16),

$$-\operatorname{Div}P^{\mathrm{sk}} = \nabla\cdot\eta_r(\operatorname{curl}\boldsymbol{v})^{\mathrm{ps}}$$

$$= (\nabla\eta_r)\cdot(\operatorname{curl}\boldsymbol{v})^{\mathrm{ps}} + \eta_r\nabla\cdot(\operatorname{curl}\boldsymbol{v})^{\mathrm{ps}}$$

$$= (\operatorname{grad}\eta_r)\times\operatorname{curl}\boldsymbol{v} + \eta_r\operatorname{curl}\operatorname{curl}\boldsymbol{v}. \tag{21}$$

Substituting eqs. (4) and (18—21) in eq. (1),

$$\rho\frac{d\boldsymbol{v}}{dt} = -\operatorname{grad}p + \left(\eta_V + \frac{\eta}{3}\right)\operatorname{grad}\operatorname{div}\boldsymbol{v} + \eta\operatorname{div}\operatorname{Grad}\boldsymbol{v}$$

$$+ \sum_k \rho_k\boldsymbol{F}_k + \eta_r\operatorname{curl}\operatorname{curl}\boldsymbol{v} + [\operatorname{grad}\eta_V(\operatorname{div}\boldsymbol{v})$$

$$+ 2\operatorname{grad}\eta\cdot\operatorname{Grad}^{0,\mathrm{sy}}\boldsymbol{v} + (\operatorname{grad}\eta_r)\times\operatorname{curl}\boldsymbol{v}]. \tag{22}$$

Now we assume that the viscosity relations are linear, i.e., the viscosity coefficients are not functions of pressure and velocity. If we further assume that the viscosity coefficients are constants, the square bracket in eq. (22) vanishes. η_r will also vanish if we assume that P is symmetric. We have

$$\rho \frac{d\boldsymbol{v}}{dt} = -\operatorname{grad} p + \sum_k \rho_k \boldsymbol{F}_k + \left(\frac{\eta}{3} + \eta_V\right) \operatorname{grad} \operatorname{div} \boldsymbol{v}$$

$$+ \eta \operatorname{Div} \operatorname{Grad} \boldsymbol{v}. \tag{23}$$

This is the Navier-Stokes equation. The equation can be further simplified in the case of dilute gases, $\eta_V = 0$; or of incompressible fluids, $\operatorname{div} \boldsymbol{v} = 0$.

References

[1] L. Tisza, *Phys. Rev.*, **61**, 531 (1942).

[2] D. R. Bland: The Theory of Linear Viscoelasticy, Pergaman Press, New York, p. 67 (1960).

§6 *CONSERVATION OF ENERGY*

The total energy is conserved in any system or in any process. The conservation of energy has been expounded in every book of thermodynamics. In this section we cast the law in such a form that irreversible processes are incorporated.

We rewrite eq. (5, 1),

$$\rho \frac{d\boldsymbol{v}}{dt} = -\operatorname{Div} \Pi + \sum_k \rho_k \boldsymbol{F}_k. \tag{1}$$

Dotted with $\boldsymbol{v}$,

$$\rho \boldsymbol{v} \cdot \frac{d\boldsymbol{v}}{dt} = -(\operatorname{Div} \Pi) \cdot \boldsymbol{v} + \sum_k \rho_k \boldsymbol{F}_k \cdot \boldsymbol{v}$$

$$= -\operatorname{div}(\Pi \cdot \boldsymbol{v}) + \Pi : \operatorname{Grad} \boldsymbol{v} + \sum_k \rho_k \boldsymbol{F}_k \cdot \boldsymbol{v}. \tag{2}$$

The last equality of eq. (2) can be seen by writing the dot product $-(\operatorname{Div} \Pi) \cdot \boldsymbol{v}$ in components,

$$-\left(\frac{\partial \Pi_{ij}}{\partial x_i}\right) v_j = -\frac{\partial(\Pi_{ij} v_j)}{\partial x_i} + \Pi_{ij} \frac{\partial v_j}{\partial x_i}$$

$$= -\frac{\partial(\Pi_{ij} v_j)}{\partial x_i} + \Pi_{ji} \frac{\partial v_j}{\partial x_i}. \tag{3}$$

In the last equality of eq. (3) we have assumed that Π is symmetric. From eqs. (2) and (4, 11),

$$\frac{\partial \frac{1}{2} \rho v^2}{\partial t} = -\text{div}\left(\frac{1}{2} \rho v^2 + \Pi\cdot\right) \boldsymbol{v} + \Pi\text{:Grad } \boldsymbol{v}$$

$$+ \sum_k \rho_k \boldsymbol{F}_k \cdot \boldsymbol{v}. \tag{4}$$

If we assume that the outside force field is stationary,

$$\frac{\partial \boldsymbol{F}_k}{\partial t} = 0, \tag{5}$$

with

$$-\text{ grad } \psi_k = \boldsymbol{F}_k, \tag{6}$$

where ψ is the potential energy, then

$$\frac{\partial \rho \psi}{\partial t} = \sum_k \frac{\partial \rho_k \psi_k}{\partial t} \qquad \text{[from eq. (4, 16)]}$$

$$= \sum_k \psi_k \frac{\partial \rho_k}{\partial t} \qquad \text{[from eqs. (5) and (6)]}$$

$$= \sum_k \psi_k \left(-\text{div } \boldsymbol{J}_k - \text{div } \rho_k \boldsymbol{v} + \sum_r \Lambda_{kr} j_r\right)$$

[from eq. (4, 13)]

$$= -\left(\text{div} \sum_k \psi_k \boldsymbol{J}_k - \sum_k \boldsymbol{J}_k \cdot \text{grad } \psi_k\right)$$

$$-\left(\text{div} \sum_k \rho_k \psi_k \boldsymbol{v} - \sum_k \rho_k \boldsymbol{v} \cdot \text{grad } \psi_k\right)$$

[from eq. (4, 47) and conservation of potential energy in chemical reactions]

$$= -\operatorname{div} \rho\phi \boldsymbol{v} - \operatorname{div} \sum_k \phi_k \boldsymbol{J}_k - \sum_k \rho_k \boldsymbol{v}_k \cdot \boldsymbol{F}_k .$$

[from eqs. (6) and (4, 37)] (7)

We split the total energy E into kinetic energy, potential energy ϕ and internal energy U,

$$\rho E = \frac{1}{2} \rho v^2 + \rho\phi + \rho U ; \tag{8}$$

and propose that

$$\frac{\partial \rho E}{\partial t} = -\operatorname{div} \boldsymbol{J}_q^r - \operatorname{div} (\Pi - p\overleftrightarrow{1}) \cdot \boldsymbol{v} - \operatorname{div} p\boldsymbol{v}^0$$

$$- \operatorname{div} \sum_k \rho_k (U_k + \phi_k) \boldsymbol{v}_k - \operatorname{div} \frac{1}{2} \rho v^2 \boldsymbol{v}, \tag{9}$$

with

$$\boldsymbol{J}_q^r = \boldsymbol{J}_q - \sum_k H_k \boldsymbol{J}_k , \tag{10}$$

where $\boldsymbol{J}_q^r$ is the reduced heat flux, $\boldsymbol{J}_q$ is the heat flux and H_k is the partial enthalpy of the component k.

To justify eq. (9), we first change the partial into the substantial derivative,

$$\rho \frac{dE}{dt} = -\operatorname{div} \boldsymbol{J}_q^r - \operatorname{div} P \cdot \boldsymbol{v} - \operatorname{div} p\boldsymbol{v}^0$$

$$-\operatorname{div} \sum_k U_k \boldsymbol{J}_k - \operatorname{div} \sum_k \phi_k \boldsymbol{J}_k . \tag{11}$$

(A) The total energy is invariant, as it should be. (B) Because the total energy is a scalar, it is invariant with respect to finite uniform rotation of the coordinate system. (C) We can attach a moving frame of the coordinate system to the center of mass so that the total energy is invariant to uniform translation of the coordinate system in the sense of the Galilean transformation. (D) The first term on the right of eq. (11)

is the inward flow of heat (In a moment we shall show why we choose the reduced heat flux). The second and the third term are the pV work done on the boundary. The second term is the change of shape due to the viscous pressure based on the barycentric velocity. The third term is a change of size due to the hydrostatic pressure based on the volume velocity. The fourth and the fifth term are the inward flow of energy. The fourth term is due to the internal energy and the fifth term is due to the potential energy, both carried by the mass flux. As the name indicates, the potential energy is characterized by the fact that there is a mechanical force derived from a potential. The kinetic energy does not appear because the coordinate system is attached to the center of mass. Although we can distinguish between heat, energy or work outside the material medium, the identity is lost once we cross the boundary. (E) The first term on the right of eq. (11) is related to the heat transfer. The second term is related to the viscous pressure. The third term is related to the hydrostatic pressure. The last two terms are related to the mass flux. The chemical reactions do not appear because the potential energy is conserved in these reactions. (F) If there is no irreversibility,

$$\rho \frac{dE}{dt} = -\operatorname{div} p\boldsymbol{v}, \tag{12}$$

the change of the total energy depends only on the flow of pV work on the boundary. If the medium in non-compressible and the pressure is uniform, the total energy is stationary and thermostatic equilibrium is reached.

From eq. (10),

$$-\operatorname{div} \boldsymbol{J}_q^r = -\operatorname{div} \boldsymbol{J}_q + \operatorname{div} \sum_k H_k \boldsymbol{J}_k. \tag{13}$$

From eq. (4, 22),

$$-\operatorname{Div}(\Pi - p\vec{1}) \cdot \boldsymbol{v} = -\operatorname{div} \Pi \cdot \boldsymbol{v} + \operatorname{div} p \sum_k \rho_k V_k \boldsymbol{v}. \tag{14}$$

From eq. (4, 35),

$$-\operatorname{div} p\boldsymbol{v}^0 = -\operatorname{div} p \sum_k \rho_k V_k \boldsymbol{v}_k. \tag{15}$$

From eqs. (4, 16) and (4, 37),

$$-\operatorname{div} \sum_k \rho_k U_k \boldsymbol{v}_k = -\operatorname{div} \sum_k U_k \boldsymbol{J}_k - \operatorname{div} \rho U \boldsymbol{v}. \tag{16}$$

From eq. (4, 37),

$$-\operatorname{div} \sum_k \rho_k \psi_k \boldsymbol{v}_k = -\operatorname{div} \sum_k \psi_k \boldsymbol{J}_k - \operatorname{div} \rho \psi \boldsymbol{v}. \tag{17}$$

Also

$$\operatorname{div} \sum_k H_k \boldsymbol{J}_k - \operatorname{div} \sum_k U_k \boldsymbol{J}_k - \operatorname{div} p \sum_k \rho_k V_k (\boldsymbol{v}_k - \boldsymbol{v}) = 0. \tag{18}$$

Substituting eqs. (13—18) in eq. (9),

$$\frac{\partial \rho E}{\partial t} = -\operatorname{div} \boldsymbol{J}_q - \operatorname{div} \rho \left(\frac{1}{2} v^2 + \psi + U \right) \boldsymbol{v} - \operatorname{div} \Pi \cdot \boldsymbol{v} - \operatorname{div} \sum_k \psi_k \boldsymbol{J}_k. \tag{19}$$

Subtracting eqs. (4) and (7) from eq. (19),

$$\frac{\partial \rho U}{\partial t} = -\operatorname{div} \rho U \boldsymbol{v} - \operatorname{div} \boldsymbol{J}_q - \Pi : \operatorname{Grad} \boldsymbol{v} + \sum_k \boldsymbol{F}_k \cdot \boldsymbol{J}_k. \tag{20}$$

The reduced heat flux is devoid of enthalpy associated with the diffused mass. Because we can associate enthalpy, entropy or free energy with the mass, there are other definitions of heat flux similar to $\boldsymbol{J}_q^r$ (See §8). Although the use of $\boldsymbol{J}_q$ simplifies the form of eq. (20), the use of $\boldsymbol{J}_q^r$ is preferred as shown below. From the definition of specific enthalpy,

$$\sum_k \rho_k H_k = \rho H = \rho U + p. \tag{21}$$

The partial differentiation of the first member of eq. (21) yields

$$\sum_k \rho_k \frac{\partial H_k}{\partial t} + \sum_k H_k \frac{\partial \rho_k}{\partial t} = \sum_k \rho_k \frac{\partial H_k}{\partial t} - \operatorname{div} \sum_k \rho_k H_k \boldsymbol{v}_k$$

$$+ \sum_k \rho_k \boldsymbol{v}_k \cdot \operatorname{grad} H_k + \sum_k H_k \sum_r \Lambda_{kr} j_r . \tag{22}$$

The partial differentiation of the third member of eq. (21) yields

$$\frac{\partial p}{\partial t} + \frac{\partial \rho U}{\partial t} = \frac{\partial p}{\partial t} - \operatorname{div} \rho U \boldsymbol{v} - \operatorname{div} \boldsymbol{J}_q - P\colon \operatorname{Grad} \boldsymbol{v}$$

$$- \operatorname{div} p\boldsymbol{v} + \boldsymbol{v} \cdot \operatorname{grad} p + \sum_k \boldsymbol{F}_k \cdot \boldsymbol{J}_k$$

$$= \frac{\partial p}{\partial t} - \operatorname{div} \sum_k \rho_k H_k \boldsymbol{v} - \operatorname{div} \boldsymbol{J}_q - P\colon \operatorname{Grad} \boldsymbol{v}$$

$$+ v \cdot \operatorname{grad} p + \sum_k \boldsymbol{F}_k \cdot \boldsymbol{J}_k . \tag{23}$$

Combining eqs. (22) and (23),

$$-\operatorname{div}\left(\boldsymbol{J}_q - \sum_k H_k \boldsymbol{J}_k\right) = \sum_k \rho_k \frac{\partial H_k}{\partial t} + \sum_k \rho_k \boldsymbol{v}_k \cdot \operatorname{grad} H_k$$

$$+ \sum_r \Delta H_r j_r - \frac{\partial p}{\partial t} - \boldsymbol{v} \cdot \operatorname{grad} p$$

$$+ P\colon \operatorname{Grad} \boldsymbol{v} - \sum_k \boldsymbol{F}_k \cdot \boldsymbol{J}_k , \tag{24}$$

where ΔH_r is the change of enthalpy for the chemical reaction r. From the defining equation (10), the left of eq. (24) is the divergence of $\boldsymbol{J}'_q$. The first term on the right is the partial differential of H_k, the second term is the spatial differentiation of H_k and the third term is the change of enthalpy of all the reactions. Thus the divergence of $\boldsymbol{J}'_q$ is invariant with the ground level of energy.

We collect now the equations for the balance of energy

derived before:

$$\frac{\partial \frac{1}{2}\rho v^2}{\partial t} = -\text{div}\left(\frac{1}{2}\rho v^2 + \Pi\cdot\right)\boldsymbol{v} + \Pi{:}\text{Grad}\,\boldsymbol{v} + \sum_k \rho_k \boldsymbol{F}_k \cdot \boldsymbol{v}, \tag{4}$$

$$\frac{\partial \rho\phi}{\partial t} = -\text{div}\,\rho\phi\boldsymbol{v} - \text{div}\sum_k \phi_k \boldsymbol{J}_k - \sum_k \rho_k \boldsymbol{v}_k \cdot \boldsymbol{F}_k, \tag{7}$$

$$\frac{\partial \rho U}{\partial t} = -\text{div}\,\rho U\boldsymbol{v} - \text{div}\,\boldsymbol{J}_q - \Pi{:}\,\text{Grad}\,\boldsymbol{v} + \sum_k \boldsymbol{F}_k \cdot \boldsymbol{J}_k. \tag{20}$$

§7 *EXAMPLES OF ENERGY BALANCE*

A test of the law of conservation of energy is the energy balance. Our confidence in the law of conservation energy is so great that, when the energy changes are not balanced, we invent other forms of energy or matter to account for the missing change. For example, the principle of relativity admits another form of energy, which can be converted from mass. As another example, the neutrino was invented to account for the missing energy change in β-decay of radioactive nuclei, although it took 30 years for the experimental verification of the neutrino. Energy balance can be applied to a system whose energy is either conservative or non-conservative, or split into linear or nonlinear factors, referred to a stationary, a uniformly rotating, or a uniformly translating frame. In conservative systems, there is no entropy-production. In systems with series connections only (§22), the energy balance is the same as the (mechanical) force balance. The Lagrangian equation, the Hamiltonian equation, and the Routhian equation are all energy balances but with different methods of splitting energy into factors. In this section we give four examples of energy balance.

Energy Balance of a Linear Conservative System

Consider the simple example of a harmonic oscillator in which a mass and a spring are connected in series. The (mechanical) force balance is

$$m\ddot{x} + kx = 0, \tag{1}$$

where m is the mass and k is the spring constant. The Hamiltonian integral is

$$\int \dot{q}dp - \int \dot{p}dq = \mathscr{H}, \tag{2}$$

where p is the generalized momentum and q is the generalized coordinate. In this example,

$$q = x \text{ and } p = m\dot{x}. \tag{3}$$

However, we should not be tempted to think that the generalized momentum is always momentum and the generalized coordinate is always a spatial coordinate. Actually they could be any physical quantity or even the role of the two may be interchanged, so long as the integral $\iint dp\, dq$ is invariant and $\dot{p}q$ has the dimensions of energy. Substituting eqs. (1) and (3) in eq. (2),

$$\mathscr{H} = \int m\dot{x}\ d\dot{x} - \int m\ddot{x}\ dx = \int m\dot{x}\ d\dot{x} + \int kx\ dx. \tag{4}$$

On integration,

$$\frac{1}{2}m\dot{x}^2 + \frac{1}{2}kx^2 = h = a \text{ constant.} \tag{5}$$

Dividing eq. (5) by h and letting $m/2h = 1/\alpha^2$ and $k/2h = 1/\beta^2$,

$$\frac{\dot{x}^2}{\alpha^2} + \frac{x^2}{\beta^2} = 1, \tag{6}$$

which is a family of ellipses with semi-axes α and β.

In (mechanical) forces of a central field, the potential

energy is a function of radius only, $\phi = f(r)$. The path lies in a plane. If we extend from a plane oscillator to a space oscillator, we recognize that the potential energy is $\frac{1}{2}kr^2$, that the path is closed and that variation of each coordinate x and y is a simple oscillation with the same frequency $\omega = \sqrt{k/m}$. We set

$$x = A\cos(\omega t + \alpha) = A\cos\phi, \tag{7}$$

$$\begin{aligned} y &= B\cos(\omega t + \beta) = B\cos(\phi - \alpha + \beta) \\ &= B\cos(\phi - \delta). \end{aligned} \tag{8}$$

where

$$\phi = \omega t + \alpha \text{ and } \delta = \alpha - \beta.$$

Solving for $\cos\phi$ and $\sin\phi$ and equating the sum of the squares to unity,

$$\frac{x^2}{A^2} + \frac{y^2}{B^2} - \frac{2xy\cos\delta}{AB} = \sin^2\delta, \tag{9}$$

which is a family of rotated ellipses with centers at the origin.

Energy Balance of a Nonlinear Conservative System[1]

Consider a linear capacitor and a saturated iron core inductor in series. Let C be the capacitance, i the current, q the charge in the condenser, ϕ the total magnetic flux through one turn of the inductance coil, and n the number of turns. It is assumed that the hysteresis loss is negligible; therefore the system is conservative.

Because of the linearity of the capacitor, the voltage through the condenser is

$$\int \frac{i}{C}\,dt = \frac{q}{C}. \tag{10}$$

Because of the nonlinearity of the inductance (The iron core is saturated), the voltage through the inductor is $n\phi(i)$. The series connection means the force balance is

$$n\dot{\phi}(i) + \frac{q}{C} = 0. \tag{11}$$

The Hamiltonian is

$$\mathscr{H} = \text{K. E.} + \psi = \int \dot{q}dp - \int \dot{p}dq. \tag{12}$$

Because $q^2/2C$ may be considered as potential energy, we recognize

$$\dot{p} = -\frac{q}{C}. \tag{13}$$

by taking account of $\partial\mathscr{H}/\partial q = -\dot{p}$. From eqs. (11) and (13),

$$p = n\phi(i). \tag{14}$$

Substituting eqs. (13) and (14) in eq. (12),

$$\mathscr{H} = n\int \dot{q}d\phi(i) + \frac{q^2}{2C}. \tag{15}$$

The problem is to find a relation between ϕ and i. In electrical engineering there exists a convenient application of the function of $\phi(i)$ due to Dregfuss, quoted by Minorsky[1],

$$\phi(i) = \alpha \arctan\frac{ni}{\gamma} + \beta\frac{ni}{\gamma}, \tag{16}$$

where α, β, and γ are positive constants by means of which one can fit the formula into any case of saturation effects. Thus

$$d\phi(i) = \left[\frac{\dfrac{\alpha\gamma}{2n}}{\dfrac{\gamma^2}{2n^2} + \dfrac{\dot{q}^2}{2}}\right] d\dot{q} + \frac{\beta n}{\gamma}\, d\dot{q}. \tag{17}$$

Substituting eq. (17) in eq. (15),

$$\mathscr{H} = \frac{q^2}{2C} + \frac{\alpha\gamma}{2}\int \frac{\dot{q}d\dot{q}}{\dfrac{\gamma^2}{2n^2} + \dfrac{\dot{q}^2}{2}} + \frac{\beta n^2}{\gamma}\int \dot{q}d\dot{q}. \tag{18}$$

On integration,

$$\frac{q^2}{2C} + \frac{\alpha\gamma}{2}\ln\left(\frac{\gamma^2}{2n^2} + \frac{\dot{q}^2}{2}\right) + \frac{\beta n^2}{2\gamma}\dot{q}^2 = \text{a constant.} \tag{19}$$

This determines a family of closed curves resembling ellipses with a singular point at the center of the family.

A similar calculation can be made for another conservative system in which the nonlinearity is in the capacity term,

$$L\ddot{q} + \int\left[\frac{i}{C\,(i)}\right]dt = 0, \tag{20}$$

where L is the inductance. It is known that certain minerals used as dielectrics in condensers result in a non-proportionality between charge and voltage across the condenser.

Energy Balance of a Uniformly Rotating Frame

Newton's law was designed to relate to coordinate axes fixed in the center of the sun. Because physical measurements are usually done on the earth, either we approximate these measurements or derive the inertia-frame measurements by taking consideration of the rotation of the earth. The formulae connecting the related velocities and accelerations are derived in many physics books. In this example, following Karnopp[2], we derive the transformation laws between these two frames in a novel way, i.e., from the energy balance.

Let $\boldsymbol{r}$ be the displacement vector. Its components are (x_1, x_2) on the inertia frame, but they are (x_1', x_2') on the rotating frame, where prime quantities refer to the rotating frame. Thus

$$r_i = \begin{bmatrix} C & -S \\ S & C \end{bmatrix} r_r. \tag{21}$$

In components,

$$\begin{bmatrix} x_1 \\ x_2 \end{bmatrix} = \begin{bmatrix} C & -S \\ S & C \end{bmatrix}\begin{bmatrix} x_1' \\ x_2' \end{bmatrix}, \tag{22}$$

where

$$\begin{bmatrix} C & -S \\ S & C \end{bmatrix} = \begin{bmatrix} \cos\theta & -\sin\theta \\ \sin\theta & \cos\theta \end{bmatrix}$$

is the matrix of coordinate transformation and θ is the angle of rotation. Differentiating eq. (21) with respect to time (not on the same frame),

$$\dot{\boldsymbol{r}}_i = \begin{bmatrix} C & -S \\ S & C \end{bmatrix} \dot{\boldsymbol{r}}_r + \begin{bmatrix} -S & -C \\ C & -S \end{bmatrix} \dot{\theta} \boldsymbol{r}_r. \tag{23}$$

In components,

$$\begin{bmatrix} \dot{x}_1 \\ \dot{x}_2 \end{bmatrix} = \begin{bmatrix} C & -S \\ S & C \end{bmatrix} \begin{bmatrix} \dot{x}'_1 \\ \dot{x}'_2 \end{bmatrix} + \begin{bmatrix} -S & -C \\ C & -S \end{bmatrix} \dot{\theta} \begin{bmatrix} x'_1 \\ x'_2 \end{bmatrix}. \tag{24}$$

For later use, eq. (24) can also be written as

$$\begin{bmatrix} \dot{x}_1 \\ \dot{x}_2 \end{bmatrix} = \begin{bmatrix} C & -S \\ S & C \end{bmatrix} \left\{ \begin{bmatrix} 1 & 0 \\ 0 & 1 \end{bmatrix} \begin{bmatrix} \dot{x}'_1 \\ \dot{x}'_2 \end{bmatrix} + \begin{bmatrix} 0 & -1 \\ 1 & 0 \end{bmatrix} \dot{\theta} \begin{bmatrix} x'_1 \\ x'_2 \end{bmatrix} \right\}. \tag{25}$$

Equation (24) can be differentiated again,

$$\begin{bmatrix} \ddot{x}_1 \\ \ddot{x}_2 \end{bmatrix} = \begin{bmatrix} C & -S \\ S & C \end{bmatrix} \begin{bmatrix} \ddot{x}'_1 \\ \ddot{x}'_2 \end{bmatrix} + 2 \begin{bmatrix} -S & -C \\ C & -S \end{bmatrix} \dot{\theta} \begin{bmatrix} \dot{x}'_1 \\ \dot{x}'_2 \end{bmatrix} + \begin{bmatrix} -C & S \\ -S & -C \end{bmatrix} \dot{\theta}^2 \begin{bmatrix} x'_1 \\ x'_2 \end{bmatrix}. \tag{26}$$

For later use, eq. (26) can also be written as

$$\begin{bmatrix} \ddot{x}_1 \\ \ddot{x}_2 \end{bmatrix} = \begin{bmatrix} C & -S \\ S & C \end{bmatrix} \left\{ \begin{bmatrix} 1 & 0 \\ 0 & 1 \end{bmatrix} \begin{bmatrix} \ddot{x}'_1 \\ \ddot{x}'_2 \end{bmatrix} + 2 \begin{bmatrix} 0 & -1 \\ 1 & 0 \end{bmatrix} \dot{\theta} \begin{bmatrix} \dot{x}'_1 \\ \dot{x}'_2 \end{bmatrix} + \begin{bmatrix} -1 & 0 \\ 0 & -1 \end{bmatrix} \dot{\theta}^2 \begin{bmatrix} x'_1 \\ x'_2 \end{bmatrix} \right\}. \tag{27}$$

It can be shown[3] that the energy balance between these two frames is

$$E_i = E_r + \boldsymbol{M} \cdot \dot{\boldsymbol{\theta}}, \tag{28}$$

where $\boldsymbol{M}$ is the angular momentum.

With no dissipation of energy, the power balance is

$$(F_1, F_2)\begin{bmatrix}\dot{x}_1\\ \dot{x}_2\end{bmatrix} = (F'_1, F'_2)\begin{bmatrix}\dot{x}'_1\\ \dot{x}'_2\end{bmatrix} + \mathscr{T}\cdot\dot{\theta}, \tag{29}$$

where $\boldsymbol{F}$ are mechanical forces and $\mathscr{T}$ is the torque. Multiplying eq. (24) on the left by (F_1, F_2),

$$(F_1, F_2)\begin{bmatrix}\dot{x}_1\\ \dot{x}_2\end{bmatrix} = (F_1, F_2)\begin{bmatrix}C & -S\\ S & C\end{bmatrix}\begin{bmatrix}\dot{x}'_1\\ \dot{x}'_2\end{bmatrix}$$

$$+ (F_1, F_2)\begin{bmatrix}-S & -C\\ C & -S\end{bmatrix}\dot{\theta}\begin{bmatrix}x'_1\\ x'_2\end{bmatrix}. \tag{30}$$

Comparing eqs. (29) and (30), we find

$$\begin{bmatrix}F'_1\\ F'_2\end{bmatrix} = \begin{bmatrix}C & S\\ -S & C\end{bmatrix}\begin{bmatrix}F_1\\ F_2\end{bmatrix} \tag{31}$$

and

$$\mathscr{T} = (x'_1, x'_2)\begin{bmatrix}-S & C\\ -C & -S\end{bmatrix}\begin{bmatrix}F_1\\ F_2\end{bmatrix}. \tag{32}$$

Equations (31) and (32) can be combined as

$$\begin{bmatrix}C & S\\ -S & C\\ -Sx'_1 - Cx'_2 & Cx'_1 - Sx'_2\end{bmatrix}\begin{bmatrix}F_1\\ F_2\end{bmatrix} = \begin{bmatrix}F'_1\\ F'_2\\ \mathscr{T}\end{bmatrix}. \tag{33}$$

Equation (24) can be rewritten as

$$\begin{bmatrix}\dot{x}_1\\ \dot{x}_2\end{bmatrix} = \begin{bmatrix}C & -S & -Sx'_1 - Cx'_2\\ S & C & Cx'_1 - Sx'_2\end{bmatrix}\begin{bmatrix}\dot{x}'_1\\ \dot{x}'_2\\ \dot{\theta}\end{bmatrix}. \tag{34}$$

Equations (33) and (34) are transformation laws of velocities and mechanical forces between the two frames.

The velocities and the mechanical forces are different not only because they are on different frames but also because the frames are moving with respect to each other. Indeed we could base velocities on the same frame by omitting

$$\begin{bmatrix} C & -S \\ S & C \end{bmatrix} \text{ in eq. (25),}$$

$$\begin{bmatrix} \dot{x}_1 \\ \dot{x}_2 \end{bmatrix}^{\mathrm{T}} = \begin{bmatrix} \dot{x}'_1 \\ \dot{x}'_2 \end{bmatrix}^{\mathrm{T}} + (-\dot{\theta}x'_2, \dot{\theta}x'_1)$$

$$= \begin{bmatrix} \dot{x}'_1 \\ \dot{x}'_2 \end{bmatrix}^{\mathrm{T}} + (0, 0, \dot{\theta}) \begin{bmatrix} 0 & 0 & x'_2 \\ 0 & 0 & -x'_1 \\ -x'_2 & x'_1 & 0 \end{bmatrix}. \tag{35}$$

Thus the vector equation of velocity is

$$\boldsymbol{v}_i = \boldsymbol{v}_r + \dot{\boldsymbol{\theta}} \times \boldsymbol{r}_r, \tag{36}$$

which is the Coriolis law.

Similarly, we could base accelerations on the same frame by omitting $\begin{bmatrix} C & -S \\ S & C \end{bmatrix}$ in eq. (27),

$$\begin{bmatrix} \ddot{x}_1 \\ \ddot{x}_2 \end{bmatrix}^{\mathrm{T}} = \begin{bmatrix} \ddot{x}'_1 \\ \ddot{x}'_2 \end{bmatrix}^{\mathrm{T}} + 2(-\dot{\theta}\dot{x}'_2, \dot{\theta}\dot{x}'_1) + (-\dot{\theta}^2 x'_1, -\dot{\theta}^2 x'_2)$$

$$= \begin{bmatrix} \ddot{x}'_1 \\ \ddot{x}'_2 \end{bmatrix}^{\mathrm{T}} + 2(0, 0, \dot{\theta}) \begin{bmatrix} 0 & 0 & \dot{x}'_2 \\ 0 & 0 & -\dot{x}'_1 \\ -\dot{x}'_2 & \dot{x}'_1 & 0 \end{bmatrix}$$

$$+ (0, 0, \dot{\theta}) \begin{bmatrix} 0 & 0 & x'_1\dot{\theta} \\ 0 & 0 & x'_2\dot{\theta} \\ -x'_1\dot{\theta} & -x'_2\dot{\theta} & 0 \end{bmatrix}. \tag{37}$$

The vector equation of acceleration is

$$\boldsymbol{a}_i = \boldsymbol{a}_r + 2\dot{\boldsymbol{\theta}} \times \boldsymbol{v}_r + \dot{\boldsymbol{\theta}} \times (-x'_2\dot{\theta}, x'_1\dot{\theta})$$

$$= \boldsymbol{a}_r + 2\dot{\boldsymbol{\theta}} \times \boldsymbol{v}_r + \dot{\boldsymbol{\theta}} \times (\dot{\boldsymbol{\theta}} \times \boldsymbol{r}_r), \tag{38}$$

where $2\dot{\boldsymbol{\theta}} \times \boldsymbol{v}_r$ is the Coriolis acceleration, and $\dot{\boldsymbol{\theta}} \times (\dot{\boldsymbol{\theta}} \times \boldsymbol{r}_r)$ is the centrifugal acceleration.

If we limit ourselves to the inertia frame but convert the

rectangular system $\hat{R}$ to the polar system $\hat{P}$, we can find accelerations in two perpendicular directions. Because

$$x_1 = \rho\cos\theta, \quad x_2 = \rho\sin\theta,$$

the differention twice on the same rectangular system yields

$$(\ddot{x}_1, \ddot{x}_2) = (A, B),$$

where

$$\begin{aligned} A &= (\ddot{\rho} - \rho\dot{\theta}^2)C - (2\dot{\rho}\dot{\theta} + \rho\ddot{\theta})S, \\ B &= (\rho\ddot{\theta} + 2\dot{\rho}\dot{\theta})C + (\ddot{\rho} - \rho\dot{\theta}^2)S. \end{aligned} \tag{39}$$

We have

$$\begin{aligned} (\ddot{x}_1, \ddot{x}_2)\hat{R} &= (A, B)\hat{R} \\ &= (A, B)\begin{bmatrix} C & -S \\ S & C \end{bmatrix}\hat{P} = (\ddot{\rho} - \rho\dot{\theta}^2, \rho\ddot{\theta} + 2\dot{\rho}\dot{\theta})\hat{P}. \end{aligned} \tag{40}$$

Thus the tangential acceleration is $\ddot{\rho} - \rho\dot{\theta}^2$, and the drag acceleration is $\rho\ddot{\theta} + 2\dot{\rho}\dot{\theta}$.

Energy Balance of a Uniformly Translating Frame

In this example we use the electromagnetic force on a charged particle to illustrate the energy balance of a uniformly translating frame.

Let us consider the electrostatic force $\boldsymbol{F}$ on a charged particle with the charge q in presence of an electric field $\boldsymbol{E}$ and a magnetic field $\boldsymbol{H}$. The particle is at rest. It only responds to the electric field,

$$\boldsymbol{F} = q\boldsymbol{E} = -q \operatorname{grad} \zeta, \tag{41}$$

where ζ is the scalar potential. The Hamiltonian is

$$\mathscr{H} = \frac{1}{2} mv^2 + q\zeta. \tag{42}$$

Now let us turn our attention to the electromagnetic force on the same particle. This time the particle is moving. It also responds to the magnetic field. The additional force density acting on an element of $\boldsymbol{I}$ is $\boldsymbol{I} \times \boldsymbol{B}$, where $\boldsymbol{B}$ is the magnetic induction. A charge in motion is equivalent to a current, or the additional force is $q(\boldsymbol{v} \times \boldsymbol{B})$. Thus

$$\boldsymbol{F} = q\boldsymbol{E} + q(\boldsymbol{v} \times \boldsymbol{B}). \tag{43}$$

We can bring $\boldsymbol{F}$ into other form by using Maxwell's electromagnetic field equations (§11),

$$\boldsymbol{B} = \text{curl}\, \boldsymbol{A}, \tag{44}$$

$$\boldsymbol{E} = -\text{grad}\, \zeta - \frac{\partial \boldsymbol{A}}{\partial t}, \tag{45}$$

and

$$\begin{aligned} \boldsymbol{v} \times \text{curl}\, \boldsymbol{A} &= \boldsymbol{v} \times (\nabla \times \boldsymbol{A}) = \nabla(\boldsymbol{A} \cdot \boldsymbol{v}) - (\boldsymbol{v} \cdot \nabla)\boldsymbol{A} \\ &= \text{grad}\, \boldsymbol{A} \cdot \boldsymbol{v} - (\boldsymbol{v} \cdot \text{grad})\, \boldsymbol{A} \\ &= \text{grad}\, \boldsymbol{A} \cdot \boldsymbol{v} + \frac{\partial \boldsymbol{A}}{\partial t} - \frac{d\boldsymbol{A}}{dt}, \end{aligned} \tag{46}$$

eq. (4, 1) having been used in the last equality of eq. (46). Substituting eq. (44—46) in eq. (43),

$$\begin{aligned} \boldsymbol{F} &= q\left(-\text{grad}\, \zeta - \frac{\partial \boldsymbol{A}}{\partial t}\right) + q(\boldsymbol{v} \times \text{curl}\, \boldsymbol{A}) \\ &= q\left(-\text{grad}\, \zeta - \frac{d\boldsymbol{A}}{dt} + \text{grad}\, \boldsymbol{A} \cdot \boldsymbol{v}\right). \end{aligned} \tag{47}$$

Let

$$\phi = q(\zeta - \boldsymbol{A} \cdot \boldsymbol{v}), \tag{48}$$

where ϕ is the potential energy. Then

$$\frac{\partial \phi}{\partial \boldsymbol{v}} = -q\boldsymbol{A}, \tag{49}$$

and

$$\frac{\partial \phi}{\partial \boldsymbol{x}} = q(\text{grad}\, \zeta - \text{grad}\, \boldsymbol{A} \cdot \boldsymbol{v}). \tag{50}$$

Substituting eqs. (49) and (50) in eq. (47),

$$\boldsymbol{F} = \frac{d}{dt}\left(\frac{\partial \psi}{\partial \boldsymbol{v}}\right) - \frac{\partial \psi}{\partial \boldsymbol{x}}. \tag{51}$$

Equation (51) is interesting because $\boldsymbol{F}$ is velocity-dependent. Forces prescribed this way are called velocity-dependent forces.

Because the Lagrangian $\mathscr{L}$ is the difference between kinetic energy and potential energy, from eq. (48),

$$\mathscr{L} = \frac{1}{2}mv^2 + q\boldsymbol{A}\cdot\boldsymbol{v} - q\zeta. \tag{52}$$

The momentum p can be found from eq. (52),

$$\boldsymbol{p} = \frac{\partial \mathscr{L}}{\partial \boldsymbol{v}} = m\boldsymbol{v} + q\boldsymbol{A}. \tag{53}$$

The Hamiltonian is

$$\mathscr{H} = \boldsymbol{p}\cdot\boldsymbol{v} - \mathscr{L} = \frac{1}{2}mv^2 + q\zeta. \tag{54}$$

This turns out to be the same as eq. (42). However, in the example of a velocity-dependent force, and therefore a velocity-dependent potential, we are not sure that $\mathscr{H}$ is still the sum of the kinetic and potential energy.

The energy balance can be better interpreted in terms of translating frames of reference. Let us have two frames. Both are inertia frames and one is translating at a uniform velocity of $q\boldsymbol{A}/m$ with respect to the other. The velocities are related as

$$\boldsymbol{v} = \boldsymbol{v}' + \frac{q}{m}\boldsymbol{A}. \tag{55}$$

The momenta are related as

$$\boldsymbol{p} = \boldsymbol{p}' + q\boldsymbol{A}. \tag{56}$$

The energies are related as

$$E = \frac{1}{2}mv^2 + \psi.$$

$$= \left(\frac{1}{2} mv'^2 + \psi\right) + q\boldsymbol{A} \cdot \boldsymbol{v}' + \frac{1}{2}\frac{q^2A^2}{m}$$

$$= E' + \frac{q}{m}\boldsymbol{A} \cdot \boldsymbol{p}' + \frac{1}{2}\frac{q^2A^2}{m}, \tag{57}$$

where $(q/m)\boldsymbol{A} \cdot \boldsymbol{p}'$ can be interpreted as an energy connected to the relative motion of frames, and $\frac{1}{2}\, q^2A^2/m$ as the internal energy. In the Galilean transformation, all frames moving in straight lines are equivalent (Note that this is correct only in a non-relativistic sense. To preserve the velocity of light, the Galilean transformation must be replaced by the Lorentz transformation). Because there is no preferential frame, we could use the frame in which $\boldsymbol{p}'$ is zero.

$$E = \frac{p^2}{2m} + E'. \tag{58}$$

We see that the partition of total energy into kinetic energy, potential energy and internal energy is a matter of choice.

References

[1] N. Minorsky: Nonlinear Oscillations, D. Van Nostrand, Princeton, pp. 62—64 (1962).

[2] D. Karnopp: *J. Franklin Inst.*, **291**, 211 (1971).

[3] L. D. Landau and E. M. Lifschitz (translated by J. B. Sykes and J. S. Bell): Mechanics, Pergamon Press, Oxford, p. 129 (1960).

§8 *THE RATE OF ENTROPY-PRODUCTION*

Except in units and sign, entropy-production is almost synonymous with dissipation of energy (although strictly speaking, energy cannot be dissipated). In this sense the rate of entropy-production is a negative power.

According to the postulate of local equilibrium, § 2, we could use the Gibbs equation in the dynamic form,

$$\rho \frac{dS}{dt} = \frac{\rho}{T}\frac{dU}{dt} + \frac{\rho p}{T}\frac{dV}{dt} - \frac{\rho}{T}\sum_k \mu_k \frac{dc_k}{dt}. \tag{1}$$

From eqs. (4, 11) and (6, 20),

$$\frac{\rho}{T}\frac{dU}{dt} = -\frac{\operatorname{div} \boldsymbol{J}_q}{T} - \frac{\Pi:\operatorname{Grad} \boldsymbol{v}}{T} + \sum_k \frac{\boldsymbol{F}_k \cdot \boldsymbol{J}_k}{T}. \tag{2}$$

From eq. (4, 53),

$$p\frac{\rho}{T}\frac{dV}{dt} = \frac{p}{T}\operatorname{div} \boldsymbol{v}. \tag{3}$$

From eq. (4, 52),

$$-\frac{\rho}{T}\sum_k \mu_k \frac{dc_k}{dt} = -\sum_k \mu_k \frac{\sum_r \Lambda_{kr} j_r - \operatorname{div} \boldsymbol{J}_k}{T}. \tag{4}$$

Substituting eqs. (2)—(4) in eq. (1),

$$\begin{aligned}\rho \frac{dS}{dt} = &-\frac{1}{T}\operatorname{div} \boldsymbol{J}_q + \frac{1}{T}\sum_k \mu_k \operatorname{div} \boldsymbol{J}_k \\ &- \frac{1}{T}[(\Pi - p\overleftrightarrow{1}):\operatorname{Grad} \boldsymbol{v}] + \frac{1}{T}\sum_k \boldsymbol{F}_k \cdot \boldsymbol{J}_k \\ &- \frac{1}{T}\sum_{kr} \mu_k \Lambda_{kr} j_r.\end{aligned} \tag{5}$$

From eq. (4, 47),

$$\begin{aligned}&-\frac{1}{T}(\operatorname{div} \boldsymbol{J}_q) + \frac{1}{T}\sum_k \mu_k \operatorname{div} \boldsymbol{J}_k \\ &\quad = -\operatorname{div} \frac{1}{T}\left(\boldsymbol{J}_q - \sum_k \mu_k \boldsymbol{J}_k\right) + \boldsymbol{J}_q \cdot \operatorname{grad}\left(\frac{1}{T}\right) \\ &\quad - \sum_k \boldsymbol{J}_k \cdot \operatorname{grad}\left(\frac{\mu_k}{T}\right).\end{aligned} \tag{6}$$

If we define

$$\boldsymbol{J}_q - \sum_k \mu_k \boldsymbol{J}_k = T\boldsymbol{J}_S, \tag{7}$$

and

$$\sum_k \mu_k \Lambda_{kr} = A_r, \tag{8}$$

where $\boldsymbol{J}_s$ is the entropy flux and A_r is the affinity for the chemical reaction r, substituting eqs. (6—8) in eq. (5), we have

$$\rho \frac{dS}{dt} = -\text{div}\, \boldsymbol{J}_S + \sigma \tag{9}$$

and

$$\sigma = -\frac{\boldsymbol{J}_q \cdot \text{grad}\, T}{T^2} - \frac{1}{T}\left\{\sum_k \boldsymbol{J}_k \cdot \left[T \,\text{grad}\, \frac{\mu_k}{T} - \boldsymbol{F}_k\right]\right\}$$

$$- \frac{1}{T}[(\Pi - p\overleftrightarrow{1}):\text{Grad}\, \boldsymbol{v}] - \sum_r \frac{A_r j_r}{T}. \tag{10}$$

The physical meaning of eq. (9) is that the rate of change of entropy is due to flow of entropy from outside and production of entropy from within. The last term on the right of eq. (10) is essentially the ordinary product of the rate of chemical reaction and the affinity. The first term is the dot product of the rate of heat flow and the gradient of the reciprocal temperature. The second term is essentially the dot product of the rate of mass flow and the chemical potential gradient modified by the external mechanical force. The third term is essentially the double contracted product of the viscous pressure tensor and the velocity gradient. In short, the (negative) dissipation function is the sum of the inner products,

$$T\sigma = \sum_r [j_r | -A_r] + [\boldsymbol{J}_q | -d \ln T]$$

$$+ \sum_k \left[\boldsymbol{J}_k \middle| -\left\{T \,\text{grad}\left(\frac{\mu_k}{T}\right) - \boldsymbol{F}_k\right\}\right]$$

$$+ [(\Pi - p\overleftrightarrow{1}) | - \text{Grad}\, \boldsymbol{v}]. \tag{11}$$

One member of each inner product is the flux $\boldsymbol{J}$ and the other is the force $\boldsymbol{X}$. Because the inner product is symmetric, the separation of force and flux is arbitrary.

As mentioned in §6, $\boldsymbol{J}_q$ is not the only heat flux which we could use. We have defined the reduced heat flux as

$$\boldsymbol{J}_q^r = \boldsymbol{J}_q - \sum_k H_k \boldsymbol{J}_k. \tag{12}$$

Also we can split grad (μ_k/T) into an isothermal and a non-isothermal part:

$$\begin{aligned} \operatorname{grad}\left(\frac{\mu_k}{T}\right) &= \left[\partial\left(\frac{\mu_k}{T}\right) \Big/ \partial\left(\frac{1}{T}\right)\right] d\left(\frac{1}{T}\right) + \left[\frac{\operatorname{grad}\mu_k}{T}\right]_{T=\text{constant}} \\ &= H_k \operatorname{grad}\left(\frac{1}{T}\right) + \frac{\operatorname{grad}_T \mu_k}{T}. \end{aligned} \tag{13}$$

If we transform the old fluxes in eq. (11) into new ones,

$$\begin{bmatrix} \boldsymbol{J}'_k \\ \boldsymbol{J}_q^r \end{bmatrix} = \begin{bmatrix} 1 & 0 \\ -H_k & 1 \end{bmatrix} \begin{bmatrix} \boldsymbol{J}_k \\ \boldsymbol{J}_q \end{bmatrix}, \tag{14}$$

the old forces are, from eqs. (11) and (13),

$$\begin{bmatrix} \boldsymbol{X}_k \\ \boldsymbol{X}_q \end{bmatrix} = \begin{bmatrix} \left(\dfrac{\boldsymbol{F}_k}{T}\right) - H_k \operatorname{grad}\left(\dfrac{1}{T}\right) - \dfrac{(\operatorname{grad}_T \mu_k)}{T} \\ \operatorname{grad}\dfrac{1}{T} \end{bmatrix}. \tag{15}$$

To preserve the inner product, the new forces are

$$\begin{aligned} \begin{bmatrix} \boldsymbol{X}'_k \\ \boldsymbol{X}'_q \end{bmatrix} &= \begin{bmatrix} 1 & -H_k \\ 0 & 1 \end{bmatrix}^{-1} \begin{bmatrix} \boldsymbol{X}_k \\ \boldsymbol{X}_q \end{bmatrix} = \begin{bmatrix} 1 & H_k \\ 0 & 1 \end{bmatrix} \begin{bmatrix} \boldsymbol{X}_k \\ \boldsymbol{X}_q \end{bmatrix} \\ &= \begin{bmatrix} \left(\dfrac{\boldsymbol{F}_k}{T}\right) - \dfrac{\operatorname{grad}_T \mu_k}{T} \\ \operatorname{grad}\left(\dfrac{1}{T}\right) \end{bmatrix}. \end{aligned} \tag{16}$$

Instead of eq. (11), the dissipation function is

$$T\sigma = \sum_r [j_r | -A_r] + [\boldsymbol{J}_q^r | -d \ln T]$$

$$+\sum_k [\boldsymbol{J}_k | -(\mathrm{grad}_T\,\mu_k - \boldsymbol{F}_k)]$$

$$+[(\Pi - p\overleftrightarrow{1}) | -\mathrm{Grad}\,\boldsymbol{v}]. \tag{17}$$

The use of $\boldsymbol{J}_q'$ as heat flux simplifies the force conjugate to $\boldsymbol{J}_k$.

$\boldsymbol{J}_s$ is the historical entropy flow vector. In eqs. (11) and (17) we use it for convection. If we use it for heat flux too, from eq. (7), the transformation is

$$\begin{bmatrix} \boldsymbol{J}_k' \\ \boldsymbol{J}_S \end{bmatrix} = \begin{bmatrix} 1 & 0 \\ -\dfrac{\mu_k}{T} & \dfrac{1}{T} \end{bmatrix} \begin{bmatrix} \boldsymbol{J}_k \\ \boldsymbol{J}_q \end{bmatrix}. \tag{18}$$

The old forces are

$$\begin{bmatrix} \boldsymbol{X}_k \\ \\ \boldsymbol{X}_q \end{bmatrix} = \begin{bmatrix} \left(\dfrac{\boldsymbol{F}_k}{T}\right) - \mathrm{grad}\left(\dfrac{\mu_k}{T}\right) \\ \\ \mathrm{grad}\left(\dfrac{1}{T}\right) \end{bmatrix}. \tag{19}$$

To preserve the inner product, the new forces are

$$\begin{bmatrix} \boldsymbol{X}_k' \\ \\ \boldsymbol{X}_q' \end{bmatrix} = \begin{bmatrix} 1 & -\dfrac{\mu_k}{T} \\ \\ 0 & \dfrac{1}{T} \end{bmatrix}^{-1} \begin{bmatrix} \boldsymbol{X}_k \\ \\ \boldsymbol{X}_q \end{bmatrix} = \begin{bmatrix} 1 & \mu_k \\ \\ 0 & T \end{bmatrix} \begin{bmatrix} \boldsymbol{X}_k \\ \\ \boldsymbol{X}_q \end{bmatrix}$$

$$= \begin{bmatrix} \left(\dfrac{\boldsymbol{F}_k}{T}\right) - \dfrac{(\mathrm{grad}\,\mu_k)}{T} \\ \\ T\,\mathrm{grad}\left(\dfrac{1}{T}\right) \end{bmatrix}. \tag{20}$$

Instead of eq. (11), the dissipation function is

$$T\sigma = \sum_r [j_r | -A_r] + [\boldsymbol{J}_S | -\mathrm{grad}\,T]$$

$$+\sum_k [\boldsymbol{J}_k | -(\mathrm{grad}\,\mu_k - \boldsymbol{F}_k)]$$

$$+[(\Pi - p\overleftrightarrow{1}) | -\mathrm{grad}\,\boldsymbol{v}]. \tag{21}$$

The use of $\boldsymbol{J}_s$ as heat flux simplifies both forces conjugate to heat flow and mass flow. If $\boldsymbol{F}_k$ is the negative gradient of a scalar potential ψ_k,

$$\text{grad}\,\mu_k - \boldsymbol{F}_k = \text{grad}\,(\mu_k + \psi_k) = \text{grad}\,\tilde{\mu}_k, \tag{22}$$

where $\tilde{\mu}_k$ is the generalized chemical potential. Thus the use of the generalized chemical potential simplifies the writing of the force conjugate to the mass flow.

We collect now the equations for the rate of entropy-production derived before:

$$\begin{aligned}\sigma = \sum_r \frac{1}{T}[j_r| - A_r] &+ \left[\boldsymbol{J}_q \middle| - \frac{\text{grad}\,T}{T^2}\right] \\ &+ \sum_k \frac{1}{T}\left[\boldsymbol{J}_k \middle| - \left\{T\,\text{grad}\left(\frac{\mu_k}{T}\right) - \boldsymbol{F}_k\right\}\right] \\ &+ \frac{1}{T}[(\Pi - p\overleftrightarrow{1})| - \text{Grad}\,\boldsymbol{v}],\end{aligned} \tag{11}$$

$$\begin{aligned}\sigma = \sum_r \frac{1}{T}[j_r| - A_r] &+ \left[\boldsymbol{J}_q^r \middle| \text{grad}\left(\frac{1}{T}\right)\right] \\ &+ \sum_k \frac{1}{T}[\boldsymbol{J}_k| - (\text{grad}_T\,\mu_k - \boldsymbol{F}_k)] \\ &+ \frac{1}{T}[(\Pi - p\overleftrightarrow{1}) - \text{Grad}\,\boldsymbol{v}],\end{aligned} \tag{17}$$

$$\begin{aligned}\sigma = \sum_r \frac{1}{T}[j_r| - A_r] &+ \frac{1}{T}[\boldsymbol{J}_s| - \text{grad}\,T] \\ &+ \sum_k \frac{1}{T}[\boldsymbol{J}_k| - (\text{grad}\,\mu_k - \boldsymbol{F}_k)] \\ &+ \frac{1}{T}[(\Pi - p\overleftrightarrow{1})| - \text{Grad}\,\boldsymbol{v}].\end{aligned} \tag{21}$$

§9 *TRANSPORT EQUATIONS*

At the end of §4, §6 and §8, we listed the transport

equations of mass, energy and entropy. In this section we shall derive the transport equation of temperature or pressure. We hasten to add that, although sometimes it is convenient to use these equations, nothing new is added physically to the important transport equation of entropy, eq. (8, 9).

By definition, $$\frac{dH}{dt} = \frac{dU}{dt} + \frac{pdV}{dt} + \frac{Vdp}{dt}. \tag{1}$$

By considering H as a function of pressure, temperature and concentration,

$$\frac{dH}{dt} = \left(\frac{\partial H}{\partial T}\right)_{pc_k} \frac{dT}{dt} + \left(\frac{\partial H}{\partial p}\right)_{Tc_k} \frac{dp}{dt} + \sum_k \left(\frac{\partial H}{\partial c_k}\right)_{Tp} \frac{dc_k}{dt}. \tag{2}$$

Because of the thermodynamical identity,

$$\left(\frac{\partial H}{\partial p}\right)_{Tc_k} = T\left(\frac{\partial S}{\partial p}\right)_{Tc_k} + V = -T\left(\frac{\partial V}{\partial T}\right)_{pc_k} + V = -T\alpha V + V = V(1 - T\alpha), \tag{3}$$

where α is the coefficient of thermal expansion at constant pressure, eq. (2) becomes

$$\frac{dH}{dt} = C_p \frac{dT}{dt} + V(1 - T\alpha)\frac{dp}{dt} + \sum_k H_k \frac{dc_k}{dt}, \tag{4}$$

where H_k is the partial enthalpy of the component k (at constant temperature and pressure). Combining eqs. (1) and (4),

$$\rho C_p \frac{dT}{dt} - \alpha T \frac{dp}{dt} = \rho \frac{dU}{dt} + \rho p \frac{dV}{dt} - \sum_k H_k \rho \frac{dc_k}{dt}. \tag{5}$$

Substituting eqs. (6, 20), (4, 52) and (4, 53) in eq. (5),

$$\rho C_p \frac{dT}{dt} - \alpha T \frac{dp}{dt} = -\operatorname{div} \boldsymbol{J}_q^r - (\Pi - p\overset{\leftrightarrow}{1}) : \operatorname{Grad} \boldsymbol{v}$$
$$- \sum_k \boldsymbol{J}_k \cdot (\operatorname{grad} H_k - \boldsymbol{F}_k) - \sum_r j_r \sum_k \Lambda_{kr} H_k, \quad (6)$$

where the last term is the decrease of enthalpy of all chemical reactions.

To separate the transport equation of temperature from pressure, we consider now the internal energy as a function of volume, temperature and concentration,

$$\frac{dU}{dt} = C_V \frac{dT}{dt} + \left(\frac{\partial U}{\partial V}\right)_{Tc_k} \frac{dV}{dt}$$
$$+ \sum_k \left(\frac{\partial U}{\partial c_k}\right)_{TV} \frac{dc_k}{dt}. \quad (7)$$

The following are thermodynamical identities:

$$\left(\frac{\partial U}{\partial V}\right)_{Tc_k} = T\left(\frac{\partial S}{\partial V}\right)_{Tc_k} - p = T\left(\frac{\partial p}{\partial T}\right)_{Vc_k} - p$$
$$= -T\left(\frac{\partial p}{\partial V}\right)_{Tc_k}\left(\frac{\partial V}{\partial T}\right)_{pc_k} - p = \frac{\alpha T}{\kappa} - p, \quad (8)$$

where κ is the coefficient of isothermal compressibility, and

$$\left(\frac{\partial U}{\partial c_k}\right)_{VT} = \left(\frac{\partial U}{\partial c_k}\right)_{pT} + \left(\frac{\partial U}{\partial p}\right)_{c_k T}\left(\frac{\partial p}{\partial c_k}\right)_{VT}$$
$$= U_k - \left[\left(\frac{\partial U}{\partial V}\right)_{Tc_k}\left(\frac{\partial V}{\partial p}\right)_{Tc_k}\right]\left[\left(\frac{\partial p}{\partial V}\right)_{Tc_k}\left(\frac{\partial V}{\partial c_k}\right)_{Tp}\right]$$
$$= U_k - \left(\frac{\alpha T}{\kappa} - p\right) V_k = H_k - \left(\frac{\alpha T V_k}{\kappa}\right). \quad (9)$$

Substituting eqs. (8) and (9) in eq. (7),

$$\rho \frac{dU}{dt} = \rho C_V \frac{dT}{dt} + \left(\frac{\alpha T}{\kappa} - p\right) \rho \frac{dV}{dt}$$
$$+ \sum_k \left(H_k - \frac{\alpha T V_k}{\kappa}\right) \rho \frac{dc_k}{dt}. \quad (10)$$

Substituting eqs. (6, 20), (4, 52) and (4, 53) in eq. (10),

$$\rho\kappa C_V \frac{dT}{dt} = -\kappa \operatorname{div} \boldsymbol{J}_q^r - \kappa(\Pi - p\overset{\leftrightarrow}{1}):\operatorname{Grad} \boldsymbol{v} - \alpha T \operatorname{div} \boldsymbol{v}$$

$$- \alpha T \sum_k V_k \operatorname{div} \boldsymbol{J}_k - \kappa \sum_k \boldsymbol{J}_k \cdot (\operatorname{grad} H_k - \boldsymbol{F}_k)$$

$$- \sum_r j_r \left(\kappa \sum_k H_k \Lambda_{kr} - \alpha T \sum_k V_k \Lambda_{kr} \right), \tag{11}$$

where the summation of $-j_r V_k \Lambda_{kr}$ is the decrease of volume for all chemical reactions.

Multiplying eq. (11) by $C_p/\kappa C_V$, subtracting it from eq. (6), and multiplying the equation of the difference by $\rho\kappa C_V/\alpha T$,

$$\rho\kappa C_V \frac{dp}{dt} = -\alpha \operatorname{div} \boldsymbol{J}_q^r - \alpha(\Pi - p\overset{\leftrightarrow}{1}):\operatorname{Grad} \boldsymbol{v}$$

$$- \alpha \sum_k \boldsymbol{J}_k \cdot (\operatorname{grad} H_k - \boldsymbol{F}_k) - \rho C_p \operatorname{div} \boldsymbol{v}$$

$$- \rho C_p \sum_k V_k \operatorname{div} \boldsymbol{J}_k - \sum_r j_r \left(\alpha \sum_k H_k \Lambda_{kr} \right.$$

$$\left. - \rho C_p \sum_k V_k \Lambda_{kr} \right), \tag{12}$$

eq. (3, 7) having been used.

CHAPTER IV
EXTENSION TO DIFFERENT FIELDS AND DIRECTIONS

§ 10 *MECHANICAL EQUILIBRIUM*

At mechanical equilibrium the barycentric velocity is constant. From eq. (5, 23),

$$dp = \sum_k \rho_k F_k . \tag{1}$$

The physical meaning of eq. (1) is that at mechanical equilibrium the mechanical force is balanced by the non-uniform pressure distribution. For example, the vertical equilibrium of the atmosphere is preserved by a uniform decrease of pressure, which is so adjusted that the pressure gradient is everywhere the same as the gravitational force per unit volume. Thus for any fluid at mechanical equilibrium, the pressure is not uniform: the only reason why the non-uniform pressure can be disregarded is that, unless the fluid is very dense or of great depth, the variation of the pressure within the fluid is small compared with the absolute value.

The importance of mechanical equilibrium is due to the fact that the mechanical equilibrium is established long before thermodynamic equilibrium.

By using

$$\mu = \sum_k \mu_k c_k , \tag{2}$$

and the Gibbs equation

$$dU = TdS - pdV + \sum_k \mu_k dc_k, \tag{3}$$

the total differential of

$$U = \mu + TS - pV \tag{4}$$

yields the Gibbs-Duhem relation

$$-SdT + Vdp - \sum_k c_k d\mu_k = 0. \tag{5}$$

Or

$$dp = \sum_k \rho_k d\mu_k + \sum_k \rho_k S_k dT. \tag{6}$$

Because the chemical potential can be split into an isothermal and a non-isothermal part,

$$d\mu_k = \frac{\partial \mu_k}{\partial T} dT + d\mu_{k,T=\text{constant}} = -S_k dT + d_T \mu_k. \tag{7}$$

From eqs. (6) and (7), eq. (1) is reduced to

$$\sum_k \rho_k (d_T \mu_k - \boldsymbol{F}_k) = 0. \tag{8}$$

Prigogine[1] proposed the following theorem: At mechanical equilibrium, the choice of reference velocity in the rate of entropy-production is immaterial. The proof is simple. In eq. (8,17) we use the barycentric velocity as the reference velocity. If we wish to change it to $\boldsymbol{v}^b$, from eq. (8),

$$\sum_k [\boldsymbol{J}_k | (\boldsymbol{F}_k - \text{grad}_T \mu_k)] - \sum_k [\boldsymbol{J}_k^b | (\boldsymbol{F}_k - \text{grad}_T \mu_k)]$$

$$= \sum_k [(\boldsymbol{v}^b - \boldsymbol{v}) \rho_k | (\boldsymbol{F}_k - \text{grad}_T \mu_k)] = 0.$$

Therefore

$$\sum_k [\boldsymbol{J}_k | (\boldsymbol{F}_k - \text{grad}_T \mu_k)] = \sum_k [\boldsymbol{J}_k^b | (\boldsymbol{F}_k - \text{grad}_T \mu_k)]. \tag{9}$$

In eq. (8, 21) with heat conduction and diffusion only, from eqs. (7) and (8),

$$T\sigma = -[\boldsymbol{J}_s|\text{grad } T] - \sum_k [\boldsymbol{J}_k|(\text{grad } \mu_k - \boldsymbol{F}_k)]$$

$$= -[\boldsymbol{J}_s + \rho S\boldsymbol{v}|\text{grad } T] - \sum_k [\rho_k \boldsymbol{v}_k|(\text{grad } \mu_k - \boldsymbol{F}_k)]$$

$$= -[\boldsymbol{J}_{S,\text{tot}}|\text{grad } T] - \sum_k [\rho_k \boldsymbol{v}_k|(\text{grad } \mu_k - \boldsymbol{F}_k)], \quad (10)$$

where $\boldsymbol{J}_{S,\text{tot}}$ is the total entropy flux,

$$\boldsymbol{J}_S + \rho S\boldsymbol{v} = \boldsymbol{J}_{S,\text{tot}}. \quad (11)$$

Equation (10) has application in thermoelectricity (§49).

Reference

[1] I. Prigogine: Étude Thermodynamique des Phénomènes irréversibles, Desoer, Paris-Liège (1947).

§11 ELECTRODYNAMICS

Maxwell's Equations

The basic Maxwell's equations of electric and magnetic fields are

$$\text{curl } \boldsymbol{E} = -\frac{\partial \boldsymbol{B}}{\partial t}, \quad (1)$$

$$\text{curl } \boldsymbol{H} = \boldsymbol{I} + \frac{\partial \boldsymbol{D}}{\partial t}, \quad (2)$$

$$\text{div } \boldsymbol{B} = 0, \quad (3)$$

$$\text{div } \boldsymbol{D} = \rho Z. \quad (4)$$

From eq. (3) we can define $\boldsymbol{B}$ in terms of a vector potential $\boldsymbol{A}$,

$$\boldsymbol{B} = \text{curl } \boldsymbol{A}. \quad (5)$$

Then eq. (1) can be written as

$$\operatorname{curl}\left(\boldsymbol{E}+\frac{\partial \boldsymbol{A}}{\partial t}\right)=0. \tag{6}$$

This means that the quantity with vanishing curl in eq. (6) can be written as the gradient of some scalar function, namely, a scalar potential ζ,

$$\boldsymbol{E}+\frac{\partial \boldsymbol{A}}{\partial t}=-\operatorname{grad} \zeta. \tag{7}$$

The meaning of symbols is listed in Table 1. We have used the rationalized International System of Units (often abbreviated as SI, for Système International d'Unités). As is seen from the Table, the four basic quantities are E, l, t, and q.

There are many systems of units in electricity and magnetism and some of their physical constants differ in magnitude and dimensions. Heaviside devised a system of units so constructed that 4π appears in formulae involving a geometrical situation of spherical symmetry, 2π in formulae involving circular symmetry and 1 in formulae involving rectangular symmetry. A system of units yielding such formulae is said to be rationalized.

The Coulomb law, the mechanical force F between two charges q at a distance r, is written as

$$F_1=\frac{q_1 q_2}{k_1 r^2}. \tag{8}$$

For an electrostatic system (not rationalized) k_1 is unity and has dimensions of $(\text{statcoulomb})^2 \cdot \text{erg}^{-1} \cdot \text{cm}^{-1}$. In a rationalized system

$$k_1=4\pi\varepsilon_0.$$

In the rationalized SI system

$$\varepsilon_0=\frac{1}{4\pi}(\text{statcoulomb})^2 \cdot \text{erg}^{-1} \cdot \text{cm}^{-1}=\frac{10^7}{4\pi c_0^2}\, \text{C}^2 \cdot \text{J}^{-1} \cdot \text{m}^{-1},$$

Table 1

Units, Symbols and Dimensions of Physical Quantities in Electricity and Magnetism

Physical Quantities	Unit	Symbol	Dimensions
work, energy	joule = N·M	W, E	E
length	meter	l	l
time	second	t	t
charge	coulomb = A·sec	q	q
mass	kilogram	m	$El^{-2}t^{2}$
force	newton = M·kg·sec^{-2}	F	El^{-1}
power	watt = J·sec^{-1}	P	Et^{-1}
capacitance	farad = C·V^{-1}	C	$E^{-1}q^{2}$
charge density	C·M^{-3}	ρZ	$l^{-3}q$
charge per unit mass	C·kg^{-1}	Z	$E^{-1}l^{2}t^{-2}q$
current	ampere = C·sec^{-1}	i	$t^{-1}q$
current density (charge flux)	A·M^{-2}	$\boldsymbol{J}_{el}, \boldsymbol{I}$	$l^{-2}t^{-1}q$
electrical displacement	C·M^{-2}	$\boldsymbol{D}$	$l^{-2}q$
electrical field	V·M^{-1}	$\boldsymbol{E}$	$El^{-1}q^{-1}$
electrical polarization	C·M^{-2}	$\mathscr{P}$	$l^{-2}q$
inductance	henry = Wb·A^{-1}	L	$Et^{2}q^{-2}$
magnetic field	ampere-turn·M^{-1}	$\boldsymbol{H}$	$l^{-1}t^{-1}q$
magnetic flux	weber = V·sec	ϕ	Etq^{-1}
magnetic induction	tesla = Wb·M^{-2}	$\boldsymbol{B}$	$El^{-2}tq^{-1}$
magnetic polarization	Wb·M^{-2}	$\mathscr{J}$	$El^{-2}tq^{-1}$
magnetic vector potential	V·sec·M^{-1}	$\boldsymbol{A}$	$El^{-1}tq^{-1}$
permeability of vacuum	H·M^{-1}	μ_0	$El^{-1}t^{2}q^{-2}$
permeability	H·M^{-1}	μ	$El^{-1}t^{2}q^{-2}$
permittivity of vacuum	F·M^{-1}	ε_0	$E^{-1}l^{-1}q^{2}$
permittivity	F·M^{-1}	ε	$E^{-1}l^{-1}q^{2}$
potential	volt = W·A^{-1}	V, ζ	Eq^{-1}
relative permeability (magnetic constant)		μ/μ_0	1
relative permittivity (dielectric constant)		$\varepsilon/\varepsilon_0$	1
resistance	ohm = V·A^{-1}	R	Etq^{-2}
resistivity	ohm·M	r	$Eltq^{-2}$

where $c_0 \approx 3 \times 10^8$, or the known value of speed of light in vacuo in m·sec^{-1}.

According to Ampere's law, the mechanical force per unit length between two infinitely long wires carrying current i at a distance d is

$$\frac{F_2}{l} = \frac{k_2 i_1 i_2}{d}. \tag{9}$$

For an electromagnetic system (not rationalized), k_2 is unity and has dimensions of $\mathrm{erg \cdot cm^{-1} \cdot (abampere)^{-2}}$. In a rationalized system $k_2 = \mu_0/4\pi$. In the rationalized SI system

$$\mu_0 = 4\pi \,\mathrm{erg \cdot cm^{-1} \cdot (abampere)^{-2}} = 4\pi \times 10^{-7} \mathrm{J \cdot m^{-1} \cdot A^{-2}}$$

From the wave equation, in any system,

$$\varepsilon_0 \mu_0 = \frac{1}{c_0^2}. \tag{10}$$

Constitutive Relations

In free space,

$$\boldsymbol{D} = \varepsilon_0 \cdot \boldsymbol{E}, \tag{11}$$

and

$$\boldsymbol{B} = \mu_0 \cdot \boldsymbol{H}, \tag{12}$$

where ε_0 and μ_0 are universal constants. If we restrict ourselves to an isotropic material medium,

$$\boldsymbol{D} = \varepsilon_0 \boldsymbol{E} + \mathscr{P} (= \varepsilon \boldsymbol{E}), \tag{13}$$

and

$$\boldsymbol{B} = \mu_0 \boldsymbol{H} + \mathscr{I} (= \mu \boldsymbol{H}). \tag{14}$$

The relative permittivity (dielectric constant) is defined as the dimensionless ratio $\varepsilon/\varepsilon_0$. The relative permeability (magnetic constant) is defined as μ/μ_0. We further define electric susceptibility κ and magnetic susceptibility χ:

$$\kappa = \mathscr{P}/\boldsymbol{E}, \tag{15}$$

and

$$\chi = \mathscr{I}/\boldsymbol{H}^*. \tag{16}$$

---

* Some authors define

$$\boldsymbol{H} = \boldsymbol{B}/\mu_0 - \mathscr{I}'$$

and

$$\chi' = \mathscr{I}'/\boldsymbol{H}.$$

Then

$$\mathscr{I}' = \mathscr{I}/\mu_0 \text{ and } \chi' = \chi/\mu_0.$$

Then $\kappa = \varepsilon - \varepsilon_0,$ (17)

$$\chi = \mu - \mu_0. \tag{18}$$

For a real material medium, the coefficients are second-rank tensors and the functional relationship is not necessarily linear or constant.

Conservation of Charge

Taking divergence of eq. (2), we have

$$\frac{\partial \rho Z}{\partial t} = -\text{div}\, \boldsymbol{I}. \tag{19}$$

This is the conservation of charge (continuity). From eq-(4, 11), eq. (19) becomes

$$\rho \frac{dZ}{dt} = -\text{div}\,(\boldsymbol{I} - \rho Z \boldsymbol{v}). \tag{20}$$

From eq. (4, 52),

$$\rho \frac{dc_k}{dt} = -\text{div}\, \boldsymbol{J}_k. \tag{21}$$

Because of

$$\sum_k c_k Z_k = Z, \tag{22}$$

and

$$\sum_k Z_k \boldsymbol{J}_k = \boldsymbol{J}_{\text{el}}, \tag{23}$$

multiplying eq. (21) by Z_k and summing in k, we have

$$\rho \frac{dZ}{dt} = -\text{div}\, \boldsymbol{J}_{\text{el}}. \tag{24}$$

By comparing eqs. (20) and (24),

$$\boldsymbol{J}_{\text{el}} = \boldsymbol{I} - \rho Z \boldsymbol{v}. \tag{25}$$

Thus the conductive part of $\boldsymbol{J}_{\text{el}}$ is $\boldsymbol{I}$ and the convective part is

$\rho Z\boldsymbol{v}$ (For conductors, because of electroneutrality, there is no difference between $\boldsymbol{J}_{el}$ and $\boldsymbol{I}$). From another point of view, $\boldsymbol{J}_{el}$ is the current flux in the moving frame of reference and $\boldsymbol{I}$ is that in the frame at rest. Equation (25) shows that the Galilean transformation is invariant, correct to the order of v^2/c_0^2.

Balance of Momentum

The Lorentz force (dimensions El^{-4}) is $\rho Z\boldsymbol{E} + \boldsymbol{I} \times \boldsymbol{B}$ Using Maxwell's equations,

$$\rho Z\boldsymbol{E} + \boldsymbol{I} \times \boldsymbol{B} = \boldsymbol{E}\,(\operatorname{div} \boldsymbol{D}) + (\operatorname{curl} \boldsymbol{H}) \times \boldsymbol{B} + \frac{\partial(\boldsymbol{B} \times \boldsymbol{D})}{\partial t} + (\operatorname{curl} \boldsymbol{E}) \times \boldsymbol{D} + \boldsymbol{H}(\operatorname{div} \boldsymbol{B}). \qquad (26)$$

Because of identities

$$\operatorname{Div} \boldsymbol{ab} = (\boldsymbol{a} \cdot \operatorname{grad})\,\boldsymbol{b} + \boldsymbol{b}\,(\operatorname{div} \boldsymbol{a}), \qquad (27)$$

$$\operatorname{Div}\,(\boldsymbol{a} \cdot \boldsymbol{b})\overleftrightarrow{1} = \operatorname{grad}\,(\boldsymbol{a} \cdot \boldsymbol{b}) = \boldsymbol{a} \cdot \operatorname{Grad} \boldsymbol{b} + \boldsymbol{b} \cdot \operatorname{Grad} \boldsymbol{a} = (\boldsymbol{a} \cdot \operatorname{grad})\,\boldsymbol{b} + \boldsymbol{a} \times (\operatorname{curl} \boldsymbol{b}) + (\boldsymbol{b} \cdot \operatorname{grad})\,\boldsymbol{a} + \boldsymbol{b} \times (\operatorname{curl} \boldsymbol{a}), \qquad (28)$$

$$\operatorname{Div}\,(\boldsymbol{ab} - \boldsymbol{a} \cdot \boldsymbol{b}\overleftrightarrow{1}) = \boldsymbol{b}\,(\operatorname{div} \boldsymbol{a}) + (\boldsymbol{a} \cdot \operatorname{grad})\,\boldsymbol{b} - \operatorname{Grad}\,(\boldsymbol{a} \cdot \boldsymbol{b}), \qquad (29)$$

$$\rho Z\boldsymbol{E} + \boldsymbol{I} \times \boldsymbol{B} = \operatorname{Div}\,(\boldsymbol{DE} + \boldsymbol{BH} - \boldsymbol{D} \cdot \boldsymbol{E}\overleftrightarrow{1} - \boldsymbol{B} \cdot \boldsymbol{H}\overleftrightarrow{1}) + \frac{\partial(\boldsymbol{B} \times \boldsymbol{D})}{\partial t} + \boldsymbol{E} \cdot \operatorname{Grad} \boldsymbol{D} + \boldsymbol{H} \cdot \operatorname{Grad} \boldsymbol{B}. \qquad (30)$$

There are many expressions for the ponderomotive force (dimensions El^{-4}). The simplest one is $\mathscr{P} \cdot \operatorname{Grad} E + \mathscr{I} \cdot \operatorname{Grad} \boldsymbol{H}$.

$$\mathscr{P} \cdot \operatorname{Grad} \boldsymbol{E} + \mathscr{I} \cdot \operatorname{Grad} \boldsymbol{H} = \operatorname{Div}\,(\mathscr{P} \cdot \boldsymbol{E}\overleftrightarrow{1}) + \operatorname{Div}\,(\mathscr{I} \cdot \boldsymbol{H}\overleftrightarrow{1}) - [\operatorname{Grad}\,(\boldsymbol{D} - \varepsilon_0\boldsymbol{E})] \cdot \boldsymbol{E} - [\operatorname{Grad}\,(\boldsymbol{B} - \mu_0\boldsymbol{H})] \cdot \boldsymbol{H}$$

$$= \operatorname{Div}\left(\mathscr{P} \cdot \boldsymbol{E} + \mathscr{I} \cdot \boldsymbol{H} + \frac{1}{2}\,\varepsilon_0 E^2 + \frac{1}{2}\,\mu_0 H^2\right)\overleftrightarrow{1}$$

$$- \boldsymbol{E} \cdot \text{Grad}\, \boldsymbol{D} - \boldsymbol{H} \cdot \text{Grad}\, \boldsymbol{B}, \tag{31}$$

where eqs. (13) and (14) have been used. Hence

the sum of the Lorentz and the ponderomotive force $= \rho \boldsymbol{F}'$

$$= \text{Div}\left(\boldsymbol{DE} + \boldsymbol{BH} - \frac{1}{2}\, \varepsilon_0 E^2 \overleftrightarrow{1} - \frac{1}{2}\, \mu_0 H^2 \overleftrightarrow{1} \right)$$

$$- \frac{\partial(\boldsymbol{D} \times \boldsymbol{B})}{\partial t}$$

$$= \text{Div}\, T - \frac{\partial(\boldsymbol{D} \times \boldsymbol{B})}{\partial t}, \tag{32}$$

where

$$T = \boldsymbol{DE} + \boldsymbol{BH} - \frac{1}{2}\, \varepsilon_0 E^2 \overleftrightarrow{1} - \frac{1}{2}\, \mu_0 H^2 \overleftrightarrow{1}$$

is called modified Maxwell's stress tensor (dimensions El^{-3}) and $\boldsymbol{D} \times \boldsymbol{B}$ is called electromagnetic momentum vector (dimensions $El^{-4}t$). These two terms can be understood as follows:

$$\frac{\partial(\boldsymbol{D} \times \boldsymbol{B})}{\partial t} = \text{Div}\, T - \rho \boldsymbol{F}', \tag{33}$$

$$\frac{\partial \rho \boldsymbol{v}}{\partial t} = -\text{Div}\, \Pi - \text{Div}\, \rho \boldsymbol{vv} + \rho \boldsymbol{F}' + \rho \boldsymbol{F}, \tag{34}$$

where $\boldsymbol{F}$ is the force other than the Lorentz and the ponderomotive force. We see that $\boldsymbol{D} \times \boldsymbol{B}$ is equivalent to the momentum vector $\rho \boldsymbol{v}$ and T is equivalent to the negative pressure tensor. Adding eqs. (33) and (34),

$$\frac{\partial \rho \boldsymbol{v}}{\partial t} + \frac{\partial(\boldsymbol{D} \times \boldsymbol{B})}{\partial t} = -\text{Div}\, (\Pi - T) - \text{Div}\, \rho \boldsymbol{vv} + \rho \boldsymbol{F}. \tag{35}$$

Thus the total momentum is not invariant. This is not surprising as momentum can be changed into force.

If $\mathscr{P} = \mathscr{I} = 0$, the Maxwell stress tensor is

$$\varepsilon_0 \boldsymbol{EE} + \mu_0 \boldsymbol{HH} - \frac{1}{2}\, \varepsilon_0 E^2 \overleftrightarrow{1} - \frac{1}{2}\, \mu_0 H^2 \overleftrightarrow{1},$$

and $\boldsymbol{E} \times \boldsymbol{H}$ is Poynting's vector, the energy flux (dimensions $El^{-2}t^{-1}$).

Conservation of Energy

Using one of Maxwell's equations,

$$\boldsymbol{I} \cdot \boldsymbol{E} = (\text{curl } \boldsymbol{H}) \cdot \boldsymbol{E} - \frac{\partial \boldsymbol{D}}{\partial t} \cdot \boldsymbol{E}. \tag{36}$$

From the identity

$$\text{div}\,(\boldsymbol{E} \times \boldsymbol{H}) = \boldsymbol{H} \cdot \text{curl } \boldsymbol{E} - \boldsymbol{E} \cdot \text{curl } \boldsymbol{H} \tag{37}$$

and the use of eqs. (13—14), eq. (36) becomes

$$\begin{aligned}\boldsymbol{I} \cdot \boldsymbol{E} &= -\text{div}\,(\boldsymbol{E} \times \boldsymbol{H}) - \boldsymbol{H} \cdot \frac{\partial \boldsymbol{B}}{\partial t} - \boldsymbol{E} \cdot \frac{\partial \boldsymbol{D}}{\partial t} \\ &= -\text{div}\,(\boldsymbol{E} \times \boldsymbol{H}) - \mu_0 \boldsymbol{H} \cdot \frac{\partial \boldsymbol{H}}{\partial t} - \boldsymbol{H} \cdot \frac{\partial \mathscr{I}}{\partial t} \\ &\quad - \varepsilon_0 \boldsymbol{E} \cdot \frac{\partial \boldsymbol{E}}{\partial t} - \boldsymbol{E} \cdot \frac{\partial \mathscr{P}}{\partial t}.\end{aligned} \tag{38}$$

If we assume that the electromagnetic energy density (dimensions El^{-3}) is

$$E_{\text{el}} = \frac{1}{2}\,(\mu_0 H^2 + \varepsilon_0 E^2), \tag{39}$$

$$\begin{aligned}\frac{\partial E_{\text{el}}}{\partial t} &= -\text{div}\,(\boldsymbol{E} \times \boldsymbol{H}) - \boldsymbol{I} \cdot \boldsymbol{E} - \boldsymbol{H} \cdot \frac{\partial \mathscr{I}}{\partial t} \\ &\quad - \boldsymbol{E} \cdot \frac{\partial \mathscr{P}}{\partial t}.\end{aligned} \tag{40}$$

From eq. (6, 20),

$$\begin{aligned}\frac{\partial \rho U}{\partial t} &= -\,\text{div}\,\boldsymbol{J}_q - \text{div}\,\rho U \boldsymbol{v} - \Pi\!:\!\text{Grad}\,\boldsymbol{v} + \sum_k \boldsymbol{F}_k \cdot \boldsymbol{J}_k \\ &\quad + \sum_k \rho_k \boldsymbol{F}'_k \cdot \boldsymbol{v}_k - \rho \boldsymbol{F}' \cdot \boldsymbol{v}.\end{aligned} \tag{41}$$

From eq. (6, 4),

$$\partial \frac{1}{2} \frac{\rho v^2}{\partial t} = -\mathrm{div}\, \Pi \cdot \boldsymbol{v} - \mathrm{div} \frac{1}{2} \rho v^2 \boldsymbol{v} + \Pi : \mathrm{Grad}\, \boldsymbol{v}$$

$$+ \sum_k \boldsymbol{F}_k \rho_k \cdot \boldsymbol{v} + \rho \boldsymbol{F}' \cdot \boldsymbol{v}. \qquad (42)$$

From eq. (6, 7),

$$\frac{\partial \rho \phi}{\partial t} = -\,\mathrm{div}\, \rho \phi \boldsymbol{v} - \mathrm{div} \sum_k \phi_k \boldsymbol{J}_k - \sum_k \boldsymbol{F}_k \rho_k \cdot \boldsymbol{v}_k. \qquad (43)$$

It is noted that we have included the Lorentz and the ponderomotive force in eqs. (41) and (42) but not in eq. (43). This is because they cannot be derived from potentials. If we define the total energy density

$$\rho E' = \frac{1}{2} \rho v^2 + \rho \phi + \rho U + E_{\mathrm{el}} = \rho E + E_{\mathrm{el}}, \qquad (44)$$

the sum of eqs. (40—43) yields

$$\frac{\partial \rho E'}{\partial t} = -\mathrm{div} \Big(\rho U \boldsymbol{v} + \frac{1}{2} \rho v^2 \boldsymbol{v} + \rho \phi \boldsymbol{v} + \boldsymbol{J}_q + \Pi \cdot \boldsymbol{v}$$

$$+ \sum_k \phi_k \boldsymbol{J}_k + \boldsymbol{E} \times \boldsymbol{H} \Big) - \boldsymbol{E} \cdot \boldsymbol{I} - \boldsymbol{E} \cdot \frac{\partial \mathscr{P}}{\partial t}$$

$$- \boldsymbol{H} \cdot \frac{\partial \mathscr{I}}{\partial t} + \sum_k \boldsymbol{F}'_k \cdot \rho_k \boldsymbol{v}_k. \qquad (45)$$

But the total energy should be invariant:

$$\sum_k \boldsymbol{F}'_k \cdot \rho_k \boldsymbol{v}_k = \boldsymbol{E} \cdot \boldsymbol{I} + \boldsymbol{E} \cdot \frac{\partial \mathscr{P}}{\partial t} + \boldsymbol{H} \cdot \frac{\partial \mathscr{I}}{\partial t}. \qquad (46)$$

We define

$$\rho H' = \rho H + P : \overleftrightarrow{1} - (\boldsymbol{E} \cdot \mathscr{P} + \boldsymbol{H} \cdot \mathscr{I}), \qquad (47)$$

and

$$\sum_k \rho_k H'_k \boldsymbol{v}_k = \sum_k \rho_k H_k \boldsymbol{v}_k + (P : \overleftrightarrow{1}) \boldsymbol{v}. \qquad (48)$$

From eqs. (47) and (48),

$$\sum_k H'_k \boldsymbol{J}_k = -\boldsymbol{J}^r_q + \boldsymbol{J}_q + (\boldsymbol{E} \cdot \mathscr{P} + \boldsymbol{H} \cdot \mathscr{I}) \boldsymbol{v}. \qquad (49)$$

Equations (46—48) are consistent with the fact that the Lorentz, the ponderomotive force and the electromagnetic energy are not conservative. The irreversible parts are reflected in the appearance of $\boldsymbol{I}$ and P. On the whole there is arbitrariness in splitting the stress tensor into viscous pressure and hydrostatic pressure and the total energy into electromagnetic and non-electromagnetic energy.

Substituting eqs. (46) and (49) in eq. (45),

$$\frac{\partial \rho E'}{\partial t} = -\text{div}\left[\frac{1}{2}\rho v^2 \boldsymbol{v} + \rho U \boldsymbol{v} + \sum_k \rho_k \psi_k \boldsymbol{v}_k + \boldsymbol{J}'_q + \sum_k H'_k \boldsymbol{J}_k - (\boldsymbol{E}\cdot\boldsymbol{\mathscr{P}} + \boldsymbol{H}\cdot\boldsymbol{\mathscr{I}})\boldsymbol{v} + \Pi\cdot\boldsymbol{v} + (\boldsymbol{E}\times\boldsymbol{H})\right]. \quad (50)$$

While we can distinguish the various forms of energy and heat outside the material medium, it is no longer distinguishable once they cross the boundary.

Our previous formulae for transport equations of kinetic energy, internal and potential energy need no change in the presence of electromagnetic field if we include the Lorentz and the ponderomotive force in the kinetic energy and the internal energy but not in the potential energy.

§ 12 *RELATIVISTIC THERMODYNAMICS*

Except in fast plasmas, disturbances in ordinary irreversible processes are too slow to warrant the relativistic treatment; furthermore, not all physical laws can be written in covariant form now, although they should be. The following relativistic (special relativity) treatment ignores the conversion between mass and energy and considers only isotropic material. Incomplete as the treatment may be, nevertheless we can write the thermodynamic equations in an elegant form.

Mechanical Laws in Special Relativity

In Galilean relativity, space and time coordinates are unconnected. Consequently, under Galilean transformations, the infinitesimal elements of the distance $dx^2 + dy^2 + dz^2$ and time dt^2 are separably invariant. For the Lorentz transformations, on the other hand, the space and time coordinates are interrelated. Thus $dx^2 + dy^2 + dz^2 - c^2dt^2$ is invariant, where c is velocity of light. We are led to consider a four-dimensional (world) displacement vector with elements (x_k, ict), $k = 1, 2, 3$. The matrix of Lorentz transformation is:

$$\begin{bmatrix} c/\sqrt{c^2 - v^2} & 0 & 0 & iv/\sqrt{c^2 - v^2} \\ 0 & 1 & 0 & 0 \\ 0 & 0 & 1 & 0 \\ -iv/\sqrt{c^2 - v^2} & 0 & 0 & c/\sqrt{c^2 - v^2} \end{bmatrix}$$

We form a four-velocity vector, $\left(\frac{dx_k}{dt'}, ic\frac{dt}{dt'}\right)$, where t' is the world time. Consider another world vector $(0, 0, 0, ic)$. We have

$$-c^2dt'^2 = dx^2 + dy^2 + dz^2 - c^2dt^2. \tag{1}$$

Or

$$\frac{dt'}{dt} = \frac{1}{c}\sqrt{c^2 - v^2}. \tag{2}$$

Thus the world velocity vector is

$$c\left(\frac{v_k}{\sqrt{c^2 - v^2}}, \frac{ic}{\sqrt{c^2 - v^2}}\right) = c(u_k, u_4), \tag{3}$$

where

$$u_k = \frac{v_k}{\sqrt{c^2 - v^2}} \quad \text{and} \quad u_4 = \frac{ic}{\sqrt{c^2 - v^2}}. \tag{4}$$

Also we form a four-momentum vector $\left(m\frac{dx_k}{dt'}, icm\frac{dt}{dt'}\right)$

and a four-force vector $\left[m \frac{d}{dt'}\left(\frac{dx_k}{dt'}, ic \frac{dt}{dt'}\right)\right]$ where m is the rest mass. Because the dot product of the world force and the world momentum is zero,

$$\sum_k^3 m^2 \frac{dx_k}{dt'} \frac{d}{dt'}\left(\frac{dx_k}{dt'}\right) - m^2c^2 \frac{dt}{dt'} \frac{d}{dt'}\left(\frac{dt}{dt'}\right)$$

$$= \frac{1}{2} m^2 \frac{d}{dt'}\left[\sum_k^3 \left(\frac{dx_k}{dt'}\right)^2 - c^2\left(\frac{dt}{dt'}\right)^2\right] = 0, \tag{5}$$

and because the spatial part of the world force is $cF_k/\sqrt{c^2 - v^2}$,

$$\sum_k^3 \left[\frac{(cF_k)}{\sqrt{c^2 - v^2}}\right]\left(\frac{v_k c}{\sqrt{c^2 - v^2}}\right) + \left(\frac{ic^2}{\sqrt{c^2 - v^2}}\right) K_4 = 0. \tag{6}$$

Thus the world force vector is

$$\left(\frac{cF_k}{\sqrt{c^2 - v^2}}, i\boldsymbol{F} \cdot \frac{\boldsymbol{v}}{\sqrt{c^2 - v^2}}\right) = (K_k, K_4), \tag{7}$$

where

$$K_k = \frac{cF_k}{\sqrt{c^2 - v^2}} \text{ and } K_4 = i\boldsymbol{F} \cdot \frac{\boldsymbol{v}}{\sqrt{c^2 - v^2}}, \tag{8}$$

and the world momentum vector is

$$\left(\frac{cmv_k}{\sqrt{c^2 - v^2}}, \frac{iT}{\sqrt{c^2 - v^2}}\right), \tag{9}$$

where T is the non-relativistic kinetic energy, $\boldsymbol{F} \cdot \boldsymbol{v} = \frac{dT}{dt}$. If the world momentum is conserved, i.e., constant in world time, then both momentum and kinetic energy are conserved. Thus the conservative laws of momentum and energy are not separate but different aspects of a single four-momentum vector.

Covariant Form of Maxwell's Equations

Einstein laid down the principle that the laws of physics have the same form in different Lorentz frames. This is equivalent to saying that the equations for physical laws must be covariant in form. By covariant we mean that the equations can be written so that both sides have the same well-defined characteristics under Lorentz transformations.

By forming a four-charge flux vector $(I_k, ic\rho Z)$, $k = 1, 2, 3$, the equation for the conservation of charge [eq. (11, 19)] is

$$\frac{\partial I_\alpha}{\partial x_\alpha} = 0, \ \alpha = 1, 2, 3, 4. \tag{10}$$

We also form a four-potential vector $(A_k, i\zeta/c)$, where $\boldsymbol{A}$ and ζ are vector and scalar potentials respectively, a four-magnetic induction (pseudo-) tensor

$$B_{\alpha\beta} = \begin{bmatrix} 0 & B_3 & -B_2 & -\frac{iE_1}{c} \\ -B_3 & 0 & B_1 & -\frac{iE_2}{c} \\ B_2 & -B_1 & 0 & -\frac{iE_3}{c} \\ \frac{iE_1}{c} & \frac{iE_2}{c} & \frac{iE_3}{c} & 0 \end{bmatrix}, \tag{11}$$

and a four-magnetic field (pseudo-) tensor

$$H_{\alpha\beta} = \begin{bmatrix} 0 & H_3 & -H_2 & -icD_1 \\ -H_3 & 0 & H_1 & -icD_2 \\ H_2 & -H_1 & 0 & -icD_3 \\ icD_1 & icD_2 & icD_3 & 0 \end{bmatrix}. \tag{12}$$

The equation

$$B_{\alpha\beta} = \frac{\partial A_\beta}{\partial x_\alpha} - \frac{\partial A_\alpha}{\partial x_\beta} \tag{13}$$

is the combination of eqs. (11, 5) and (11, 7). If we interpret the four columns of $H_{\alpha\beta}$ as four components,

$$\frac{\partial H_{\alpha\beta}}{\partial x_\beta} = I_\alpha \tag{14}$$

is the inhomogeneous pair of Maxwell's equations, eqs. (11, 2) and (11, 4). The equation

$$\frac{\partial B_{\alpha\beta}}{\partial x_\gamma} + \frac{\partial B_{\beta\gamma}}{\partial x_\alpha} + \frac{\partial B_{\gamma\alpha}}{\partial x_\beta} = 0 \tag{15}$$

is the homogeneous pair of Maxwell's equations, eqs. (11, 1) and (11, 3).

Finally we form a traceless momentum-energy tensor

$$W_{\alpha\beta} = B_{\alpha\gamma}H_{\gamma\beta} - (B_{\alpha\gamma}H_{\gamma\beta})\frac{\delta_{\alpha\beta}}{4}$$

$$= \left[\begin{array}{c|c} E_jD_k + H_jB_k - \frac{1}{2}\boldsymbol{D}\cdot\boldsymbol{E}\delta_{jk} - \frac{1}{2}\boldsymbol{B}\cdot\boldsymbol{H}\delta_{jk} & -ic(\boldsymbol{D}\times\boldsymbol{B}) \\ \hline \dfrac{i(\boldsymbol{H}\times\boldsymbol{E})}{c} & \dfrac{\boldsymbol{D}\cdot\boldsymbol{E}+\boldsymbol{H}\cdot\boldsymbol{B}}{2} \end{array}\right], \tag{16}$$

$$\alpha, \beta = 1, 2, 3, 4; \quad j, k = 1, 2, 3.$$

For $\alpha = 1, 2, 3,$ by using eqs. (16), (11, 13), (11, 14) and (11, 30),

$$\sum_{\alpha}^{3}\sum_{\beta}^{4}\frac{\partial W_{\alpha\beta}}{\partial x_\beta} = \boldsymbol{E}(\text{div}\,\boldsymbol{D}) + (\boldsymbol{D}\cdot\text{grad})\,\boldsymbol{E} - \frac{1}{2}\text{grad}\,\boldsymbol{D}\cdot\boldsymbol{E}$$

$$+ \boldsymbol{H}(\text{div}\,\boldsymbol{B}) + (\boldsymbol{B}\cdot\text{grad})\,\boldsymbol{H} - \frac{1}{2}\text{grad}\,\boldsymbol{B}\cdot\boldsymbol{H}$$

$$-\frac{\partial(\boldsymbol{D}\times\boldsymbol{B})}{\partial t} = \text{Div}\,(\boldsymbol{DE} + \boldsymbol{BH} - \boldsymbol{D}\cdot\boldsymbol{E}\overleftrightarrow{1} - \boldsymbol{B}\cdot\boldsymbol{H}\overleftrightarrow{1})$$

$$+ \frac{1}{2}\,\text{grad}\,(\boldsymbol{D}\cdot\boldsymbol{E} + \boldsymbol{B}\cdot\boldsymbol{H}) - \frac{\partial(\boldsymbol{D}\times\boldsymbol{B})}{\partial t}$$

$$= \rho Z\boldsymbol{E} + \boldsymbol{I}\times\boldsymbol{B} + \mathscr{P}\cdot\text{Grad}\,\boldsymbol{E}$$

$$+ \mathscr{I}\cdot\text{Grad}\,\boldsymbol{H} - \frac{1}{2}\text{grad}\,(\mathscr{P}\cdot\boldsymbol{E} + \mathscr{I}\cdot\boldsymbol{H}). \tag{17}$$

If we assume that the energy density $\mathscr{P}\cdot\boldsymbol{E} + \mathscr{I}\cdot\boldsymbol{H}$ is uniform,

$$\sum_{\alpha}^{3}\sum_{\beta}^{4}\frac{\partial W_{\alpha\beta}}{\partial x_{\beta}}=\rho\boldsymbol{F}'. \tag{18}$$

For $\alpha=4$, by using eqs. (16), (11, 13), (11, 14), (11, 38) and (11, 46),

$$\sum_{\beta}^{4}\frac{\partial W_{4\beta}}{\partial x_{\beta}}=\frac{i}{c}\left[\operatorname{div}(\boldsymbol{H}\times\boldsymbol{E})-\frac{1}{2}\frac{\partial(\boldsymbol{D}\cdot\boldsymbol{E}+\boldsymbol{B}\cdot\boldsymbol{H})}{\partial t}\right]$$

$$=\frac{i}{c}\left[\sum_{k}^{3}\rho_{k}\boldsymbol{F}'_{k}\cdot\boldsymbol{v}_{k}-\frac{1}{2}\frac{\partial(\mathscr{P}\cdot\boldsymbol{E}+\mathscr{I}\cdot\boldsymbol{H})}{\partial t}\right]. \tag{19}$$

If we assume that the energy density $\mathscr{P}\cdot\boldsymbol{E}+\mathscr{I}\cdot\boldsymbol{H}$ is also steady,

$$\sum_{\beta}^{4}\frac{\partial W_{4\beta}}{\partial x_{\beta}}=\frac{i}{c}\sum_{k}^{3}\rho_{k}\boldsymbol{F}'_{k}\cdot\boldsymbol{v}_{k}. \tag{20}$$

Combining eqs. (18) and (20) and using eq. (8),

$$\sum_{\alpha\beta}^{4}\frac{\partial W_{\alpha\beta}}{\partial x_{\beta}}\approx\sum_{k=1}^{4}\rho_{k}\boldsymbol{F}'_{k}. \tag{21}$$

[Here we neglect the difference between partial and substantial derivatives. For the more formal expression, see eq. (32)]

Covariant form of Many Components

Eckart[1] has recast the formulae for irreversible processes in covariant form for pure substances. Kluitenberg and de Groot[2] modified them for many components. We shall follow closely their treatment.

We use the superscripts to designate the jth component ($j=1, 2, \cdots, n$). We lose the freedom of distinguishing between covariant and contravariant tensors. This does not matter. Because we use orthogonal Lorentz coordinate systems, there is no difference between covariance and contravariance.

The momentum vector can be considered as the total (rest)

mass vector,

$$m_\alpha = [\rho v_k (k = 1, 2, 3.), i\rho c]. \tag{22}$$

The partial mass vector is defined as

$$m_\alpha^{(j)} = (\rho^{(j)} v_k^{(j)}, i\rho^{(j)} c). \tag{23}$$

Evidently

$$\sum_j^n m_\alpha^{(j)} (\alpha = 1, 2, 3, 4.) = m_\alpha. \tag{24}$$

The equation for the continuity of mass, eq. (4, 45),

$$\frac{\partial \rho^{(j)}}{\partial t} = -\operatorname{div} \sum_\alpha \rho^{(j)} v_\alpha^{(j)} + \sum_r \Lambda^{(j)(r)} j^{(r)}$$

can be written as

$$\sum_\alpha \frac{\partial m_\alpha^{(j)}}{\partial x_\alpha} = \sum_r \Lambda^{(j)(r)} j^{(r)}. \tag{25}$$

Summing in j,

$$\sum_\alpha \frac{\partial m_\alpha}{\partial x_\alpha} = 0. \tag{26}$$

The relative flow of mass, eq. (4, 37),

$$J_\alpha^{(j)} = \rho^{(j)} (v_\alpha^{(j)} - v_\alpha)$$

can be modified to

$$I_\alpha^{(j)} = m_\alpha^{(j)} - c^{(j)} m_\alpha, \tag{27}$$

where

$$c^{(j)} = -\sum_\beta \frac{m_\beta m_\beta^{(j)}}{\rho^2 (c^2 - v^2)} = \frac{\rho^{(j)}}{\rho}. \tag{28}$$

We see, from eqs. (27) and (28),

$$\sum_j I_\alpha^{(j)} = 0. \tag{29}$$

Furthermore, from eqs. (4) and (27),

$$\sum_{\alpha} u_\alpha I_\alpha^{(j)} = 0. \tag{30}$$

Equation (30) shows that all relative flows are perpendicular to the four-velocity u_α representing the barycentric velocity

$$v = \sum_j c^{(j)} v^{(j)}.$$

Three directions perpendicular to u_α are the local axis of the proper space.

Covariant Form of Rate of Entropy-production

We write the energy-momentum tensor $W_{\alpha\beta}$ in the form

$$\left[\begin{array}{c|c} \Pi_{jk} + g_j v_k & icg_j \\ \hline \left(\dfrac{i}{c}\right) I_k^{(q)} & -E \end{array}\right], \qquad j, k = 1, 2, 3, \tag{31}$$

where $\boldsymbol{g}$ is the momentum density vector, $\boldsymbol{gv}$ is the convective part of the momentum flow and the remaining part is the pressure tensor or the negative stress tensor. Of the equation,

$$\sum_{\beta}^{4} \frac{\partial W_{\alpha\beta}}{\partial x_\beta} = \sum_j \rho_0^{(j)} K_\alpha^{(j)}, \tag{32}$$

$\rho_0^{(j)}$ is the density of rest mass measured with the observer moving with the component,

$$\rho_0^{(j)} / \rho^{(j)} = \sqrt{c^2 - \boldsymbol{v}^{(j)} \cdot \frac{\boldsymbol{v}}{c}}. \tag{33}$$

On the other hand,

$$\rho'^{(j)} / \rho^{(j)} = \frac{\sqrt{c^2 - v^2}}{c}. \tag{34}$$

The prime refers to the measurement with the observer moving with the barycentric velocity. In principle $\rho_0^{(j)}$ is different from $\rho'^{(j)}$, but in practice the difference in value is small.

Usings eqs. (8) and (31), for $\alpha = 1, 2, 3$, eq. (32) is

$$\operatorname{Div}(\Pi + \boldsymbol{g}\boldsymbol{v}) + \frac{\partial \boldsymbol{g}}{\partial t} = \sum_j \frac{\rho_0^{(j)} c \boldsymbol{F}^{(j)}}{\sqrt{c^2 - v^2}}$$

$$\approx \sum_j \frac{\rho_0^{(j)} c \boldsymbol{F}^{(j)}}{\sqrt{c^2 - \boldsymbol{v}^{(j)} \cdot \boldsymbol{v}}} = \sum_j \rho^{(j)} \boldsymbol{F}^{(j)}, \tag{35}$$

which is eq. (6, 1); for $\alpha = 4$, eq. (32) is

$$\left(\frac{i}{c}\right)\left[\operatorname{div} I_{\alpha}^{(q)} + \frac{\partial \rho U}{\partial t}\right] = \left(\frac{i}{c}\right)\left[\sum_j \rho_0^{(j)} c \boldsymbol{F}^{(j)} \cdot \frac{\boldsymbol{v}^{(j)}}{\sqrt{c^2 - v^2}}\right]$$

$$\approx \left(\frac{i}{c}\right)\left[\sum_j \rho_0^{(j)} c \boldsymbol{F}^{(j)} \cdot \frac{\boldsymbol{v}^{(j)}}{\sqrt{c^2 - \boldsymbol{v}^{(j)} \cdot \boldsymbol{v}}}\right]$$

$$= \left(\frac{i}{c}\right) \sum_j \rho^{(j)} \boldsymbol{F}^{(j)} \cdot \boldsymbol{v}^{(j)}, \tag{36}$$

which is eq. (6, 20) except that the ground level of U and $\boldsymbol{J}_q$ is different.

$$\text{Let } S_{\alpha\beta} = \delta_{\alpha\beta} + u_\alpha u_\beta, \quad \alpha, \beta = 1, 2, 3, 4. \tag{37}$$

S is evidently symmetric. Because of eq. (4),

$$\sum_\alpha u_\alpha u_\alpha = -1, \tag{38}$$

and

$$\sum_\alpha \frac{u_\alpha \partial u_\alpha}{\partial x_\beta} = 0. \tag{39}$$

Also

$$\sum_\alpha u_\alpha S_{\alpha\beta} = 0, \quad \sum_\beta u_\beta S_{\beta\alpha} = 0. \tag{40}$$

Now $W_{\alpha\beta}$ can be resolved into

$$W_{\alpha\beta} = \left(\sum_{\alpha\beta} u_\alpha W_{\alpha\beta} u_\beta\right) u_\alpha u_\beta + \left(-\sum_{\beta\gamma} S_{\alpha\beta} W_{\beta\gamma} u_\gamma\right) u_\beta$$

$$+ \left(-\sum_{\alpha\gamma} S_{\beta\alpha} W_{\alpha\gamma} u_\gamma\right) u_\alpha + \left(\sum_{\gamma\delta} S_{\alpha\delta} W_{\delta\gamma} S_{\gamma\beta}\right). \tag{41}$$

The interpretation of different terms in the preceding equation is faciliated if we notice that

$$\mu'_\alpha = (0, 0, 0, i), \tag{42}$$

$$S'_{\alpha\beta} = \begin{bmatrix} 1 & 0 & 0 & 0 \\ 0 & 1 & 0 & 0 \\ 0 & 0 & 1 & 0 \\ 0 & 0 & 0 & 0 \end{bmatrix}, \tag{43}$$

and that scalars are invariant in any Lorentz frame.

The first bracket in eq. (41) is a scalar, which we interpret as internal energy density, whose ground level is not specified,

$$\sum_{\alpha\beta} W_{\alpha\beta} u_\alpha u_\beta = \rho' U'. \tag{44}$$

The second and third bracket are vectors, which we interpret as heat fluxes,

$$-\sum_{\beta\gamma} S_{\alpha\beta} W_{\beta\gamma} u_\gamma = \frac{1}{c} I_\alpha^{(q)}, \tag{45}$$

$$-\sum_{\alpha\gamma} S_{\beta\alpha} W_{\alpha\gamma} u_\gamma = \frac{1}{c} I_\beta^{(q)}. \tag{46}$$

The fourth bracket is a second-rank tensor, which we interpret as a pressure or a negative stress tensor,

$$\sum_{\delta\gamma} S_{\alpha\delta} W_{\delta\gamma} S_{\gamma\beta} = \Pi_{\alpha\beta}. \tag{47}$$

For $I^{(q)}$ and Π, because of eqs. (40), (45) and (47),

$$\sum_\alpha u_\alpha I_\alpha^{(q)} = 0, \quad \sum_\alpha u_\alpha \Pi_{\alpha\beta} = 0. \tag{48}$$

Or all four elements of $I^{(q)}$ or all four components of $\Pi_{\alpha\beta}$ are perpendicular to the four vectors of u_α. Furthermore

$$I_4^{(q)} = 0, \tag{49}$$

and the number of independent components of Π is reduced to six.

The counterpart of d/dt is the new operator, d/dt', which is abbreviated as D. By using eqs. (2) and (4),

$$D = \left(\frac{c}{\sqrt{c^2 - v^2}}\right)\left(\frac{\partial}{\partial t} + \boldsymbol{v} \cdot \text{grad}\right) = \sum_{\alpha}^{4} c u_\alpha \frac{\partial}{\partial x_\alpha}. \tag{50}$$

From eq. (32),

$$\sum_{\alpha\beta}\left(\frac{\partial u_\alpha W_{\alpha\beta}}{\partial x_\beta} - W_{\alpha\beta}\frac{\partial u_\alpha}{\partial x_\beta}\right) = \sum_{\alpha}\sum_{j} \rho_0^{(j)} u_\alpha K_\alpha^{(j)}. \tag{51}$$

Substituting eqs. (41) and (44—47) in the left side of eq. (51) and using eqs. (38—40),

$$\begin{aligned}\sum_{\alpha\beta}&\left(\frac{\partial u_\alpha W_{\alpha\beta}}{\partial x_\beta} - W_{\alpha\beta}\frac{\partial u_\alpha}{\partial x_\beta}\right) \\ &= \sum_{\beta}\left(-\frac{\partial \rho' u_\beta U'}{\partial x_\beta} - \frac{1}{c}\frac{\partial I_\beta^{(q)}}{\partial x_\beta}\right) \\ &\quad - \sum_{\alpha\beta}\left(\frac{1}{c} I_\alpha^{(q)} u_\beta \frac{\partial u_\alpha}{\partial x_\beta} + \Pi_{\alpha\beta}\frac{\partial u_\alpha}{\partial x_\beta}\right)\end{aligned} \tag{52}$$

The right side of eq. (51), after using eqs. (4), (5), (22), (27), and (34), is

$$\begin{aligned}\sum_{\alpha}\sum_{j} \rho_0^{(j)} u_\alpha K_\alpha^{(j)} &= \sum_{\alpha}\sum_{j} \rho_0^{(j)}\left(\frac{c^{(j)}}{\rho^{(j)}}\right)\left(\frac{m_\alpha}{\sqrt{c^2 - v^2}}\right) K_\alpha^{(j)} \\ &= \sum_{\alpha}\sum_{j}\left(\frac{\rho_0^{(j)}}{c\rho'^{(j)}}\right)(-m_\alpha^{(j)} + m_\alpha c^{(j)}) K_\alpha^{(j)} \\ &= -\left(\frac{1}{c}\right)\sum_{\alpha}\sum_{j} \omega^{(j)} I_\alpha^{(j)} K_\alpha^{(j)},\end{aligned} \tag{53}$$

where

$$\omega^{(j)} = \frac{\rho_0^{(j)}}{\rho'^{(j)}}. \tag{54}$$

Equating eqs. (52) and (53) and using eqs. (4), (22), (26),

(34), and (50),

$$\rho' DU' = -\sum_{\alpha} \frac{\partial I_{\alpha}^{(q)}}{\partial x_{\alpha}} - \frac{1}{c} \sum_{\alpha} I_{\alpha}^{(q)} Du_{\alpha} - c \sum_{\alpha\beta} \Pi_{\alpha\beta} \frac{\partial u_{\beta}}{\partial x_{\alpha}}$$

$$+ \sum_{\alpha} \sum_{j} \omega^{(j)} I_{\alpha}^{(j)} K_{\alpha}^{(j)}. \tag{55}$$

By using eqs. (4), (22), (26), (34), (37), (39), and (50),

$$\rho' D\left(\frac{1}{\rho'}\right) = \sum_{\alpha} m_{\alpha} \frac{\partial V'}{\partial x_{\alpha}} = \sum_{\alpha} \frac{\partial m_{\alpha} V'}{\partial x_{\alpha}}$$

$$= c \sum_{\alpha} \frac{\partial u_{\alpha}}{\partial x_{\alpha}} = c \sum_{\alpha\beta} S_{\alpha\beta} \frac{\partial u_{\beta}}{\partial x_{\alpha}}. \tag{56}$$

By using eqs. (4), (22), (25), (26), (27), (34), and (50),

$$\rho' Dc'^{(j)} = \sum_{\alpha} m_{\alpha} \frac{\partial c^{(j)}}{\partial x_{\alpha}} = \sum_{\alpha} \left(\frac{\partial m^{(j)}}{\partial x_{\alpha}} - \frac{\partial m^{(j)}}{\partial x_{\alpha}} \right.$$

$$\left. + \frac{\partial m_{\alpha} c^{(j)}}{\partial x_{\alpha}} \right) = \sum_{\alpha} \left(\sum_{r} \Lambda^{(j)(r)} j^{(r)} - \frac{\partial I_{\alpha}^{(j)}}{x_{\alpha}} \right). \tag{57}$$

Substituting eqs. (55—57) in

$$\rho' DS' = \rho' \frac{DU'}{T'} + \rho' p' \frac{DV'}{T'} - \sum_{j} \rho' \mu'^{(j)} \frac{Dc'^{(j)}}{T'}, \tag{58}$$

$$\rho' DS' = -\sum_{\alpha} \frac{\partial}{\partial x_{\alpha}} \left(I_{\alpha}^{(q)} - \sum_{j} \mu'^{(j)} I_{\alpha}^{(j)} \right) \Big/ T'$$

$$- \frac{1}{T'} \sum_{\alpha} I_{\alpha}^{(q)} \left(\frac{1}{T'} \frac{\partial T'}{\partial x_{\alpha}} + \frac{1}{c} Du_{\alpha} \right)$$

$$+ \frac{1}{T'} \sum_{\alpha} \sum_{j} I_{\alpha}^{(j)} \left(\omega^{(j)} K_{\alpha}^{(j)} - T' \frac{\partial}{\partial x_{\alpha}} \frac{\mu'^{(j)}}{T'} \right)$$

$$- \frac{c}{T'} \sum_{\alpha\beta} (\Pi_{\alpha\beta} - p' S_{\alpha\beta}) \frac{\partial u_{\beta}}{\partial x_{\alpha}}$$

$$- \frac{1}{T'} \sum_{r} j^{(r)} \sum_{j} \Lambda^{(j)(r)} \mu'^{(j)}. \tag{59}$$

Comparing eq. (59) with eqs. (8, 7), (8, 9), and (8, 10), we see that the first bracket is the entropy flux. The second bracket is the force for heat flow, although the second term within the bracket has no classic meaning. It may be interpreted as a flow of heat through accelerated matter. The third bracket is the modified chemical potential. The fourth bracket is the viscous pressure. The fifth bracket is the chemical affinity. Thus the rate of entropy-production is covariant.

References

[1] Eckart, C.: *Phys. Rev.*, **53**, 919 (1940).
[2] Kluitenberg, G. A. and de Groot, S. R.: *Physica*, **19**, 689 (1953); *ibid.*, **19**, 1079 (1953); *ibid.*, **20**, 109 (1954).

§13 *THE CURIE THEOREM AND ISOTROPY*

The Curie theorem dates back to 1894[1]. As was pointed out by Fitts[2], Curie neither stated nor proved the theorem. The Curie theorem is important in irreversible thermodynamics. It limits the possible ways of the coupling of fluxes and forces in an isotropic system. It may be stated as follows: In an isotropic system, fluxes and forces of different tensorial ranks do not couple.

The proof[3] is simple. Because the contracted inner product of any coupled fluxes and forces yields the rate of entropy-production and because contraction occurs in pairs of the indices, the ranks of tensors of fluxes and forces will differ by an indeterminate even number (zero included). Similarly, the phenomenological coefficients, which are ratios of fluxes and forces, are tensors of ranks of an indeterminate even number. Geometrically, the inner product is a projection of a flux on a force (or vice versa). That the tensor of the rate of entropy-production or of phenomenological coefficients is indeterminate in rank, is only relevant to anisotropic substances, for which the properties in one direction are not necessarily the same as those in another direction. Also, the coupling of nonzero fluxes with nonzero forces, to give zero rate of entropy-production in certain directions but nonzero (positive) rate in certain other directions, is not excluded. For isotropic substances, such directions cannot be found. Thus for isotropic substances the phenomenological coefficients are scalars.

Although there are no isotropic crystals, in the sense that there are cubic crystals, for example, some physical properties of single crystals crystallized in certain systems are isotropic. Furthermore, fluids and powdered specimens can be considered isotropic. To find the number of independent components of an isotropic tensor is just as important as to find the independent components of a tensor corresponding to a physical property of single crystals crystallized in the 32 systems (for example, see the book by Nye, reference [4]).

In a straightforward manner the restraints of an isotropic tensor can be found from the transformation law of tensors. However, for a high-ranked tensor, say of rank 8, hundreds of equations must be processed. Furthermore, each equation involves hundreds of variables. To trim such a large number of dependent equations to independent ones is extremely tedious; other approaches can be made. Group theory can be used to find the number of independent components only[5,6]. The restraints on the components of a fourth-rank tensor in three dimensions and undergoing proper orthogonal transformation have been given by Jeffreys[7], Thomas[8] and de Groot and Mazur[9] from tensor analysis. These treatments vary in the degree of elegance but none of them is short even in the simple case. The simplest treatment takes one printed page. The results only of a simple treatment[10] are given below. For details the original reference should be consulted.

(A) Let the rank of a tensor be p and the number of dimensions be n. The splitting of p into a number of integers like $[1^{\alpha}, 2^{\beta}, 3^{\gamma}, \cdots]$, with $\alpha + 2\beta + 3\gamma + \cdots = p$, to show the composition of the tensor is called partitioning. For example, for $p = 4, n = 4$, there are $4^4 = 256$ components. The components of $[3, 1]$ are permutations of 1112, 1113, 1114, 2221, 2223, 2224, 3331, 3332, 3334, 4441, 4442, and 4443, or 48 in all.

(B) For proper orthogonal transformations, all components of a tensor are identically zero if p is an odd integer. This is because the n-proper orthogonal matrix may be considered as a product of n, or less than n, reflection matrices. The reflection matrices must appear in pairs or not at all; otherwise the value of the determinant would not be positive unity.

(C) For the same reason, the partitions of an isotropic tensor are limited to those, each entry of which is an even number. For example, for $p = 6$, $[6]$, $[4, 2]$, and $[2^3]$ are the allowed partitions.

(D) For the partition $[2^{\alpha}, 4^{\beta}, 6^{\gamma}, \cdots]$, let

$$\pi = \frac{p!}{(2!)^{\alpha}(4!)^{\beta}(6!)^{\gamma}\cdots\alpha!\beta!\gamma!\cdots}, \tag{1}$$

where π is simply the number of modes in which p distinguishable objects may be assorted into piles, with α piles of 2 objects each, β piles of 4 objects each, and so on, such that $2\alpha + 4\beta + 6\gamma + \cdots = p$. For our purposes we interpret π as the number of apparently independent components, for each permutation sequence in the particular partition. We say apparently independent because restraints are not considered yet. For example, for $p = 6, n = 3$, the permutation sequence 112233 of $[2^3]$ has 15 apparently independent components, namely 112233, 112323, 112332, 121233, 121323, 121332, 122133, 123123, 123132, 122313, 123213, 123312, 122331, 123231, and 123321. π for all possible partitions with $p = 2$—10 are listed on the left half of Table 1.

(E) Because the matrix of an orthogonal transformation can be reduced to a special orthogonal matrix,

$$\begin{bmatrix} 0 & 1 & 0 & \cdots \\ 0 & 0 & 1 & \cdots \\ \cdots & \cdots & \cdots & \cdots \\ 1 & 0 & 0 & \cdots \end{bmatrix},$$

the transformation turns x_1 into x_2, x_2 into x_3, $\cdots$ and x_n into x_1. Or the permutation may be represented by $(1, 2, \cdots, n)$. This is called cyclic symmetry. For the example in (D), the 112233 sequence is equal to the 223311 or the 331122 sequence. Because of inherent symmetry, the 112233 sequence is equal to the 113322 sequence. Thus the 112233 sequence is the only sequence possible in the partition [2, 2, 2].

(F) Let m be the number of entries in the partition.

Consider the example of the partition [4, 2] of an isotropic tensor with $p = 6$. From Table 1 we see that $\pi = 15$. We also note that $m = 2$.

If $n = 2$, the 112222 sequence namely, 112222, 121222, 122122, $\cdots$, has 15 apparently independent components. The other sequence 221111 is equal to the 112222 sequence due to inherent symmetry.

If $n = 3$, $\binom{n}{m} = 3$, there are three sequences, each of which has 15 apparently independent components. The sequence 223333 can be derived from the 112222 sequence by cyclic symmetry. Due to inherent symmetry, the sequence 113333 is equal to the sequence 331111, which can be derived from the sequence 112222 by cyclic symmetry.

If $n = 4$, $\binom{n}{m} = 6$, the sequences 112222 and 113333 give 30 apparently independent components. The sequences

Table 1

π for Different Partitions of a Tensor

Rank of an isotropic tensor	Partition	π	Rank of an ordinary tensor	Partition	π
2	[2]	1	1	[1]	1
4	[4]	1	2	[2]	1
	$[2^2]$	3		$[1^2]$	1
6	[6]	1	3	[3]	1
	[4,2]	15		[2,1]	3
	$[2^3]$	15		$[1^3]$	1
8	[8]	1	4	[4]	1
	[6,2]	28		[3,1]	4
	$[4^2]$	35		$[2^2]$	3
	$[4,2^2]$	210		$[2,1^2]$	6
	$[2^4]$	105		$[1^4]$	1
10	[10]	1	5	[5]	1
	[8,2]	45		[4,1]	5
	[6,4]	210		[3,2]	10
	$[6,2^2]$	630		$[3,1^2]$	10
	$[4^2,2]$	1,575		$[2^2,1]$	15
	$[4,2^3]$	3,150		$[2,1^3]$	10
	$[2^5]$	945		$[1^5]$	1

223333 and 334444 can be derived from the sequence 112222 and the sequence 224444 from the sequence 113333 by cyclic symmetry. The sequence 114444 is equal to the sequence 441111, which can be derived from the sequence 112222 by cyclic symmetry.

It is clear that $\binom{n}{m}$ is the number of permutation sequences possible and $\binom{n}{m}\Big/n$ is the number of permutation sequences realized. Thus the number of apparently independent components S is

$$S = \pi\binom{n}{m}\Big/n\,. \tag{2}$$

We understand that $\binom{n}{m}\Big/ n$ is the next integer if the ratio is not an integer and that it has no meaning if $n < m$.

For $p = 2$,

$$S = 1. \tag{3}$$

For $p = 4$,

$$S = 1 + 3\binom{n}{2}\Big/ n. \tag{4}$$

For $p = 6$,

$$S = 1 + 15\binom{n}{2}\Big/ n + 15\binom{n}{3}\Big/ n. \tag{5}$$

For $p = 8$,

$$S = 1 + 63\binom{n}{2}\Big/ n + 210\binom{n}{3}\Big/ n + 105\binom{n}{4}\Big/ n. \tag{6}$$

For $p = 10$,

$$S = 1 + 255\binom{n}{2}\Big/ n + 2{,}205\binom{n}{3}\Big/ n + 3{,}150\binom{n}{4}\Big/ n + 945\binom{n}{5}\Big/ n. \tag{7}$$

(G) A tensor of the pth rank may be represented by a polynomial of the pth degree occupying n–field space. For example, an ordinary tensor of the third rank may be represented by

$$a_{[3]}\sum x_1^3 + \pi_{[2,1]}a_{[2,1]}\sum x_1^2 x_2 + \pi_{[1,1,1]}a_{[1,1,1]}\sum x_1 x_2 x_3,$$

where the a's are multinomial coefficients. The coefficients are introduced to take care of non-homogeneous (not symmetric) variables. If the variables are homogeneous (e.g. in an isotropic tensor), or the polynomial is a perfect nth power, the coefficients are unities.

The partitions of an isotropic tensor are the same as all possible partitions of an ordinary tensor of the $p/2$ rank except for a factor of 2. For example, an isotropic tensor with $p=6$ has partitions [6], [4, 2], and [2, 2, 2]. But [3], [2, 1], and [1, 1, 1] are exactly the partitions of an ordinary tensor with $p=3$. Thus the isotropic tensor may be represented by

$$a_{[6]}\sum x_1^6+\pi_{[2,1]}a_{[4,2]}\sum x_1^4x_2^2+\pi_{[1,1,1]}a_{[2,2,2]}\sum x_1^2x_2^2x_3^2.$$

Because fewer variables are required for the 6-ic form, the Hessian for the 6-ic form is zero. Or the 6-ic form is a perfect 3rd power. The 6-ic form is represented by

$$\sum x_1^6+\pi_{[2,1]}\sum x_1^4x_2^2+\pi_{[1,1,1]}\sum x_1^2x_2^2x_3^2,$$

the multinomial coefficients disappear but π's are those of the ordinary tensor. To find the number of terms for summation, we could set x's unity and count the terms. Then

$$\frac{\text{number of terms of the partition [4, 2] in the isotropic tensor}}{\text{number of terms of the partition [2, 2, 2] in the isotropic tensor}}$$

$$=\frac{\pi\text{ of the partition [2, 1] in the ordinary tensor}}{\pi\text{ of the partition [1, 1, 1] in the ordinary tensor}}.$$

Herein lies the key for finding restraints, which is illustrated below. The π's for an ordinary tensor with $p=1$—5 are listed on the right half of Table 1.

Now we examine the isotropic tensors with $p=2$—10 in detail.

$p=2$

The number of independent components is 1 irrespective of n.

$p=4$

For $n=1$, only the partition [4] is allowed. There is only one independent component.

For $n=2$, from the second and third row of Table 1, we write

$$1\text{ term of }[4]=\frac{1}{1}\times\text{the sum of 3 terms of }[2,2].\quad(8)$$

There is only one way of writing eq. (8),

$$1111 = 1122 + 1212 + 1221. \tag{9}$$

The other way of writing it, namely,

$$1111 = 2112 + 2121 + 2211 \tag{10}$$

is redundant because of symmetry. There are $\left[1 + \frac{3\binom{2}{2}}{2}\right] - 1$ $= 3$ independent components.

For $n = 3$, equation (9) is still the only restraint. There are still

$$\left[1 + \frac{3\binom{3}{2}}{3}\right] - 1 = 3 \text{ independent components.}$$

For $n = 4$,

$$1111 = 1122 + 1212 + 2211 \tag{11}$$

and

$$1111 = 1133 + 1313 + 3311 \tag{12}$$

are restraints. There are $\left[1 + \frac{3\binom{4}{2}}{4}\right] - 2 = 5$ independent components.

In general, the number of independent components t is

$$t = 1 + \frac{3\binom{n}{2}}{n} - \frac{\binom{n}{2}}{n} = 1 + \frac{2\binom{n}{2}}{n}, \text{ for } n > 3. \tag{13}$$

$p = 6$

For $n = 1$, only partition [6] is allowed. There is only 1 independent component.

For $n = 2$, the partitions [6] and [4, 2] only are allowed.

From the fourth and fifth row of Table 1, we write

$$1 \text{ term of } [6] = \frac{1}{3} \times \text{the sum of 15 terms of } [4, 2]. \quad (14)$$

There is only one way of writing this,

$$\begin{aligned} 111111 = \frac{1}{3}(&221111 + 212111 + 211211 + 211121 \\ &+ 211112 + 122111 + 121211 + 121121 \\ &+ 121112 + 112211 + 112121 + 112112 \\ &+ 111221 + 111212 + 111122). \end{aligned} \quad (15)$$

There are $\left[1 + \frac{15\binom{2}{2}}{2}\right] - 1 = 15$ independent components.

For $n = 3$, there is still one way of writing eq. (14). From the fifth and sixth row of Table 1, we write

$$\text{the sum of 15 terms of } [4, 2] = 3 \times \text{the sum of 15 terms of } [2^3]. \quad (16)$$

Or $$1 \text{ term of } [4, 2] = 3 \text{ terms of } [2^3]. \quad (17)$$

There are 15 ways of writing eq. (17) such as

$$111122 = 331122 + 313122 + 311322, \quad (18)$$

$$111212 = 331212 + 313212 + 311232, \quad (19)$$

$$\cdots\cdots\cdots\cdots.$$

Each term on the left side of the equations like eq. (18) is an apparently independent component of [4, 2]. Each term on the right side of the equations like eq. (18) belongs to one of the sequences 331122, 332211 and 223311. However, because of inherent symmetry, these 3 sequences are equal. In other words, equations like eq. (18) give 15 restraints.

There are $31 - 16 = 15$ independent components.

For $n = 4$, because $\dfrac{\binom{4}{2}}{4} = 2$, we have restraints

$$111111 = \frac{1}{3} \times \text{the sum of 15 terms like } 111122 \tag{20}$$

and

$$111111 = \frac{1}{3} \times \text{the sum of 15 terms like } 111133. \tag{21}$$

However, eqs. (20) and (21) are equivalent due to eqs. like eq. (18).

Because $\dfrac{\binom{4}{3}}{4} = 1$, we have 15 restraints like

$$111122 = 331122 + 313122 + 311322. \tag{22}$$

There are $46 - 16 = 30$ independent components.

For $n = 5$, because $\dfrac{\binom{5}{2}}{5} = 2$, there are restraints

$$111111 = \frac{1}{3} \times \text{the sum of 15 terms like } 112222 \tag{23}$$

and

$$111111 = \frac{1}{3} \times \text{the sum of 15 terms like } 113333. \tag{24}$$

However, eqs. (23) and (24) are counted twice due to eqs. (25—26).

Because $\dfrac{\binom{5}{3}}{5} = 2$, there are 30 restraints like

$$112222 = 113322 + 113232 + 113223 \tag{25}$$

and

$$113333 = 114433 + 114343 + 114334. \tag{26}$$

There are $61 - 31 = 30$ independent components.

For $n = 6$, because $\frac{\binom{6}{2}}{6} = 3$, there are three restraints:

$$111111 = \frac{1}{3} \times \text{sum of 15 terms like } 112222, \tag{27}$$

$$111111 = \frac{1}{3} \times \text{sum of 15 terms like } 113333, \tag{28}$$

$$111111 = \frac{1}{3} \times \text{sum of 15 terms like } 114444. \tag{29}$$

However, eqs. (27—29) are counted thrice due to eqs. (30—32).

Because $\frac{\binom{6}{3}}{6} = 3$, there are 45 restraints like

$$112222 = 113322 + 113232 + 113223, \tag{30}$$

$$113333 = 114433 + 114343 + 114334, \tag{31}$$

$$114444 = 115544 + 115454 + 115445. \tag{32}$$

There are $106 - 46 = 60$ independent components.

In general, let the components of [4, 2] be independent, the components of [6] are eliminated. The components of $[2^3]$ are also eliminated except if

$$\frac{\binom{n}{3}}{n} > \frac{\binom{n}{2}}{n}.$$

Each unit of difference introduces 15 independent components. Thus

$$t = 15\frac{\binom{n}{2}}{n} + 15\left[\frac{\binom{n}{3}}{n} - \frac{\binom{n}{2}}{n}\right] = 15\frac{\binom{n}{3}}{n} \quad \text{for } n \geqslant 5. \tag{33}$$

The work involved in $p = 8$ is already formidable. The independent components of an isotropic tensor with $p = 2—10$ are listed in Table 2.

Table 2

Number of Independent Components of an Isotropic Tensor under Proper Orthogonal Transformations

Dimensions	Rank				
	2	4	6	8	10
1	1	1	1	1	1
2	1	3	15	7	135
3	1	3	15	91	603
4	1	5	30	98	738
5	1	5	30	182	1,836
6	1	7	60	483	4,644
7	1	7	75	693	8,109
8	1	9	105	1,204	19,737
9	1	9	150	2,044	37,062
10	1	11	180	2,870	64,755

References

[1] P. Curie: *J. Physique*, (Ser. 3), **3**, 393 (1894).
[2] D. D. Fitts: Nonequilibrium Thermodynamics, McGraw-Hill, New York, p. 35 (1962).
[3] Y. L. Yao: *J. Chem. Phys.*, **48**, 537 (1968).
[4] J. F. Nye: Physical Properties of Crystals, Oxford Univ. Press, London (1957).
[5] V. Heine: Group Theory in Quantum Mechanics, Pergamon Press, Oxford, p. 67 (1960).
[6] M. Hamermesh: Group Theory, Addison-Wesley, Reading, Mass., p. 331 (1962).
[7] H. Jeffreys: Cartesian Tensors, Cambridge Univ. Press, London, p. 66 (1931).
[8] T. Y. Thomas: Concepts from Tensor Analysis and Differential Geometry, Academic Press, New York, p. 65 (1961).
[9] S. R. de Groot and P. Mazur: Non-equilibrium Thermodynamics, Interscience, New York, p. 62 (1962).
[10] Y. L. Yao: Physical Metallurgy Division Internal Report PM-M-67-11, Mines Branch, Department of Energy, Mines and Resources, Ottawa (1967).

CHAPTER V
LINEAR THERMODYNAMICS

§14 *HYPOTHESIS OF LINEARITY BETWEEN FLUXES AND FORCES*

Once the inner product of fluxes and forces is formulated and fluxes and forces are chosen, it remains to find the relation between fluxes and forces. One possibility is that fluxes are linear homogeneous functions of forces,

$$\boldsymbol{J} = -\frac{L\boldsymbol{X}}{T}, \tag{1}$$

where L are phenomenological coefficients.

For one irreversible process, this hypothesis has foundation from a large number of phenomenological laws, e.g., Fourier's law between the flow of heat and temperature gradient, Fick's first law between the flow of material and concentration gradient, Ohm's law between the flow of electricity and potential gradient, and Newton's law between the shear stress and velocity gradient.

For more than one irreversible process, it is expected that each flux $\boldsymbol{J}_i$ is not only related to its principal force X_i, but may be also related to all other forces X_j, $i \neq j$. The cross effects are described by non-diagonal terms in the matrix of phenomenological coefficients. For example, thermal diffusion is described by L_{ij}, where i refers to diffusion current and j to temperature gradient. The reciprocal phenomenon, the Dufour effect, is described by L_{ji}, where j refers to heat current and

i concentration gradient. In discontinuous systems with a semi-permeable membrane, the thermal osmosis corresponds to the thermal diffusion and the osmotic Dufour effect corresponds to the Dufour effect. The introduction of the cross effects is not arbitrary but necessary because these effects exist. The ratio of the non-diagonal coefficient to the diagonal coefficient gives a measure of selectivity of the cross effect. For example, in the cross effect of osmotic flow induced by osmotic pressure, the membrane may be so selective that no solute can pass or may be so non-selective that solute and solvent move at equal velocities. However, the sign of the ratio is not predictable. In the previous example cases are known where the solute can move at a higher speed than the solvent.

Mathematically, the linear relationship is the simplest one between two sets of variables. A linear relationship has two characteristics: one is superposition mentioned in the previous paragraph and the other is the predictability of extrapolation from the particular to the general case. For example, the formal writing of eq. (1, 12) with large volume V is more plausible if eq. (1) is valid.

There is another way to look at the hypothesis. Macroscopically, there are frictional forces opposing the motion (If the movement is random and atomic, there is no friction. The reasons underlying this are complicated and controversial[1,2]. Suffice it to say that we can ignore this random movement, no matter how fast is them ovement). Rayleigh's dissipation function $\mathscr{F}$[3] is defined as

$$\mathscr{F} = -T\sigma = -\frac{1}{2}\sum_{ik}\alpha_{ik}\dot{x}_i\dot{x}_k, \tag{2}$$

where α's are proportional coefficients. It is seen that $\mathscr{F}$ is a homogeneous function of the degree 2 in velocity. $\partial\mathscr{F}/\partial\dot{x}$, being a homogeneous function of the degree 1, is interpreted as frictional force, provided that the coefficient α_{ik} is symmetric in the suffixes i and k. Of all the frictional forces depending on velocity, the Rayleigh frictional force may be justified as the one having the least positive degree. Although, except for viscous fluids, friction is not explicitly expressed, the inverse of the matrix L is related to the matrix of frictional coefficients.

By substituting eq. (1) in eq. (1, 6), the matrix L is positive definite. Provided that the matrix is not singular, i.e., both the fluxes and the forces are independent, all diagonal terms of the matrix are positive. Also the determinant of the matrix or any submatrix is positive.

The hypothesis is of limited validity, either because the measurements are so refined that linearity is only a first approximation, or because the conditions are so severe that irreversible processes are no longer close to steady equilibrium. That phenomenological coefficients are constants independent of fluxes and forces, although they may be functions of other state variables, is of more limited validity. However, in Chapter V (§ 14—20) and in the entire portion of Part II, the hypothesis of linearity is supposed to be valid. Just like massless springs and frictionless levers, plane waves and linear resistors serve as simple physical concepts. The justification is that many unrelated phenomena could be explained in a unified way. The Onsager law permits us to thread our way through the maze. Of course, if the hypothesis of linearity is in conflict with experiment, it must be modified accordingly.

References

[1] J. von Neumann (translated by R. T. Beyer): Mathematical Foundations of Quantum Mechanics, Princeton University Press, Princeton, N. J. (1955).

[2] J. M. Jauch and C. Piron: *Hel. Phys. Acta*, **36**, 827 (1963).

[3] J. W. Strutt (Lord Rayleigh): *Proc. Math. Soc.*, London, **4**, 357 (1873).

§ 15 SIGNIFICANCE OF THE ONSAGER LAW

The Onsager law states that the proportional matrix L between fluxes and forces is symmetric. Before the law is proved in § 16, it is pointed out here that the reciprocal relationship is unlike any other form of macroscopic symmetry. While certain symmetries in chemical reactions can be explained by either statistical theory or the Onsager law, certain symmetries in crystallography cannot be explained in any other way than by the Onsager law.

Many authors consider the symmetry pertaining to the law to be the same as that pertaining to Maxwell's relations

$$\left(\frac{\partial T}{\partial V}\right)_S = -\left(\frac{\partial p}{\partial S}\right)_V, \tag{1}$$

$$\left(\frac{\partial T}{\partial p}\right)_S = \left(\frac{\partial V}{\partial S}\right)_p, \tag{2}$$

$$\left(\frac{\partial S}{\partial V}\right)_T = \left(\frac{\partial p}{\partial T}\right)_V, \tag{3}$$

$$\left(\frac{\partial S}{\partial p}\right)_T = -\left(\frac{\partial V}{\partial T}\right)_p. \tag{4}$$

However, the Onsager law extends beyond a 2-dimensional system. The integrability and the inaccessibility associated with the rate of entropy-production in more than 2 dimensions bear no resemblance to the symmetry in question.

The second-rank tensor A, the thermal or electrical conductivity tensor for example, is transformed according to the following law:

$$A'_{ij} = B_{ik}B_{jl}A_{kl}, \tag{5}$$

where B's are direction cosines between the old and the new axes. For crystals with just one n-fold symmetry axis, there is a plane perpendicular to the Z-axis if the symmetry axis is taken as the Z-axis. Merely rotating the axis will not change a physical quantity on the plane. Thus,

$$A'_{ij} = A_{ij}. \tag{6}$$

In matrix form,

$$\begin{bmatrix} A_{11} & A_{12} & A_{13} \\ A_{21} & A_{22} & A_{23} \\ A_{31} & A_{32} & A_{33} \end{bmatrix} = \begin{bmatrix} \cos\alpha & \sin\alpha & 0 \\ -\sin\alpha & \cos\alpha & 0 \\ 0 & 0 & 1 \end{bmatrix} \begin{bmatrix} A_{11} & A_{12} & A_{13} \\ A_{21} & A_{22} & A_{23} \\ A_{31} & A_{32} & A_{33} \end{bmatrix}$$

$$\begin{bmatrix} \cos\alpha & -\sin\alpha & 0 \\ \sin\alpha & \cos\alpha & 0 \\ 0 & 0 & 1 \end{bmatrix} = \begin{bmatrix} A_{11} & A_{12} & 0 \\ -A_{12} & A_{11} & 0 \\ 0 & 0 & A_{33} \end{bmatrix}, \tag{7}$$

where $\alpha = 2\pi/n$ is the angle of the plane rotation with respect to the Z-axis. For $n \geqq 4$, α is the angle of any amount of rotation. It is advantageous to choose α as a small angle, i.e., $\sin\alpha \approx \alpha$, $\cos\alpha \approx 1$. Then the multiplication of the matrices is simplified. From eq. (7), or crystal symmetry,

$$A_{21} = -A_{12}. \tag{8}$$

But A is the proportional matrix between current and potential gradient. From the Onsager law,

$$A_{21} = A_{12}. \tag{9}$$

Therefore,

$$A_{21} = A_{12} = 0. \tag{10}$$

We conclude that the matrix

$$\begin{bmatrix} A_{11} & 0 & 0 \\ 0 & A_{11} & 0 \\ 0 & 0 & A_{33} \end{bmatrix}$$

written for the components of the conductivity tensor for crystals having at least one n-fold symmetry axis, $n \geqq 4$, is derived from crystal symmetry and the Onsager law.

Let us assume that a substance exists as a homogeneous phase in n tautomeric forms, $Z_1, Z_2, \cdots, Z_n$, which can be converted into each other in pairs. Also let N_i be the number of moles of Z_i per unit volume, k_{ij} be the velocity constant of transforming Z_i into Z_j and k_{ji} be that of the reverse transformation.

For $n = 2$,

$$Z_1 \underset{k_{21}}{\overset{k_{12}}{\rightleftarrows}} Z_2.$$

According to the kinetic part of the law of mass action,

$$\frac{\partial N_1}{\partial t} = -N_1 k_{12} + N_2 k_{21}. \tag{11}$$

At equilibrium,

$$\frac{\partial N_1}{\partial t} = 0. \tag{12}$$

Or

$$\frac{k_{12}}{k_{21}} = \frac{N_2^{\text{eq}}}{N_1^{\text{eq}}} = K_{12}, \tag{13}$$

where K_{12} is the equilibrium constant.

For $n = 3$, there are three possible mechanisms:

mechanism A,

mechanism B,

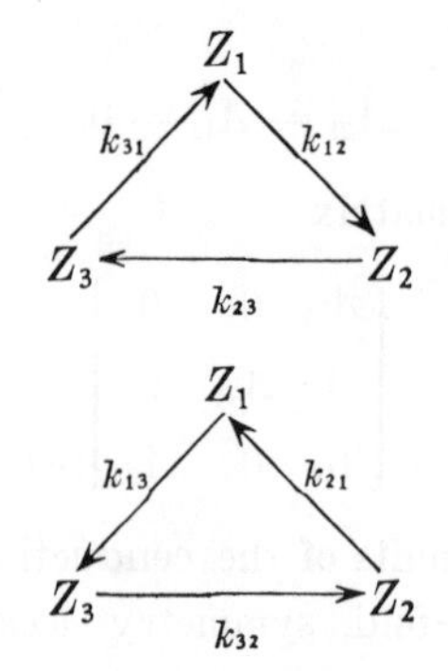

mechanism C, $Z_1 \underset{k_{21}}{\overset{k_{12}}{\rightleftarrows}} Z_2, \quad Z_2 \underset{k_{32}}{\overset{k_{23}}{\rightleftarrows}} Z_3, \quad Z_3 \underset{k_{13}}{\overset{k_{31}}{\rightleftarrows}} Z_1.$

At equilibrium, for mechanism A,

$$\begin{bmatrix} \dfrac{\partial N_1}{\partial t} \\ \dfrac{\partial N_2}{\partial t} \\ \dfrac{\partial N_3}{\partial t} \end{bmatrix} = \begin{bmatrix} -k_{12} & 0 & k_{31} \\ k_{12} & -k_{23} & 0 \\ 0 & k_{23} & -k_{31} \end{bmatrix} \begin{bmatrix} N_1 \\ N_2 \\ N_3 \end{bmatrix} = \begin{bmatrix} 0 \\ 0 \\ 0 \end{bmatrix}, \tag{14}$$

for mechanism B,

$$\begin{bmatrix} \dfrac{\partial N_1}{\partial t} \\ \dfrac{\partial N_2}{\partial t} \\ \dfrac{\partial N_3}{\partial t} \end{bmatrix} = \begin{bmatrix} -k_{13} & k_{21} & 0 \\ 0 & -k_{21} & k_{32} \\ k_{13} & 0 & -k_{32} \end{bmatrix} \begin{bmatrix} N_1 \\ N_2 \\ N_3 \end{bmatrix} = \begin{bmatrix} 0 \\ 0 \\ 0 \end{bmatrix}, \tag{15}$$

for mechanism C,

$$\begin{bmatrix} \dfrac{\partial N_1}{\partial t} \\ \dfrac{\partial N_2}{\partial t} \\ \dfrac{\partial N_3}{\partial t} \end{bmatrix} = \begin{bmatrix} -k_{12} - k_{13} & k_{21} & k_{31} \\ k_{12} & -k_{23} - k_{21} & k_{32} \\ k_{13} & k_{23} & -k_{31} - k_{32} \end{bmatrix} \cdot \begin{bmatrix} N_1 \\ N_2 \\ N_3 \end{bmatrix}$$

$$= \begin{bmatrix} 0 \\ 0 \\ 0 \end{bmatrix}. \tag{16}$$

Because the square matrices in eqs. (14—16) are singular, the equilibrium concentrations have non-trivial solutions: for mechanism A,

$$\frac{N_1^{eq}}{N_2^{eq}} = \frac{k_{23}}{k_{12}}, \quad \frac{N_2^{eq}}{N_3^{eq}} = \frac{k_{31}}{k_{23}}, \tag{17}$$

for mechanism B,

$$\frac{N_1^{eq}}{N_2^{eq}} = \frac{k_{21}}{k_{13}}, \quad \frac{N_2^{eq}}{N_3^{eq}} = \frac{k_{32}}{k_{21}}, \tag{18}$$

for mechanism C,

$$\left.\begin{aligned} \frac{N_1^{eq}}{N_2^{eq}} &= \frac{k_{23}k_{31} + k_{21}k_{31} + k_{21}k_{32}}{k_{12}k_{31} + k_{12}k_{32} + k_{32}k_{13}} \\ \frac{N_2^{eq}}{N_3^{eq}} &= \frac{k_{12}k_{31} + k_{12}k_{32} + k_{32}k_{13}}{k_{12}k_{23} + k_{23}k_{13} + k_{13}k_{21}} \end{aligned}\right\}. \tag{19}$$

To summarize, mechanisms A and B involve cyclic reactions but mechanism C involves "detailed balancing", i.e., the balancing of the reactions in pairs. Each of eqs. (17)—(19) satisfies the requirement for equilibrium. Then it is difficult to see why chemists insist in detailed balancing and why they insist in equilibrium constants being the ratio of the velocity constants in pairs.

While the answer to the first question cannot be given at present (See §27), the answer to the second question may be found in the Onsager law. Subtracting the equilibrium concentrations from the non-equilibrium ones in eq. (16),

$$\begin{bmatrix} \dfrac{\partial N_1}{\partial t} - \dfrac{\partial N_1^{eq}}{\partial t} \\ \dfrac{\partial N_2}{\partial t} - \dfrac{\partial N_2^{eq}}{\partial t} \\ \dfrac{\partial N_3}{\partial t} - \dfrac{\partial N_3^{eq}}{\partial t} \end{bmatrix} = \begin{bmatrix} -k_{12} - k_{13} & k_{21} & k_{31} \\ k_{12} & -k_{23} - k_{21} & k_{32} \\ k_{13} & k_{23} & -k_{31} - k_{32} \end{bmatrix} \begin{bmatrix} N_1 - N_1^{eq} \\ N_2 - N_2^{eq} \\ N_3 - N_3^{eq} \end{bmatrix}, \tag{20}$$

Now

$$N_i - N_i^{eq} = \Delta N_i \approx N_i^{eq} \ln\left(1 + \frac{\Delta N_i}{N_i^{eq}}\right) = N_i^{eq} \ln \frac{N_i}{N_i^{eq}}, \qquad (21)$$

because not far from equilibrium, ΔN_i is a small quantity. Assuming that the homogeneous phase is an ideal solution,

$$N_i - N_i^{eq} = \frac{N_i^{eq}(\mu_i - \mu_i^{eq})}{RT}, \qquad (22)$$

where μ is the chemical potential and R is the gas constant. Let

$$\frac{\partial(N_i - N_i^{eq})}{\partial t} = \frac{dj_i}{dt}, \qquad (23)$$

where j_i is the rate of chemical reaction. Substituting eqs. (22) and (23) in eq. (20),

$$\begin{bmatrix} \frac{dj_1}{dt} \\ \frac{dj_2}{dt} \\ \frac{dj_3}{dt} \end{bmatrix} = \frac{1}{RT} \begin{bmatrix} -k_{12} - k_{13} & k_{21} & k_{31} \\ k_{12} & -k_{23} - k_{21} & k_{32} \\ k_{13} & k_{23} & -k_{31} - k_{32} \end{bmatrix} \begin{bmatrix} (\mu_1 - \mu_1^{eq})N_1^{eq} \\ (\mu_2 - \mu_2^{eq})N_2^{eq} \\ (\mu_3 - \mu_3^{eq})N_3^{eq} \end{bmatrix}. \qquad (24)$$

From the hypothesis of linearity, §14,

$$\begin{bmatrix} \frac{dj_1}{dt} \\ \frac{dj_2}{dt} \\ \frac{dj_3}{dt} \end{bmatrix} = -\frac{1}{T} \begin{bmatrix} L_{11} & L_{12} & L_{13} \\ L_{21} & L_{22} & L_{23} \\ L_{31} & L_{32} & L_{33} \end{bmatrix} \begin{bmatrix} \mu_1 - \mu_1^{eq} \\ \mu_2 - \mu_2^{eq} \\ \mu_3 - \mu_3^{eq} \end{bmatrix}. \qquad (25)$$

Comparing eqs. (24) and (25),

$$\frac{N_2^{eq}}{N_1^{eq}} = \frac{k_{12}}{k_{21}} = K_{12} \text{ if } L_{12} = L_{21}, \tag{26}$$

$$\frac{N_3^{eq}}{N_1^{eq}} = \frac{k_{13}}{k_{31}} = K_{13} \text{ if } L_{31} = L_{13}, \tag{27}$$

$$\frac{N_3^{eq}}{N_2^{eq}} = \frac{k_{23}}{k_{32}} = K_{23} \text{ if } L_{23} = L_{32}. \tag{28}$$

The equation of compatibility is

$$k_{21}k_{13}k_{32} = k_{12}k_{31}k_{23}. \tag{29}$$

We conclude that the Onsager law is a strong condition for detailed balancing.

§ 16 *PROOF OF THE ONSAGER LAW*

We proceed to prove the Onsager law by (1) seeking a relation between macroscopic variables and entropy, (2) using the Boltzmann law for the connection between macroscopic and microscopic variables, and (3) applying the principle of microscopic reversibility and the ergodic theorem in microscopic ensembles.

Let us consider a well-insulated and isolated system, not far from equilibrium, the state of which is determined by a number of variables $\boldsymbol{\alpha}$. The deviation of entropy from its equilibrium value can be described by

$$\Delta S = S - S^{eq} = -\frac{1}{2}\,[g\boldsymbol{\alpha}|\boldsymbol{\alpha}], \tag{1}$$

where S^{eq} refers to equilibrium state and g is a positive-definite symmetric matrix. The quadratic form of eq. (1) is justified as a Taylor's series expansion terminated at the second order. The first order is zero and the second order is negative, because the entropy of an isolated system is at its maximum value at equilibrium. The matrix g is positive-definite because ΔS is necessarily negative, and it is symmetric because the matrix of a quadratic form can be so arranged. From eqs. (1, 6) and (14, 1), the rate of entropy-production is

$$\Delta \dot{S} = \frac{1}{T}[\boldsymbol{J}|-\boldsymbol{X}] = \frac{1}{T^2}[L'\boldsymbol{X}|\boldsymbol{X}] = [L\boldsymbol{X}|\boldsymbol{X}]. \tag{2}$$

We have set $L = L'/T^2$. In an isolated system there is no difference between $\Delta \dot{S}$ and $\Delta \dot{S}_i$. From eq. (1),

$$\Delta \dot{S} = [\dot{\boldsymbol{\alpha}}|-g\boldsymbol{\alpha}]. \tag{3}$$

Comparing eqs. (2) and (3),

$$\dot{\boldsymbol{\alpha}} = \frac{d\boldsymbol{\alpha}}{dt} = L\boldsymbol{X}, \tag{4}$$

$$-g\boldsymbol{\alpha} = \frac{\partial \Delta S}{\partial \boldsymbol{\alpha}} = \boldsymbol{X}. \tag{5}$$

According to the Boltzmann law,

$$S = k \ln f + \text{constant}, \tag{6}$$

where k is the Boltzmann constant and f is the thermodynamical probability. Or

$$\Delta S(\boldsymbol{\alpha}) = k\Delta \ln f(\boldsymbol{\alpha}).$$

The integrated form is

$$f(\boldsymbol{\alpha}) = f^{eq}(\boldsymbol{\alpha}) \exp\left[\frac{\Delta S(\boldsymbol{\alpha})}{k}\right], \tag{7}$$

where f^{eq} is probability at equilibrium. The random process of fluctuations of $\boldsymbol{\alpha}$ can be handled by approximating it as a Markov-Gaussian process. If the development of a process is independent of history prior to zero time, we call the process Markovian. An oft-quoted statement is that a Markov process has no memory of the past. A very general class of Markov process can be approximated by the Gaussian distribution, which is easy to manipulate mathematically. For example, if R is a symmetric and positive-definite matrix and R^{-1} denotes its inverse, the

Gaussian distribution with zero mean and covariance R in n dimensions has a probability density[1],

$$\phi(\boldsymbol{x}) = \frac{|R|^{1/2}}{(2\pi)^{n/2}} \exp\left(-\frac{1}{2}\boldsymbol{x}^T R \boldsymbol{x}\right). \tag{8}$$

We have

$$\int_{-\infty}^{\infty} \exp\left(-\frac{1}{2}\boldsymbol{x}^T R \boldsymbol{x}\right) d\boldsymbol{x} = \frac{(2\pi)^{n/2}}{|R|^{1/2}}, \tag{9}$$

$$\int_{-\infty}^{\infty} \boldsymbol{x} \exp\left(-\frac{1}{2}\boldsymbol{x}^T R \boldsymbol{x}\right) d\boldsymbol{x} = 0, \tag{10}$$

$$\int_{-\infty}^{\infty} \boldsymbol{x}^T A \boldsymbol{x} \exp\left(-\frac{1}{2}\boldsymbol{x}^T R \boldsymbol{x}\right) d\boldsymbol{x} = \frac{(2\pi)^{n/2}}{|R|^{1/2}} \operatorname{tr}(AR^{-1}), \tag{11}$$

where A is a symmetric matrix, not necessarily positive-definite. The use of eqs. (9)—(11) may be illustrated by the following example:

$$\int_{-\infty}^{\infty} (x_1^2 - 2x_1x_4) \exp\Big[-\frac{1}{2}(3x_1^2 + 2x_2^2 + 2x_3^2 + x_4^2 + 2x_1x_2$$

$$+ 2x_3x_4 - 6x_1 - 2x_2 - 6x_3 - 2x_4 + 8)\Big] dx_1dx_2dx_3dx_4$$

$$= \int_{-\infty}^{\infty} \Bigg\{(x_1 - 1, x_2, x_3 - 2, x_4 + 1)$$

$$\begin{bmatrix} 1 & 0 & 0 & -1 \\ 0 & 0 & 0 & 0 \\ 0 & 0 & 0 & 0 \\ -1 & 0 & 0 & 0 \end{bmatrix} \begin{bmatrix} x_1 - 1 \\ x_2 \\ x_3 - 2 \\ x_4 + 1 \end{bmatrix} + (4, 0, 0, -2)$$

$$\begin{bmatrix} x_1 - 1 \\ x_2 \\ x_3 - 2 \\ x_4 + 1 \end{bmatrix} + 3\Bigg\}$$

$$\exp\left\{-\frac{1}{2}(x_1-1, x_2, x_3-2, x_4+1)\begin{bmatrix}3 & 1 & 0 & 0\\ 1 & 2 & 0 & 0\\ 0 & 0 & 2 & 1\\ 0 & 0 & 1 & 1\end{bmatrix}\begin{bmatrix}x_1-1\\ x_2\\ x_3-2\\ x_4+1\end{bmatrix}\right\} d(x_1-1)dx_2d(x_3-2)d(x_4+1)$$

$$= \frac{(2\pi)^2}{\begin{vmatrix}3 & 1 & 0 & 0\\ 1 & 2 & 0 & 0\\ 0 & 0 & 2 & 1\\ 0 & 0 & 1 & 1\end{vmatrix}^{1/2}}\left\{\operatorname{tr}\begin{bmatrix}1 & 0 & 0 & -1\\ 0 & 0 & 0 & 0\\ 0 & 0 & 0 & 0\\ -1 & 0 & 0 & 0\end{bmatrix}\frac{1}{5}\begin{bmatrix}2 & -1 & 0 & 0\\ -1 & 3 & 0 & 0\\ 0 & 0 & 5 & -5\\ 0 & 0 & -5 & 10\end{bmatrix}+3\right\} = \frac{68\sqrt{5}\,\pi^2}{25}.$$

In the present example, substituting eq. (1) in eq. (7) and comparing with eq. (8),

$$R = \frac{g}{k}, \tag{12}$$

$$f(\boldsymbol{a}) = \left[\frac{|R|^{1/2}}{(2\pi)^{n/2}}\right]\exp\left[-\frac{1}{2}\boldsymbol{a}^T R\boldsymbol{a}\right]. \tag{13}$$

Let the ensemble be E. From eq. (10),

$$E(\boldsymbol{a}) = \int_{-\infty}^{\infty}\boldsymbol{a}f(\boldsymbol{a})d\boldsymbol{a} = 0. \tag{14}$$

From eq. (11),

$$E[\boldsymbol{a}|\boldsymbol{a}] = \int_{-\infty}^{\infty}\boldsymbol{a}^T\boldsymbol{a}f(\boldsymbol{a})d\boldsymbol{a} = \operatorname{tr}\frac{k}{g}. \tag{15}$$

From eqs. (5) and (11),

$$E[\boldsymbol{a}|\boldsymbol{X}] = -\int_{-\infty}^{\infty} \boldsymbol{a}^T g \boldsymbol{a} f(\boldsymbol{a}) d\boldsymbol{a} = -\mathrm{tr}(k\overleftrightarrow{\mathrm{I}}). \tag{16}$$

The assumption of microscopic reversibility requires that if α_i and α_j are two quantities which depend on the configuration of molecules and atoms, the event α_i followed Γ seconds later by α_j will occur just as often as α_j followed Γ seconds later by α_i,

$$\overline{\alpha_i(t)\alpha_j(t+\Gamma)} = \overline{\alpha_j(t)\alpha_i(t+\Gamma)}, \tag{17}$$

where the bar represents a time average, defined as

$$\overline{F(t)} = \lim_{r\to\infty} \frac{1}{2r} \int_{-r}^{r} f(t) dt. \tag{18}$$

The equation

$$\overline{\alpha_i(t)\alpha_j(t)} = \overline{\alpha_j(t)\alpha_i(t)} \neq 0 \tag{19}$$

merely indicates the interaction between fluctuating parameters. Subtracting eq. (19) from eq. (17),

$$\overline{\alpha_i(t)[\alpha_j(t+\Gamma) - \alpha_j(t)]} = \overline{\alpha_j(t)[\alpha_i(t+\Gamma) - \alpha_i(t)]}. \tag{20}$$

Dividing both sides of eq. (20) by Γ and allowing Γ to approach zero,

$$\overline{\alpha_i(t)\frac{d\alpha_j(t)}{dt}} = \overline{\alpha_j(t)\frac{d\alpha_i(t)}{dt}}. \tag{21}$$

It must be emphasized that Γ is a short time much greater than the collision time to establish a statistical average but much shorter than the relaxation time so as not to establish equilibrium.

From the ergodic theorem, time averages are equal to ensemble averages. Equation (21) becomes

$$E[\alpha_i(t)|\dot{\alpha}_j(t)] = E[\alpha_j(t)|\dot{\alpha}_i(t)]. \tag{22}$$

From eqs. (4), (16) and (22),

$$-k \sum_l L_{jl}\delta_{li} = -k \sum_l L_{il}\delta_{lj}. \tag{23}$$

Therefore,

$$L_{ji} = L_{ij}. \tag{24}$$

Q. E. D.

In spite of some misgivings about the proof, the postulate of the reciprocal relationship has been overwhelmingly verified experimentally[2]. It is justified to elevate the status of the postulate to that of a law.

It has been shown in §15 that a strong condition of detailed balancing is the Onsager law. It will be shown in §27 that a weak condition is the ergodic hypothesis. The question naturally arises of the connection between the detailed balancing and the ergodic hypothesis. Thomsen[3], from a microscopic point of view, has considered the transitions between discrete quantum levels of an isolated system. Let $\pi_i(t)$ be the initial and unconditional probability of a state i and R_{ij} be the conditional transition probability per unit time for the transition $i \to j$ (R_{ij} independent of t). Let the entropy of the system be defined by

$$S = k \ln f = -k \sum_i \pi_i \ln \pi_i. \tag{25}$$

Thomsen considered four propositions, defined as follows:

(A) microscopic reversibility (M),

$$R_{ij} = R_{ji}; \tag{26}$$

(B) detailed balancing (D),

$$\pi_i R_{ij} = \pi_j R_{ji} \text{ in equilibrium}; \tag{27}$$

(C) ergodic hypothesis (E),

$$\pi_i = \pi_j = \frac{1}{f} = \text{constant in equilibrium;} \tag{28}$$

(D) second law (S),

$$\dot{S} > 0. \tag{29}$$

He is able to derive the following interesting results: The ergodic hypothesis and the second law are equivalent; when combined with the principle of detailed balancing, the second law is equivalent to the principle of microscopic reversibility. Symbolically,

$$(S) \rightleftarrows (E); \quad (S) + (D) \rightleftarrows (M).$$

References

[1] F. A. Graybill: Introduction to Matrices, with Application to Statistics Wardsworth, Belmont, Calif. (1969).

[2] D. G. Miller: *Chem. Rev.*, **60**, 16 (1960); The Onsager Relations-Experimental Evidence, Symposium on the Foundations of Continuum Thermodynamics (1973).

[3] J. S. Thomsen: *Phys. Rev.*, **91**, 1263, 1953.

§17 *MODIFICATIONS OF THE ONSAGER LAW*

Let $\boldsymbol{J}$ and $\boldsymbol{X}$ be column vectors, with the linear relation holding,

$$\boldsymbol{J} = -\frac{L'\boldsymbol{X}}{T} = -L\boldsymbol{X}. \tag{1}$$

If we transform $\boldsymbol{J}$ and $\boldsymbol{X}$ such that

$$\boldsymbol{J}' = \boldsymbol{J}, \tag{2}$$

$$\boldsymbol{X}' = (1 + KL)\boldsymbol{X}, \tag{3}$$

where K is any skew-symmetric matrix. Because the quadratic form of a skew-symmetric matrix vanishes, the inner product is invariant,

$$[\boldsymbol{J}'|\boldsymbol{X}'] = [\boldsymbol{J}|\boldsymbol{X}]. \tag{4}$$

Let Q be any orthogonal matrix being a function of θ. From the definition of an orthogonal matrix,

$$[Q|Q] = 1. \tag{5}$$

Therefore,

$$\frac{d[Q|Q]}{dt} = \{[\dot{Q}|Q] + [Q|\dot{Q}]\}\dot{\theta} = 0. \tag{6}$$

$$\text{If } \dot{\theta} \neq 0, \quad \dot{Q}^{\mathrm{T}}Q = -Q^{\mathrm{T}}\dot{Q} = -(\dot{Q}^{\mathrm{T}}Q)^{\mathrm{T}}. \tag{7}$$

Thus $\dot{Q}^{\mathrm{T}}Q$ is a skew-symmetric matrix. Replacing K in eq. (3) by $\dot{Q}^{\mathrm{T}}Q\dot{\theta}$,

$$\boldsymbol{X}' = \boldsymbol{X} - \dot{Q}^{\mathrm{T}}Q\dot{\theta}\boldsymbol{J} = \boldsymbol{X} + Q\dot{Q}^{\mathrm{T}}\dot{\theta}\boldsymbol{J} \tag{8}$$

[Equation (8) can be also written as

$$\boldsymbol{X}' = \boldsymbol{X} + \boldsymbol{K} \times \boldsymbol{J} \tag{9}$$

by considering the skew-symmetric matrix K as a vector. The cross product can be extended to any dimensions[1]].

If $\dot{\theta} = 0$, the transformation in eq. (8) is trivial, i.e., the identity transformation.

If $\dot{\theta} \neq 0$, we can identify $\dot{\theta}$ as the angular velocity and $\boldsymbol{X}$ is an odd function of $\boldsymbol{\omega}$.

Because $\boldsymbol{J}$ is not directly related to $\boldsymbol{\omega}$, $\boldsymbol{J}$ is an even function of $\boldsymbol{\omega}$. Because

$$L = \frac{1}{2}(L^{\mathrm{sy}} + L^{\mathrm{sk}}), \tag{10}$$

and more specifically, L^{sk} is an odd function of $\boldsymbol{\omega}$,

$$L^{\mathrm{sk}}(\boldsymbol{\omega}) = -L^{\mathrm{sk}}(-\boldsymbol{\omega}); \tag{11}$$

but L^{sy} is not directly related to $\boldsymbol{\omega}$,

$$L^{\mathrm{sy}}(\boldsymbol{\omega}) = L^{\mathrm{sy}}(-\boldsymbol{\omega}). \tag{12}$$

Thus

$$L(\boldsymbol{\omega}) = \frac{1}{2}\,[L^{\mathrm{sy}}(-\boldsymbol{\omega}) - L^{\mathrm{sk}}(-\boldsymbol{\omega})] = L^{\mathrm{T}}(-\boldsymbol{\omega}). \qquad (13)$$

If a magnetic field $\boldsymbol{B}$ is applied in the direction of the Z-axis, for example, the Z-axis is a unique direction, even for an isotropic material. However, the XY-plane is still isotropic. We have the same state of affairs with angular velocity. While the symmetric part of L in XY-plane is not affected by the reverse of the sign $\boldsymbol{B}$, the skew-symmetric part of L in the direction of the Z-axis is reversed in sign by the reverse of the sign of $\boldsymbol{B}$. Or

$$L(\boldsymbol{B}) = L^{\mathrm{T}}(-B). \qquad (14)$$

Equations (13) and (14) can be combined,

$$L(\boldsymbol{\omega}, \boldsymbol{B}) = L^{\mathrm{T}}(-\boldsymbol{\omega}, -\boldsymbol{B}) \qquad (15)$$

In the derivation of the Onsager law (§16), from microscopic reversibility, $\boldsymbol{X}$ is assumed to be an even function of time (as it is true for all physical laws). The Onsager law must be modified if $\boldsymbol{X}$ is an odd function of linear velocity $\boldsymbol{v}$, such as in viscous fluids. If ε_α is defined as $+1$ for $\boldsymbol{X}_i$ being an even function of $\boldsymbol{v}$ and ε_β as -1 for $\boldsymbol{X}_j$ being an odd function of $\boldsymbol{v}$,

$$L_{ij}(\boldsymbol{\omega}, \boldsymbol{B}) = \varepsilon_\alpha \varepsilon_\beta L_{ji}(-\boldsymbol{\omega}, -\boldsymbol{B}), \qquad (16)$$

which is a complete statement of the Onsager law, the symmetry and parity relations being combined[2].

References

[1] R. L. Eisenman and D. R. Barr: *Am. Math. monthly*, **70**, 82 (1963).
[2] H. B. G. Casimir: *Rev. Mod. Phys.*, **17**, 343 (1945).

§18 THEOREMS DEDUCED FROM THE ONSAGER LAW

Theorem 1 The Onsager law holds for the new sets of $\boldsymbol{J}'$ and $\boldsymbol{X}'$, if $\boldsymbol{X}'$ is transformed contragrediently to $\boldsymbol{J}'$ and if elements of both $\boldsymbol{J}$ and $\boldsymbol{X}$ are independent.

Proof: Given $\boldsymbol{J}' = C\boldsymbol{J}$ and $\boldsymbol{X}' = (C^{\mathrm{T}})^{-1}\boldsymbol{X}$.

From eq. (1, 6),

$$\sigma = -\frac{1}{T}[\boldsymbol{J}'|\boldsymbol{X}'] = -\frac{1}{T}[C\boldsymbol{J}|\boldsymbol{X}'] = \frac{1}{T^2}[CL\boldsymbol{X}|\boldsymbol{X}']$$

$$= \frac{1}{T^2}[CLC^{\mathrm{T}}\boldsymbol{X}'|\boldsymbol{X}']. \tag{1}$$

Let L' be the transformed matrix of the phenomenological coefficients,

$$\sigma = -\frac{1}{T}[\boldsymbol{J}'|\boldsymbol{X}'] = \frac{1}{T^2}[L'\boldsymbol{X}'|\boldsymbol{X}']. \tag{2}$$

Comparing eqs. (1) and (2),

$$L' = CLC^{\mathrm{T}}, \tag{3}$$

or L is transformed congruently. Transposing eq. (3),

$$(L')^{\mathrm{T}} = CLC^{\mathrm{T}}. \tag{4}$$

Comparing eqs. (3) and (4),

$$(L')^{\mathrm{T}} = L'. \tag{5}$$

Q. E. D.

This theorem can be easily understood from group theory. If a congruent group contains a symmetric matrix, then all matrices in the group are symmetric.

Theorem 2 The Onsager law holds for the sets of $\boldsymbol{J}$ and $\boldsymbol{X}$ if elements of $\boldsymbol{J}$ are linearly dependent but those of $\boldsymbol{X}$ are independent.

Proof: Because elements of $\boldsymbol{J}$ are linearly dependent,

$$\sum_{i}^{n} a_i J_i = 0. \tag{6}$$

The rate of entropy-production is

$$\sigma = -\frac{1}{T}\sum_{i}^{n} J_i X_i = -\frac{1}{T}\sum_{i}^{n-1} J_i\left[X_i - \frac{a_i}{a_n}X_n\right]. \tag{7}$$

For the third member of eq. (7),

$$J_i = \sum_{j}^{n-1} L_{ij}X_j - \sum_{j}^{n-1}\frac{a_i}{a_n}L_{ij}X_n; \tag{8}$$

for the second member of eq. (7),

$$J_i = \sum_{j}^{n-1} L'_{ij}X_j + L'_{in}X_n, \tag{9}$$

where L and L' are the matrices of phenomenological coefficients of $n-1$ and n dimensions respectively (the factor $1/T^2$ having been absorbed in L or L'). Comparing eqs. (8) and (9),

$$L'_{ij} = L_{ij}, \quad i, j = 1, 2, \cdots, (n-1), \tag{10}$$

and

$$L'_{in} = -\sum_{i}^{n-1}\frac{a_i}{a_n}L_{ij}. \tag{11}$$

Setting $i = n$ in eq. (9),

$$J_n = \sum_{j}^{n-1} L'_{nj}X_j + L'_{nn}X_n. \tag{12}$$

Multiplying each term in eq. (8) by a_i and summing in i,

$$-a_nJ_n = \sum_{i}^{n-1}\sum_{j}^{n-1} L_{ij}\left[a_iX_j - \frac{a_i^2}{a_n}X_n\right]. \tag{13}$$

Comparing eqs. (12) and (13),

$$L'_{ni} = -\sum_{j}^{n-1}\frac{a_j}{a_n}L_{ji}. \tag{14}$$

Because L is symmetric, from eqs. (10), (11), and (14), L' is also symmetric.

Q. E. D.

This theorem can be easily understood from matrix theory. Because independent elements of $\boldsymbol{J}$ are expressed as a column matrix of $(n-1) \times 1$, and those of $\boldsymbol{X}$ as a column matrix of $n \times 1$, L is a matrix of $(n-1) \times n$. Because L is singular, the nth column of L is a linear combination of all the previous columns. If we add another row, of which the first $(n-1)$ elements are the same as those in the nth column and the last element is set zero or any positive number, the new square matrix is symmetric and singular or non-singular. It is noted that in this case the (negative) dissipation function is positive semi-definite.

Theorem 3 The Onsager law holds for the sets of $\boldsymbol{J}$ and $\boldsymbol{X}$ if elements of $\boldsymbol{X}$ are linearly dependent but those of $\boldsymbol{J}$ are independent.

Proof: Write $\boldsymbol{X} = -L\boldsymbol{J}/T$ and proceed as in the proof of theorem 2 except that the role of $\boldsymbol{X}$ and $\boldsymbol{J}$ is interchanged.

Theorem 4 The Onsager law does not necessarily hold for the sets of $\boldsymbol{J}$ and $\boldsymbol{X}$ if the elements of both $\boldsymbol{J}$ and $\boldsymbol{X}$ are linearly dependent. However, the law holds if a matrix of L' is constructed with the conditions: (A) L'_{ij} bearing a certain relationship to L_{ij}, i, $j = 1, 2, \cdots, (n-1)$; (B) $L'_{in} = L'_{ni}$, $i = 1, 2, \cdots, (n-1)$; and (C) L'_{nn} arbitrary except that the determinant of L' is not negative-definite

Proof: Because the elements of $\boldsymbol{J}$ are linearly dependent,

$$\sum_{i}^{n} a_i J_i = 0. \qquad (15)$$

Because the elements of $\boldsymbol{X}$ are linearly dependent,

$$\sum_{i}^{n} b_i X_i = 0. \qquad (16)$$

The rate of entropy-production is

$$\sigma = -\frac{1}{T}\sum_{i}^{n} J_i X_i = -\frac{1}{T}\sum_{i}^{n-1}\left[J_i X_i + \frac{a_i}{a_n} J_i \sum_{j}^{n-1}\frac{b_j}{b_n} X_j\right]. \tag{17}$$

We denote L for the coefficients of independent $\boldsymbol{J}$ and $\boldsymbol{X}$,

$$\begin{aligned} J_i &= \sum_{k}^{n-1} L_{ik}\left[X_k + \frac{a_k}{a_n}\sum_{j}^{n-1}\frac{b_j}{b_n} X_j\right] \\ &= \sum_{k}^{n-1}\left[L_{ik} + \frac{b_k}{b_n}\sum_{j}^{n-1} L_{ij}\frac{a_j}{a_n}\right] X_k. \end{aligned} \tag{18}$$

In the preceding line we use the fact, that in the double summation, the dummy indices can be interchanged. Let us denote L' for the coefficients of dependent $\boldsymbol{J}$ and $\boldsymbol{X}$,

$$J_i = \sum_{k}^{n-1}\left[L'_{ik} - \frac{b_k}{b_n} L'_{in}\right] X_k. \tag{19}$$

By comparing eqs. (18) and (19), the condition (A) stated in the theorem is

$$L'_{ik} - \frac{b_k}{b_n} L'_{in} = L_{ik} + \frac{b_k}{b_n}\sum_{j}^{n-1}\frac{a_j}{a_n} L_{ij} \tag{20}$$

Multiplying each term of eq. (18) by a_i/a_n and summing in i,

$$-J_n = \sum_{i}^{n-1}\sum_{k}^{n-1}\frac{a_i}{a_n}\left[L_{ik} + \frac{b_k}{b_n}\sum_{j}^{n-1}\frac{a_i a_j}{a_n^2} L_{ij}\right] X_k. \tag{21}$$

Comparing eqs. (19) and (21), after setting $i = n$ in eq. (19),

$$L'_{nk} - \frac{b_k}{b_n} L'_{nn} = -\sum_{i}^{n-1}\frac{a_i}{a_n}\left[L_{ik} + \frac{b_k}{b_n}\sum_{j}^{n-1}\frac{a_i a_j}{a_n^2} L_{ij}\right]. \tag{22}$$

If L'_{nn} is known, L'_{nk} can be found. L'_{nn} is arbitrary except that the determinant of L' is not negative.

Q. E. D.

In other words, to construct the symmetric L' matrix, eq. (20) furnishes $(n-1)^2$ relations, and eq. (22) and $L'_{in} = L'_{ni}$ furnish $2(n-1)$ relations. A total of (n^2-1) relations of the elements of L' is known for its n^2 elements. There is one relation short. By arbitraily setting the value of L'_{nn}, except that det $L' \geqq 0$, all n^2 elements of L' can be found.

We do not need theorems 2—4 if we trim off the dependent element of L'. However, the new elements of $\boldsymbol{J}$ and $\boldsymbol{X}$ are usually unwieldy in form.

§19 *THE DISSIPATION OF ENERGY*

Although the rate of entropy-production is discriminative in deciding the direction of an irreversible process, the evaluation of the dissipated energy is useful for keeping account of the process.

The Curie law (§13) is valid for an isotropic substance. If we insert $\boldsymbol{J} = -L\boldsymbol{X}/T$ in eq. (8, 17), $T\sigma$ for fluids can be written as the sum of quadratic forms. Before insertion, the rate of dissipation of energy for the symmetric viscous pressure tensor can be simplified:

$$
\begin{aligned}
&-[(\Pi - p\overleftrightarrow{1})^{sy} | \text{Grad}\, \boldsymbol{v}] \\
&\quad = -\left[\frac{1}{3}(\text{tr}\, p)\overleftrightarrow{1} | \text{Grad}\, \boldsymbol{v}\right] - \left[p^{sy} - \frac{1}{3}(\text{tr}\, p)\overleftrightarrow{1} | \text{Grad}\, \boldsymbol{v}\right] \\
&\quad = \eta_V [(\text{div}\, \boldsymbol{v})\overleftrightarrow{1} | \text{Grad}\, \boldsymbol{v}] + 2\eta [\text{Grad}^{0,sy}\, \boldsymbol{v} | \text{Grad}\, \boldsymbol{v}] \\
&\quad = \eta_V \left[(\text{div}\, \boldsymbol{v})\overleftrightarrow{1} | \frac{1}{3}(\text{div}\, \boldsymbol{v})\overleftrightarrow{1}\right] + 2\eta\, [\text{Grad}^{sy}\, \boldsymbol{v} | \text{Grad}^{sy}\, \boldsymbol{v}] \\
&\qquad - 2\eta \left[\frac{1}{3}(\text{div}\, \boldsymbol{v})\overleftrightarrow{1} | \frac{1}{3}(\text{div}\, \boldsymbol{v})\overleftrightarrow{1}\right]
\end{aligned}
$$

$$= \left(\eta_V - \frac{2}{3}\eta\right)(\operatorname{div} \boldsymbol{v})^2 + 2\eta(\text{Grad}^{sy}\, \boldsymbol{v})^2. \tag{1}$$

In the first equality of eq. (1), we split p into the trace and a symmetric tensor without a trace, in the second equality we use the defining eqs. (5, 5) and (5, 6), in the third equality we use the symmetry of the inner product, and in the last equality we make use of

$$\overleftrightarrow{1}:\overleftrightarrow{1} = \left\{\begin{bmatrix} 1 & 0 & 0 \\ 0 & 1 & 0 \\ 0 & 0 & 1 \end{bmatrix}\begin{bmatrix} 1 & 0 & 0 \\ 0 & 1 & 0 \\ 0 & 0 & 1 \end{bmatrix}\right\} = 3. \tag{2}$$

Thus

$$T\sigma = \sum_{ik}\left(\frac{\text{grad}\, T}{T}, \text{grad}_T\, \mu_i - F_i\right)\begin{bmatrix} L_{qq} & L \\ L & L_{ik} \end{bmatrix}\begin{bmatrix} \dfrac{\text{grad}\, T}{T} \\ \text{grad}_T\, \mu_k - F_k \end{bmatrix}$$

$$+ \sum_{ik}\frac{1}{2}\left[\frac{\partial v_i}{\partial x_k} + \frac{\partial v_k}{\partial x_i}\right](2\eta)\frac{1}{2}\begin{bmatrix} \dfrac{\partial v_i}{\partial x_k} \\ + \\ \dfrac{\partial v_k}{\partial x_i} \end{bmatrix}$$

$$+ \sum_{ik}(\operatorname{div} \boldsymbol{v}, A_i)\begin{bmatrix} \eta_V - \dfrac{2}{3}\eta & L' \\ -L' & L'_{ik} \end{bmatrix}\begin{bmatrix} \operatorname{div} \boldsymbol{v} \\ A_k \end{bmatrix}, \tag{3}$$

where the matrices in the quadratic forms are those of phenomenological coefficients or physical properties or both.

Because both heat flow and mass flow are vectorial fluxes, the interactions are allowed by the Curie law. Because both the force of the viscous flow due to div $\boldsymbol{v}$ and that of chemical reactions are scalars, the interactions are also allowed by the Curie law. Because div $\boldsymbol{v}$ is the force of the second kind, i.e., $\varepsilon = -1$ in eq. (17, 16), in contrast to the symmetric matrix L, the matrix L' is anti-symmetric. However, its effects are cancelled out in the non-diagonal part of the last quadratic form. Thus we have an example of two irreversible processes which interact but without any contribution to the dissipation of energy.

For historical reasons, we evaluate Rayleigh's dissipation function, i.e., the dissipation of energy due to viscous flow alone. Now

$$\operatorname{div} \boldsymbol{v} = \frac{\partial v_i}{\partial x_i}, \quad i = 1, 2, 3. \tag{4}$$

$$\operatorname{Grad}^{sy} \boldsymbol{v} = \frac{1}{2}\left(\frac{\partial v_i}{\partial x_j} + \frac{\partial v_j}{\partial x_i}\right), \quad i, j = 1, 2, 3. \tag{5}$$

From eq. (1),

$$\begin{aligned} T\sigma = \eta \Big\{ 2 \Big[\left(\frac{\partial v_1}{\partial x_1}\right)^2 + \left(\frac{\partial v_2}{\partial x_2}\right)^2 + \left(\frac{\partial v_3}{\partial x_3}\right)^2 \Big] \\ + \left(\frac{\partial v_1}{\partial x_2} + \frac{\partial v_2}{\partial x_1}\right)^2 + \left(\frac{\partial v_2}{\partial x_3} + \frac{\partial v_3}{\partial x_2}\right)^2 + \left(\frac{\partial v_3}{\partial x_1} + \frac{\partial v_1}{\partial x_3}\right)^2 \Big\} \\ + \left(\eta_V - \frac{2}{3}\eta\right)\left(\frac{\partial v_1}{\partial x_1} + \frac{\partial v_2}{\partial x_2} + \frac{\partial v_3}{\partial x_3}\right)^2 \\ = \eta \Big\{ \frac{2}{3} \Big[\left(\frac{\partial v_1}{\partial x_1} - \frac{\partial v_2}{\partial x_2}\right)^2 + \left(\frac{\partial v_2}{\partial x_2} - \frac{\partial v_3}{\partial x_3}\right)^2 \\ + \left(\frac{\partial v_3}{\partial x_3} - \frac{\partial v_1}{\partial x_1}\right)^2 \Big] + \left(\frac{\partial v_1}{\partial x_2} + \frac{\partial v_2}{\partial x_1}\right)^2 + \left(\frac{\partial v_2}{\partial x_3} + \frac{\partial v_3}{\partial x_2}\right)^2 \\ + \left(\frac{\partial v_3}{\partial x_1} + \frac{\partial v_1}{\partial x_3}\right)^2 \Big\} + \eta_V \left(\frac{\partial v_1}{\partial x_1} + \frac{\partial v_2}{\partial x_2} + \frac{\partial v_3}{\partial x_3}\right)^2. \end{aligned} \tag{6}$$

In hydrodynamics, η_V is usually ignored, but we see that, from eq. (6), the shear and the volume viscosity play different roles in the rate of dissipation of energy.

If

$$\frac{\partial v_i}{\partial x_j} + \frac{\partial v_j}{\partial x_i} = 0, \quad i \neq j, \tag{7}$$

$$\frac{\partial v_1}{\partial x_1} = \frac{\partial v_2}{\partial x_2} = \frac{\partial v_3}{\partial x_3}, \tag{8}$$

the shear viscosity coefficient does not contribute but the volume viscosity coefficient does. If, in addition to the conditions (7) and (8),

$$\operatorname{div} \boldsymbol{v} = 0, \tag{9}$$

the flow is a reversible process.

The integration[1] of the partial differential equations (7) and (8) yields

$$\boldsymbol{v} = \boldsymbol{a} + \boldsymbol{b} \times \boldsymbol{r} + c\boldsymbol{r} + 2\boldsymbol{r}(\boldsymbol{d} \cdot \boldsymbol{r}) - \boldsymbol{d}r^2, \tag{10}$$

where $\boldsymbol{a}$, $\boldsymbol{b}$, c and $\boldsymbol{d}$ are functions of time but independent of space coordinates, and $\boldsymbol{r}$ is the position vector. This is easily checked from the matrix of Grad $\boldsymbol{v}$,

$$\operatorname{Grad} \boldsymbol{v} = c\overleftrightarrow{1} + 2(\boldsymbol{d} \cdot \boldsymbol{r})\overleftrightarrow{1} + \boldsymbol{b} \times \overleftrightarrow{1} + 2\overleftrightarrow{1} \times (\boldsymbol{r} \times \boldsymbol{d}). \tag{11}$$

The first two matrices on the right of eq. (11) are scalar; thus eqs. (7) and (8) are satisfied. The last two matrices are skew-symmetric, thus they are also satisfied. The diagonal terms of Grad $\boldsymbol{v}$ are not zero, but the non-diagonal terms are equal and opposite in sign. In the example of the non-contribution of the shear viscosity coefficient, we can set

$$\boldsymbol{b} = \boldsymbol{d} = 0, \quad \text{or } c = 0, \boldsymbol{d} = 0.$$

Thus

$$\boldsymbol{v} = \boldsymbol{a} + c\boldsymbol{r}, \quad \text{or } \boldsymbol{v} = \boldsymbol{a} + \boldsymbol{b} \times \boldsymbol{r}. \tag{12}$$

The flow is characterized by the dilation or the rotation.

Reference

[1] J. Meixner: *Ann. Phys.*, **41**, (5) 409 (1942).

§20 THE LAW OF MINIMUM ENTROPY-PRODUCTION

For an irreversible process in an isolated system, the law

of maximum entropy in classical thermodynamics prevails. By using a quadratic function of two variables, Prigogine[1] proved that the stationary state of linear irreversible processes, in which the Onsager law is valid, is a state of minimum entropy-production per unit volume per unit time, σ. Under the same restrictions de Groot and Mazur[2] proved that the stationary state is a state of minimum entropy-production per unit time, P, by using a variational principle. By a stationary state we mean that, as time tends to infinity, the irreversibility-determining function is finite and constant, or periodic, or almost periodic (This is a more general definition than that of steady equilibrium in §2). We can easily extend the quadratic function to many variables[3].

From eq. (1, 6),

$$\sigma = -\frac{1}{T}[\boldsymbol{J}(x)|\boldsymbol{X}(x)], \tag{1}$$

in the interval $a_i \geqq x_i \geqq b_i$, where the coordinates x_i are bounded. From the Onsager law and letting $L'/T^2 = L$,

$$\sigma = [L\boldsymbol{X}|\boldsymbol{X}] = [\boldsymbol{X}|L\boldsymbol{X}]. \tag{2}$$

Because σ is invariant with respect to coordinate systems, the admissible transformation is orthogonal, which preserves the inner product,

$$\sigma = [Q\boldsymbol{Y}|LQ\boldsymbol{Y}]. \tag{3}$$

Because Q is an orthogonal matrix, we have

$$\sigma = [\boldsymbol{Y}|Q^T LQ\boldsymbol{Y}] = [\boldsymbol{Y}|\lambda\boldsymbol{Y}], \tag{4}$$

where λ are characteristic roots of the matrix L. The roots are real because L is symmetric. The roots are positive because L is positive definite.

$$\sigma = \lambda_1 y_1^2 + \lambda_2 y_2^2 + \cdots + \lambda_n y_n^2, \quad \lambda_1 \geqq \lambda_2 \geqq \cdots \geqq \lambda_n \geqq 0. \tag{5}$$

Because σ is positive for all values of $\boldsymbol{J}$ and $\boldsymbol{X}$ distinct from zero, the quadratic form $[X|LX]$ has minimum values. The problem of finding minimum values of the quadratic form $[X|LX]$ on the unit sphere $[X|X]=1$ is equivalent to the problem of finding minimum values of $[Y|\lambda Y]$ on $[Y|Y]$. Because the largest characteristic root is λ_1 and the smallest λ_n, we easily establish two local minima,

$$\lambda_1 \geqq \frac{[X|LX]}{[X|X]} \geqq \lambda_n. \tag{6}$$

By using the Courant-Fisher mini-maxi theorem[4], we can establish all local minima,

$$\lambda_1 = \max_X \frac{[X|LX]}{[X|X]}, \tag{7}$$

$$\lambda_2 = \min_{[Y|Y]=1} i \max_{[X|Y]=0} i \frac{[X|LX]}{[X|X]}, \tag{8}$$

....

The geometrical meaning is clear. The problem of finding minimum values of the ellipsoid σ on the unit sphere is equivalent to finding the semiaxis of the unit ellipsoid of the largest length. We rotate the unit ellipsoid until the axes of the unit ellipsoid and the coordinate axes are aligned. The major axis assumes the largest value. To find the semiaxis of the unit ellipsoid of the second largest length we pass a plane through the origin and orthogonal to the semiaxis of the largest length. Thus we have a unit ellipsoid of $(n-1)$ dimensions. By repeating the procedure we can find a sequence of semi-axes of decreasing length. In case the characteristic roots of the matrix L are not all distinct, two or more semi-axes may be equal in length.

We have not dealt with the case $\lambda_i \to \infty$. Although the case is excluded from the bounded coordinate, we could follow Mikhlin's[5] treatment to get an interpretation. Let $\mathscr{L}$ be an arbitrary operator and λ be some number, real or complex. If $(\mathscr{L}-\lambda E)^{-1}$ does not exist, where E is the identity operator, then λ is called a characteristic root of $\mathscr{L}$. The discrete spectrum of a self-adjoined operator lies on the real axis of the plane and has a superior and an inferior limit. If λ_i is less than the inferior limit, the operator $(\mathscr{L} - \lambda E)$ is positive definite. If λ_i is greater then the superior limit, the operator $(\lambda E - \mathscr{L})$ is positive definite. In either case λ_i is not a characteristic root of $\mathscr{L}$, but a part of the continuous spectrum of the operator.

In integral version, the inner product over some region R assumes the form

$$P = \int_R \sigma dV = \int_R L X^2(x) dV = \int_R L \left[\text{grad} \frac{\theta(x)}{T} \right]^2 dV, \qquad (9)$$

where $\boldsymbol{X} = -\,\text{grad}[\theta(x)/T]$ and θ is a potential.

Let us find what the function of θ should be in order that P has an extreme value. We can ignore the constant matrix L and the constant temperature T (If T is not constant, we have to consider θ/T) and use the calculus of variations. Here the integrand I is $(\partial\theta/\partial x)^2$ and the corresponding Euler's equations are

$$\frac{\partial I}{\partial \theta} - \frac{\partial}{\partial x} \frac{\partial I}{\partial \theta_x} = 0. \qquad (10)$$

Or

$$\theta_{xx} = \nabla^2 \theta = 0. \qquad (11)$$

Now we find a scalar point function θ for which $\nabla^2\theta = 0$, subject to boundary conditions. This is the Dirichlet problem: to reconstruct θ where θ is known at boundaries and $\nabla^2\theta = 0$ everywhere. An elaborate mathematical set up is necessary to handle this problem. Fortunately, from the probability theory of potential we can travel a shorter route. In a homogeneous material, θ tends to fall if the average value nearby is lower. If θ does not change with time, then θ at any point is equal to average θ over a not too steep straight line in one dimension, a small circle in two dimensions, a small sphere in three dimensions, and so on. Therefore θ is a homogeneous function of coordinates. We see that the distribution of σ will follow the same pattern. In other words, P is proportional to σ at the midpoint of a straight line in one dimension. If one is at a minimum, both are at minima. Therefore we establish the following theorem: During the evolution of linear irreversible processes with time, (1) if the matrix of coefficients connecting fluxes and forces is symmetric and constant, and (2) if the system is isotropic and there are many irreversible processes, or (2′) if the system is anisotropic and there is only one irreversible process, the entropy-production per unit time is a minimum corresponding to a stationary state, under the restrictions at the boundaries. If one restraint is removed, the entropy-production decreases to a minimum, smaller than, or at least equal to the previous one. If all restraints are removed, the entropy-production decreases to the smallest minimum, i.e., zero entropy-production or static equilibrium.

In the statement of the theorem it is implied that the stationary state is stable. de Groot established the stability of σ by considering the first-order-approximation pertubations[6], and stability of P by examining individual examples[2]. We can establish the stability of σ and therefore P by

$$\boldsymbol{\sigma} = \frac{d\boldsymbol{S}_i}{dt} > 0 \tag{12}$$

and

$$\frac{d\boldsymbol{\sigma}}{dt} = \frac{d^2\boldsymbol{S}_i}{dt^2} = \frac{\partial\boldsymbol{\sigma}}{\partial t} + (\boldsymbol{v}\cdot\text{grad})\,\boldsymbol{\sigma}. \tag{13}$$

Equation (12) is the defining equation (1, 6). Equation (13) is eq. (4, 1), where $\boldsymbol{v}$ is velocity of motion of entropy. From the analogy in hydrodynamics, the flow of entropy cannot be sustained unless $\boldsymbol{v}$ and grad σ have opposite signs. In the stationary state $\partial\boldsymbol{\sigma}/\partial t = 0$; from eq. (13) $d^2\boldsymbol{S}_i/dt^2 < 0$. But from eq. (12) $d\boldsymbol{S}_i/dt > 0$. Because the velocity and the acceleration of motion of entropy are opposite in sign, clearly the motion is stable.

In the integral version we try to find the total value of σ in a certain region of space. Because of restrictions on boundaries, the variables in coordinates are not all independent. Instead of eliminating dependent variables, we supply restraints as the final answer. That is, the Laplacian of the potential is zero. If one restraint is removed, one irreversible process comes to an end for an isotropic system or one degree of freedom is eliminated for an anisotropic system.

As to the total value of σ, P, we know that by analogy to the Rayleigh dissipation function, the average time rate of P is equal to twice the rate of P in the stationary state. Perhaps the following idealized example is illuminating. Suppose we wish to display a fountain in a market square. We assume no evaporation or spillage. The only frictional force is inside the pipe and is proportional to the velocity of the flow. Furthermore, the temperature of the water remains constant. The smallest size of pump to be installed is such that it furnishes twice the power to overcome the frictional force at a steady rate of flow.

References

[1] I. Prigogine: Introduction to Thermodynamics of Irreversible Processes, Interscience, New York, Chaps. 6 and 7 (1961).

[2] S. R. de Groot and P. Mazur: Non-Equilibrium Thermodynamics, Interscience, New York, Chap. 5 (1962).

[3] Y. L. Yao: *J. Chem. Phys.*, **53**, 1876 (1970).

[4] R. Bellman: Introduction to Matrix Analysis, McGraw-Hill, New York, p. 198 (1952).

[5] S. G. Mikhlin (translated by A. Feinstein): The Problem of the Minimum of a Quadratic Functional, Holden-Day, San Francisco, p. 29 (1965).

[6] S. R. de Groot: Thermodynamics of Irreversible Processes, North-Holland, Amsterdam, p. 198 (1952).

CHAPTER VI

NONLINEAR THERMODYNAMICS

§ 21 *SLIGHTLY NONLINEAR IRREVERSIBLE PROCESSES AND STEADY STATE*

The assumptions underlying linear irreversible processes, § 14, are too restricted. If the linearity is removed, the Onsager law is not necessarily valid. It is not certain whether there is a steady state. Even if there is a steady state, it is not certain whether the extreme state is a state of minimum or maximum entropy-production.

For general irreversible processes, we still maintain the model based on dynamic motion of σ, the entropy-production per unit volume per unit time (§ 1). The processes amenable geometrically to this model, so far as steady state is concerned, are slightly nonlinear ones (to be explained later) and have only one degree of freedom.

Just as in classical mechanics, for which the mechanical state is a function of x, $\dot{x}$, and t, we assume that σ is also a function of generalized coordinates, generalized velocity and time. In the further assumption that there is only one degree of freedom, we deal with a first-degree second-order differential equation.

A necessary and sufficient condition for the differential equation to represent an irreversible process is that the equation contains an $\dot{x}$ term no matter how small its coefficient may be.

For example, a Mathieu-type equation in the form of

$$\ddot{x} + (1 + \cos 2t)x = 0 \tag{1}$$

is linear and reversible; but

$$\ddot{x} + b\dot{x} + (1 + \cos 2t)x + cx^3 = 0 \tag{2}$$

is nonlinear and irreversible.

The general second-order differential equation is

$$\ddot{x} + \dot{x} + qF(t, x, \dot{x}) = 0, \tag{3}$$

where q is a parameter and F is an analytical function in x and $\dot{x}$ and periodic in t with period 2π or a rational fraction of 2π. we assume that q is small, i.e., approximately linear. Although "linearity" here is different from our usage in the inner-product model, this linearity attached to a differential equation gives us a qualitative measure such as very nonlinear or slightly nonlinear. Poincaré[1] has justified the solution in a convergent series of q for small q.

Mathematically, the exhibition of all solutions of a differential equation may be difficult. Physically it may not be necessary. In autonomous cases, where the differential equation does not contain t explicitly, topographical representation in phase space is especially simple. Here the trajectory surfaces are reduced to trajectory curves. Because two trajectories cannot cross each other, this is equivalent to saying that the distance between all pairs of points on the phase plane remain constant throughout the motion. Stated briefly, it is a rigid-body motion.

Let us define a vector $\boldsymbol{R}$ having components x and $y = \dot{x}$, which are generalized coordinates and velocities of motion respectively in the phase plane. Then the differential equation prescribes the flow of each point on the phase plane. The velocity of the flow, which is quite different from the velocity of the system, can be represented by a vector $\boldsymbol{V}$ having components $\dot{x}$ and $\dot{y}$. In terms of polar coordinates, $\boldsymbol{R}$ and $\boldsymbol{V}$ are (r, θ) and $(\dot{r}, \dot{\theta})$ respectively.

The most general displacement in a rigid body is translation and rotation. The admissible transformations are translation and orthogonal transformation. By translation it is possible to shift the origin at the mass center to an arbitrary point. An orthogonal transformation of infinitesmal rotation is almost identical with a finite rotation. σ at a point on the phase plane varies with time as the point moves, but the change will depend on the coordinate system in which it is observed. An observer situated on the body set of axes, with the mass center at the origin, will observe no change, but an observer situated on the space set of axes, with an arbitrary point as the origin, will observe a change. The relation between the two set of axes is, from eq. (7, 36),

$$\left(\frac{d\sigma}{dt}\right)_{\text{space}} = \left(\frac{d\sigma}{dt}\right)_{\text{body}} + \omega \times \sigma, \tag{4}$$

where $\omega = \dot{\theta}$ is the angular velocity of rotation. In order that $d\sigma/dt$ in space axes be zero, r, ω, and $(d\sigma/dt)_{\text{body}}$ are not necessarily zero, but $\dot{r}$ and $\dot{\omega}$ must be zero.

In terms of rectangular coordinates,

$$\frac{dr^2}{dt} = \frac{d\rho}{dt} = 2r\dot{r} = 2(x\dot{x} + y\dot{y}), \tag{5}$$

$$\frac{d\theta}{dt} = \frac{d\left(\arctan \dfrac{y}{x}\right)}{dt} = \frac{x\dot{y} - y\dot{x}}{x^2 + y^2}, \tag{6}$$

where ρ is proportional to the stored energy. In terms of vectors,

$$\frac{dr^2}{dt} = 2\boldsymbol{R} \cdot \boldsymbol{V}, \tag{7}$$

$$\frac{d\theta}{dt} = \frac{\left\| (x, y, 0) \begin{bmatrix} 0 & 0 & \dot{y} \\ 0 & 0 & -\dot{x} \\ -\dot{y} & \dot{x} & 0 \end{bmatrix} \right\|}{r^2} = \frac{\|\boldsymbol{R} \times \boldsymbol{V}\|}{r^2}. \tag{8}$$

If $dr^2/dt = 0$, $\boldsymbol{R}$ is orthogonal to $\boldsymbol{V}$. If $d\theta/dt =$ constant, the sign and the magnitude determine the plane rotation. Stated briefly, to search for a closed cycle or a periodic trajectory, we look for $dr^2/dt = 0$ and $d^2\theta/dt^2 = 0$, but $\boldsymbol{R}$ is not necessarily a constant vector.

The search for a closed cycle is a complicated matter, because it does not flow from the differential equation but from its solutions. We take advantage of criteria[2] already established for this search.

One negative criterion is paraphrased from the Bendixson theorem. If $x\dot{x} + y\dot{y}$ never changes sign, there will be no stationary state of σ. Consider the van der Pol equation

$$\ddot{x} + q(x^2 - 1)\dot{x} + x = 0, \tag{9}$$

in which q is a small positive number. Let

$$\dot{x} = y. \tag{10}$$

Then

$$\dot{y} = -x + q(1 - x^2)y. \tag{11}$$

In matrix form,

$$\begin{bmatrix} \dot{x} \\ \dot{y} \end{bmatrix} = \begin{bmatrix} 0 & 1 \\ -1 & q(1 - x^2) \end{bmatrix} \begin{bmatrix} x \\ y \end{bmatrix}. \tag{12}$$

Because $x\dot{x} + y\dot{y} = q(1 - x^2)y^2$ may change sign, no conclusion can be drawn.

One positive criterion is paraphrased from the Poincaré-Bendixson theorem. In a ring-shaped domain D with boundaries c_1 and c_2, if there is no singular point in D and trajectories enter through every point of c_1 and c_2, there will be a stationary σ. In the Nemitzky theorem, trajectories entering through every point of c_1 and c_2 are replaced by

$$x\dot{x} + y\dot{y} \geqq 0 \text{ for } c_1, \tag{13}$$

and

$$x\dot{x} + y\dot{y} \leqq 0 \text{ for } c_2. \tag{14}$$

Consider again the van der Pol equation. Because the determinant of the square matrix in eq. (12) is not zero, there is no singular point in the domain between c_1 and c_2 (but there is a singular point at the origin).

In terms of polar coordinates,

$$x\dot{x} + y\dot{y} = qr^2 \sin^2 \theta(1 - r^2 \cos^2 \theta). \tag{15}$$

If $r^2 < 1$,

$$x\dot{x} + y\dot{y} > 0, \text{ or } \frac{d\sigma}{dt} < 0. \tag{16}$$

If $r^2 > 1$, the sign of $x\dot{x} + y\dot{y}$ or $d\sigma/dt$ cannot be determined. No conclusion can be drawn about the existence of a stationary σ at $r^2 = 1$.

Another positive criterion is paraphrased from Liénard's method. An oscillary phenomenon becomes periodical when $\int (x\dot{x} + y\dot{y})dt$ over one period 2π vanishes. Consider again the van der Pol equation. For $q \to 0$,

$$\frac{dr^2}{dt} = x\dot{x} + y\dot{y} = 0, \tag{17}$$

and

$$\frac{d\theta}{dt} = -1. \tag{18}$$

By integration,

$$r^2 = \rho_0, \tag{19}$$

and

$$\theta = -t + C = -t, \tag{20}$$

if we set the constant of integration C zero.

According to this method, we substitute eqs. (19) and (20) in eq. (17) and integrate,

$$q\rho_0\left(\int_0^{2\pi} \sin^2 t\ dt - \rho_0 \int_0^{2\pi} \sin^2 t \cos^2 t\ dt\right) = 0. \qquad (21)$$

We find $\rho_0 = 4$. Thus at $r = 2$ a stationary state of σ exists.

As another example of Liénards method, we consider the Liénard equation,

$$\ddot{x} + f(x)\dot{x} + g(x) = 0. \qquad (22)$$

Equation (22) differs from a nonlinear oscillation by the extra term $f(x)\dot{x}$. In order that the system does not "collapse", which means that the system is not spiralled to static equilibrium, energy must be supplied into the system. On the other hand, too much supplied energy will initiate instability. The effective force is $\ddot{x} + f(x)\dot{x}$. The effective momentum is $\dot{x}+F(x)$, where $dF(x)/dt = f(x)\dot{x}$. Because eq. (22) is already normalized with respect to mass, the effective velocity is also $\dot{x} + F(x)$. The kinetic energy is $\frac{1}{2}[\dot{x} + F(x)]^2$. The potential energy is $\int g(x)dx$. Hence the total energy is

$$E = \frac{1}{2}[\dot{x} + F(x)]^2 + \int g(x)dx. \qquad (23)$$

The negative power is

$$-T\frac{d\sigma}{dt} = [\dot{x} + F(x)][\ddot{x} + f(x)\dot{x}] + g(x)\dot{x}. \qquad (24)$$

Substituting eq. (22) in eq. (24),

$$-T\frac{d\sigma}{dt} = F(x)[\ddot{x} + f(x)\dot{x}] = F(x)\frac{d[\dot{x} + F(x)]}{dt}. \qquad (25)$$

Depending on sign of $F(x)$, $d\sigma/dt$ may be positive or negative. At steady state,

$$\frac{d\sigma}{dt} = 0,$$

Or

$$\oint F(x)dy = 0, \tag{26}$$

where

$$y = \dot{x} + F(x). \tag{27}$$

This means that the curve $F(x)$ has a re-entrant path to correspond to a closed trajectory and, hence, to a periodic solution. If we write $\dot{x} = v$, eq. (22) becomes

$$\frac{dv}{dx} + f(x) + \frac{g(x)}{v} = 0. \tag{28}$$

Substituting eq. (27) in eq. (28),

$$\frac{dy}{dx} + \frac{g(x)}{y - F(x)} = 0. \tag{29}$$

For $g(x) = x$, eq. (29) becomes

$$\frac{dy}{dx} + \frac{x}{y - F(x)} = 0, \tag{30}$$

which can be written as

$$xdx + [y - F(x)]dy = 0, \tag{31}$$

in the form of $XdX + YdY = 0$ as of eq. (5).

For $g(x)$ being x, and $f(x)$ being $u\, f(x)$, where u is the parameter, eq. (22) becomes

$$\ddot{x} + u\; f(x)\dot{x} + x = 0. \tag{32}$$

For $u \ll 1$, eq. (32) is not much different from a linear oscillator. When $u \to \infty$, one obtains an entirely new conclusion. For $u \neq 1$, eq. (30) becomes

$$\frac{dy}{dx} + \frac{x}{y - uF(x)} = 0. \tag{33}$$

Multiplying eq. (33) by u,

$$u\frac{dy}{dx}+x\left[\frac{y}{u}-F(x)\right]=0. \tag{34}$$

If we let

$$z=\frac{y}{u}, \tag{35}$$

$$\frac{dz}{dx}=-\frac{\frac{x}{u}}{u[z-F(x)]}. \tag{36}$$

Or

$$\frac{dz}{dt}=-\frac{x}{u},\ \frac{dx}{dt}=u[z-F(x)]. \tag{37}$$

As we have assumed that u is very large, dx/dt is very large everywhere except $z=F(x)$ and dz/dt is very small when the curve $F(x)$ is followed. Thus on the arc AB (Figure 1) the motion is slow. At the point B the represented point ceases to follow the curve $F(x)$ and the horizontal velocity becomes very large. At point C a slow motion begins on the arc CD. At point D there is another jump DA, and so on.

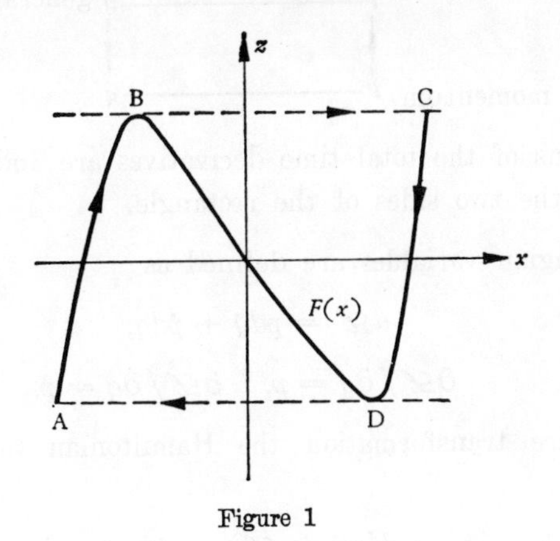

Figure 1

References

[1] H. Poincaré: Les Méthodes nouvelles de la mécanique céleste, T. 1, Gau-

therier-Villars, Paris (1892).
[2] N. Minorsky: Nonlinear Oscillations, Van Nostrand, New York, Chaps. 3, 4, and 16 (1962).

§22 *CHOICE OF FLUXES AND FORCES*

From eq. (1, 6),

$$\sigma = -\frac{1}{T}[\boldsymbol{J}|\boldsymbol{X}] = \frac{dS_i}{dt}. \tag{1}$$

Because this equation is quite general, there exists considerable freedom of choice in the specification of fluxes and forces so long as the inner product is invarient. In this section we make some suggestions about the choice.

Conservative Systems

For a conservative system, we split energy according to the following scheme,

$\dot{p}$ ———————— q generalized coordinate

↑ ↓

generalized momentum p ———————— $\dot{q}$

The relations of the total time derivatives are indicated by the arrows on the two sides of the rectangle.

The Lagrangian variables are defined as

$$d\mathscr{L} = pd\dot{q} + \dot{p}dq, \tag{2}$$

$$\partial\mathscr{L}/\partial\dot{q} = p, \quad \partial\mathscr{L}/\partial q = \dot{p}. \tag{3}$$

By Legendre transformation the Hamiltonian variables are defined as

$$d\mathscr{H} = d(p\dot{q} - \mathscr{L}) = \dot{q}dp - \dot{p}dq, \tag{4}$$

$$\partial\mathscr{H}/\partial p = \dot{q}, \quad \partial\mathscr{H}/\partial q = -\dot{p}. \tag{5}$$

If q is cyclic, i.e., q is missing, or $\partial\mathscr{L}/\partial q = 0$, $\dot{q}$ may still be present in $\mathscr{L}$. But if $\partial\mathscr{H}/\partial q = 0$, p is certainly constant in $\mathscr{H}$. Thus p can be replaced by a constant of integration and the degree of freedom of $\mathscr{H}$ is decreased by 1. On the other hand, because of the "symmetry" exhibited in eq. (3), it is convenient to use $\mathscr{L}$ variables. To take advantage of both, the Routhian variables are defined as

$$d\mathscr{R} = d(p_c\dot{q}_c - \mathscr{L}) = \dot{q}_c dp_c - \dot{p}_c dq_c - pd\dot{q} - \dot{p}dq, \qquad (6)$$

where the subscript c refers to the cyclic variables.

Thus in $\mathscr{R}$ cyclic variables are eliminated as in $\mathscr{H}$, but non-cyclic variables are retained as in $\mathscr{L}$.

As a concluding remark, $p\dot{q}$ indicates kinetic energy and $\dot{p}q$ indicates potential energy (see the first two examples of §7) in spite of the fact that the assignment of kinetic energy and potential energy is arbitrary (in radiation it is difficult to distinguish between the two). The invariance of an area of the phase plane of the Hamiltonian variables

$$\iint dp\, dq$$

does not carry over to the Lagrangian variables. As a matter of fact we have to introduce the Jacobian of the transformation $(p, q) \to (\dot{q}, q)$ to express the integral invariance of the Lagrangian variables,

$$\iint dp\, dq = \iint \begin{bmatrix} \dfrac{\partial p}{\partial \dot{q}} & \dfrac{\partial p}{\partial q} \\ \dfrac{\partial q}{\partial \dot{q}} & \dfrac{\partial q}{\partial q} \end{bmatrix} d\dot{q}dq = \iint \frac{\partial^2 \mathscr{L}}{\partial \dot{q}^2} d\dot{q}dq. \qquad (7)$$

We may imagine that the Hamiltonian variables are like a certain amount of incompressible fluid moving along in phase plane and area is conserved, but that the Lagrangian variables are like a certain amount of compressible fluid and area, weighted by "density", $\partial^2\mathscr{L}/\partial\dot{q}^2$, is conserved. Although $\dot{q}$ is like flux and $\dot{p}$ is like force, the expression for either the Hamiltonian or the Lagrangian variables contains no $\dot{q}\dot{p}$ term because there is no dissipation of energy.

Bond Graphs[1,2]

Bond graphs are devised to describe the power change in a system. They are comprised of ideal elements arranged in blocks and connected by bonds, with arrows to indicate the direction of flow of energy and strokes to select causality. Similar to chemical bonds, each bond is represented by a line with the force-like variables written above (or on the left) and the flux-like variables written below (or on the right). Because much of the terminology used in bond graphs is the generalization of electrical circuit theory, in the following, we exemplify the force-like variables as potentials ζ and the flux-like variables as currents i.

The general scheme exemplified by the linear electrical system is

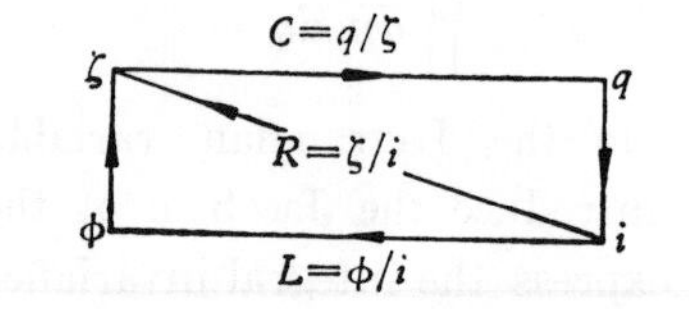

As in conservative systems, the two vertical arrows indicate the total time derivatives, while the other three arrows indicate the constitutive relations. In Table 1 the symbols and the units used in the constitutive relations are given. The SI units, some of which are unfamiliar, are enclosed in brackets. The confusion of the symbols is unavoidable because in each category the preferred symbols have been established. It is noticed that the product of R and C always has unit of sec.

The basic elements frequently used in the bond graphs are junction structures 0 and 1, transformers TF, transducers TD,

Table 1 **Symbols and Units Used in the Constitutive Relations**

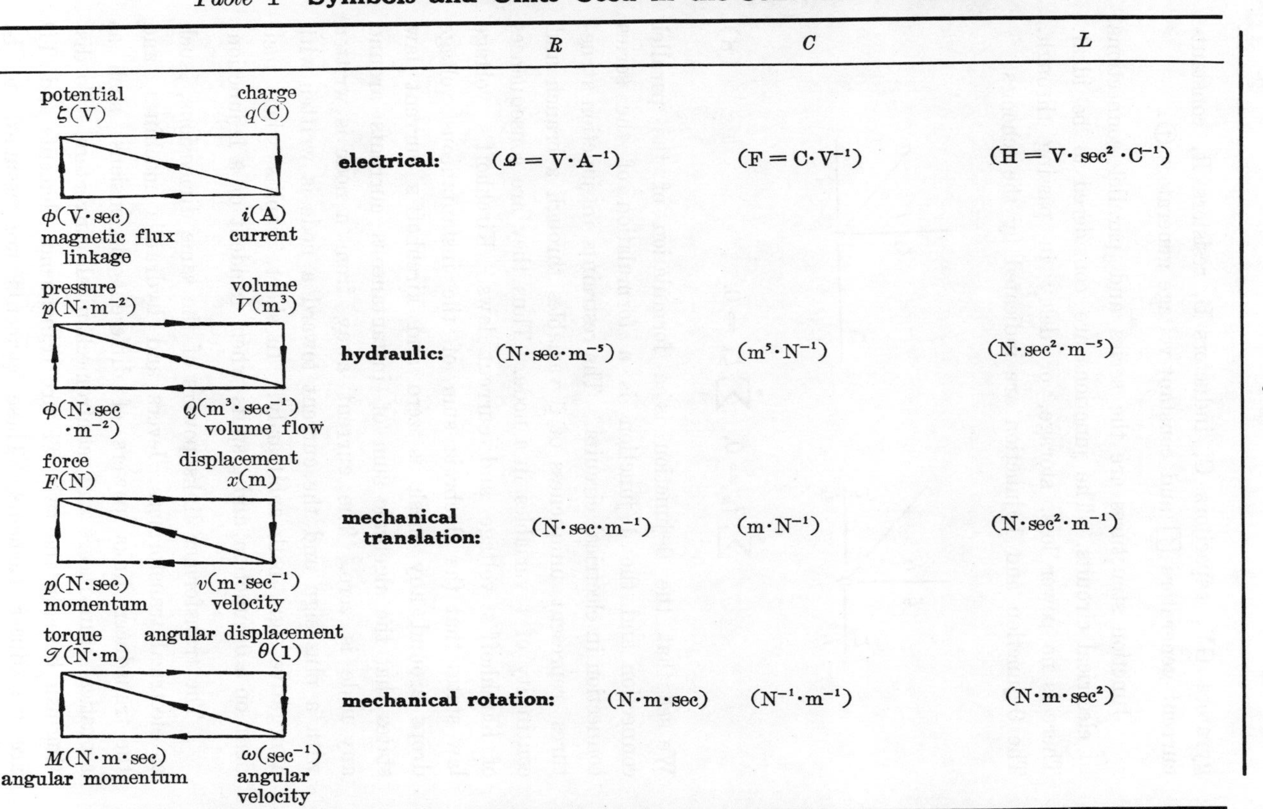

		R	C	L
potential ξ(V), charge q(C), ϕ(V·sec) magnetic flux linkage, i(A) current	**electrical:**	$(\Omega = \mathrm{V \cdot A^{-1}})$	$(\mathrm{F = C \cdot V^{-1}})$	$(\mathrm{H = V \cdot sec^2 \cdot C^{-1}})$
pressure $p(\mathrm{N \cdot m^{-2}})$, volume $V(\mathrm{m^3})$, $\phi(\mathrm{N \cdot sec \cdot m^{-2}})$, $Q(\mathrm{m^3 \cdot sec^{-1}})$ volume flow	**hydraulic:**	$(\mathrm{N \cdot sec \cdot m^{-5}})$	$(\mathrm{m^5 \cdot N^{-1}})$	$(\mathrm{N \cdot sec^2 \cdot m^{-5}})$
force F(N), displacement x(m), p(N·sec) momentum, $v(\mathrm{m \cdot sec^{-1}})$ velocity	**mechanical translation:**	$(\mathrm{N \cdot sec \cdot m^{-1}})$	$(\mathrm{m \cdot N^{-1}})$	$(\mathrm{N \cdot sec^2 \cdot m^{-1}})$
torque $\mathscr{T}$(N·m), angular displacement θ(1), M(N·m·sec) angular momentum, $\omega(\mathrm{sec^{-1}})$ angular velocity	**mechanical rotation:**	$(\mathrm{N \cdot m \cdot sec})$	$(\mathrm{N^{-1} \cdot m^{-1}})$	$(\mathrm{N \cdot m \cdot sec^2})$

gyrators GY, capacitors C, inductors L, resistors R, constant-current generators [↑] and constant-voltage generators ⓘ.

Junction structures are the series and parallel connections in electrical circuits. The junctions are considered to be ideal. There is no power loss, storage, or delay in passing through. The 0-junction and 1-junction are indicated by the schemes

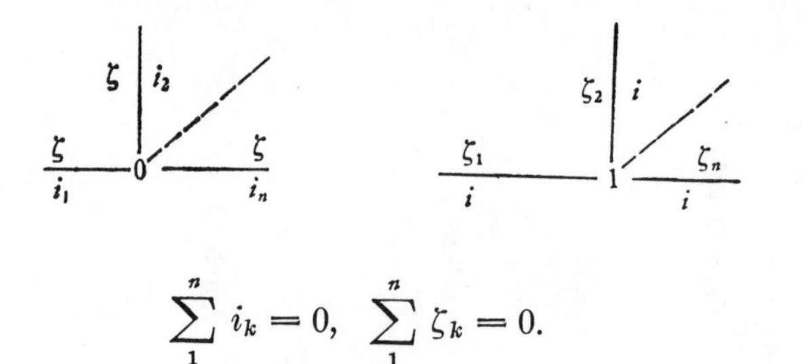

$$\sum_{1}^{n} i_k = 0, \quad \sum_{1}^{n} \zeta_k = 0. \tag{8}$$

We see that the 0-junction is a formulation of the parallel connection and the 1-junction is a formulation of the series connection in electrical circuits. The restraints in junction structures represent uniqueness of ζ variables through a branch and continuity of i variables in a node. Thus they are consequences of Kirchhoff's voltage and current laws. Kirchhoff's voltage law states that the algebraic sum of the instantaneous voltage drops around any mesh is zero and kirchhoff's current law states that the algebraic sum of instantaneous currents around any node is zero. The current away from a node is written with a plus sign and the current toward a node is written with a negative sign, both arbitrarily. In short, i splits when incident on a 0-junction and ζ splits when incident on a 1-junction.

The transformer links powers of the same dimensions, such as electrical transformers, levers and hydraulic machines; and the transducer links powers of different dimensions, such as acoustical transducers and electromechanical converters. To distinguish between the two, TF are transformer elements and TD are transducer elements. These elements are assumed to be ideal. They have no resistance in windings or friction in links and therefore dissipate no energy. For a two-terminal transformer,

$$\xrightarrow[i_1]{\zeta_1} \mathrm{TF} \xrightarrow[i_2]{\zeta_2}$$

$$\begin{bmatrix} \zeta_1 \\ i_1 \end{bmatrix} = \begin{bmatrix} r & 0 \\ 0 & \dfrac{1}{r} \end{bmatrix} \begin{bmatrix} \zeta_2 \\ i_2 \end{bmatrix}, \tag{9}$$

where r is the transformer modulus. Of course we could have $2n$ terminals and the number of primary (input) terminals need not equal to that of secondary (output) terminals. For example, the transformer in the third example of §7 is represented by

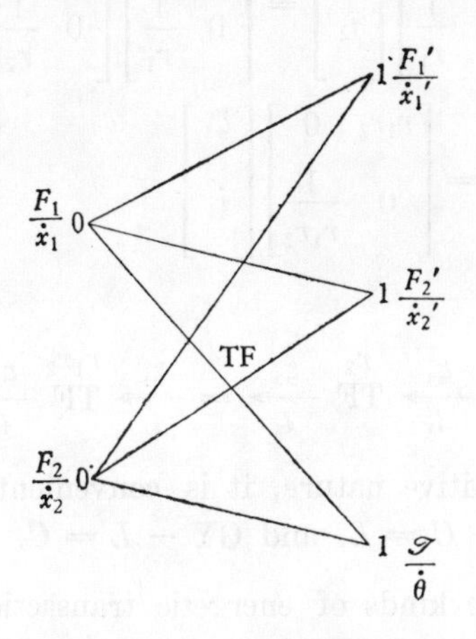

and the transformation matrix is singular.

Gyrators denoted by GY differ from transformers in that the former have the effect of interchanging the role of potentials and currents. According to Karnopp[3], the gyrator is more primitive than the transformer in the sense that gyrators cascade exactly like one transformer but transformers cascade only into a new transformer. This can be seen as follows:

$$\begin{bmatrix}\zeta_1\\ i_1\end{bmatrix}=\begin{bmatrix}0 & r_1\\ \dfrac{1}{r_1} & 0\end{bmatrix}\begin{bmatrix}\zeta_2\\ i_2\end{bmatrix}=\begin{bmatrix}0 & r_1\\ \dfrac{1}{r_1} & 0\end{bmatrix}\begin{bmatrix}0 & r_2\\ \dfrac{1}{r_2} & 0\end{bmatrix}\begin{bmatrix}\zeta_3\\ i_3\end{bmatrix}$$

$$=\begin{bmatrix}\dfrac{r_1}{r_2} & 0\\ 0 & \dfrac{r_2}{r_1}\end{bmatrix}\begin{bmatrix}\zeta_3\\ i_3\end{bmatrix}, \tag{10}$$

or

$$\xrightarrow[i_1]{\zeta_1}\overset{r_1}{\text{GY}}\xrightarrow[i_2]{\zeta_2}\overset{r_2}{\text{GY}}\xrightarrow[i_3]{\zeta_3}=\xrightarrow[i_1]{\zeta_1}\overset{r_1/r_2}{\text{TF}}\xrightarrow[i_3]{\zeta_3};$$

and

$$\begin{bmatrix}\zeta_1\\ i_1\end{bmatrix}=\begin{bmatrix}r_1 & 0\\ 0 & \dfrac{1}{r_1}\end{bmatrix}\begin{bmatrix}\zeta_2\\ i_2\end{bmatrix}=\begin{bmatrix}r_1 & 0\\ 0 & \dfrac{1}{r_1}\end{bmatrix}\begin{bmatrix}r_2 & 0\\ 0 & \dfrac{1}{r_2}\end{bmatrix}\begin{bmatrix}\zeta_3\\ i_3\end{bmatrix}$$

$$=\begin{bmatrix}r_1r_2 & 0\\ 0 & \dfrac{1}{r_1r_2}\end{bmatrix}\begin{bmatrix}\zeta_3\\ i_3\end{bmatrix}, \tag{11}$$

or

$$\xrightarrow[i_1]{\zeta_1}\overset{r_1}{\text{TF}}\xrightarrow[i_2]{\zeta_2}\overset{r_2}{\text{TF}}\xrightarrow[i_3]{\zeta_3}=\xrightarrow[i_1]{\zeta_1}\overset{r_1r_2}{\text{TF}}\xrightarrow[i_3]{\zeta_3}.$$

In spite of its primitive nature, it is convenient to retain GY. For example, GY $-\ C = L$, and GY $-\ L = C$.

There are three kinds of energetic transactions, i.e., three ways of integrating the equation power $= \zeta i$:

$$E_C(t)=\int_0^t \zeta i dt=\int_{q(0)}^{q(t)} \zeta dq, \tag{12}$$

$$E_L(t)=\int_0^t i\zeta dt=\int_{\phi(0)}^{\phi(t)} i d\phi, \tag{13}$$

$$E_R(t)=\int_0^t i\zeta dt. \tag{14}$$

To perform the integrations, we require the capacitive constitution relation

$$q = f_C(\zeta), \tag{15}$$

the inductive constitution relation

$$\phi = f_L(i), \tag{16}$$

and the resistive constitution relation

$$f_R(i, \zeta) = 0. \tag{17}$$

In the special case of linearity, eqs. (15—17) are reduced to equations written on the general scheme of p. 132. The capacitor C is an ideal element for storing capacitive (potential) energy. The inductor L is an ideal element for storing inductive (kinetic) energy. The resistor R is an ideal element for dissipating energy.

The two active elements are the constant current generator and the constant voltage generator. In the constant-current generator, the generated current is independent of the generator terminal voltage. In the constant-voltage generator, the generated voltage is independent of the generator current.

The sign convention in bond graphs is arbitrary. A half-arrowhead (A full-arrowhead is retained for the uni-directional flow.) on each bond indicates the direction of flow of energy. It is usually matched with the normal operation of the system being modelled. Except for active elements the direction of flow of energy into all ideal elements is arbitrarily assigned to be positive.

The causal stroke has nothing to do with the sign convention but signifies dependency. As individual field, the capacitor is always current-controlled, the inductor is always voltage-controlled, and resistor could be either. Because constitutive relations of capacitors and inductors are

$$i = C \frac{d\zeta}{dt} \tag{18}$$

and

$$\zeta = L \frac{di}{dt} \tag{19}$$

respectively, a voltage-controlled capacitor or a current-controlled inductor would impose infinite "velocity" on the element, which requires the supply of an infinite amount of energy. On the other hand, because of the constitutive relation of resistors, no differentiation with respect to time is required. For example, $R|\leftharpoondown$ indicates independent voltage and the constitutive relation is written as

$$i = f_R^{-1}(\zeta).$$

$R\leftharpoondown|$ indicates independent current and the constitutive relation is written as $\zeta = f_R(i)$. $C\leftharpoondown|$ indicates independent current and the constitutive relation is written as

$$\zeta = f_C^{-1} \int i dt.$$

$L|\leftharpoondown$ indicates independent voltage and the constitutive relation is written as

$$i = f_L^{-1} \int \zeta dt.$$

However, $C|\rightharpoonup 1 \ |\rightharpoonup C = C \rightharpoonup C$ indicates a causal conflict between two capacitors because for the 1-junction, i's on all the bonds are independent. The correct bond graph would be

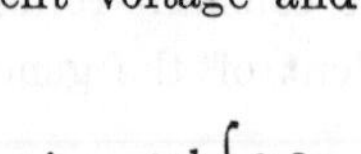

$$\begin{array}{c} R \\ \uparrow \\ C| \rightharpoonup \overline{1}| \rightharpoonup C. \end{array}$$

As another example, the bond graph

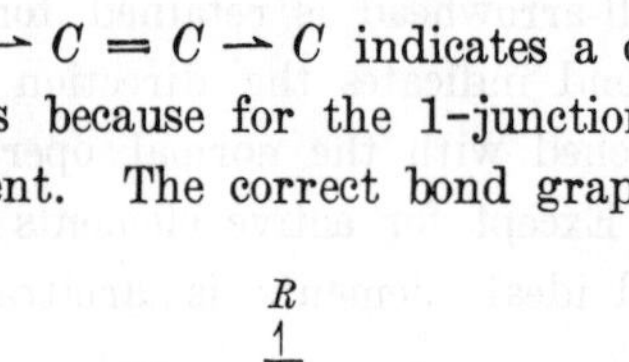

$$\begin{array}{ccccccc} & \overline{R_1} & & & \overline{R_2} & & \overline{L_1} \\ & \uparrow & & & \uparrow & & \uparrow \\ E \rightharpoonup & |\,1\,| & \rightharpoonup & 0 \rightharpoonup & |\,\overline{1}\,| & \rightharpoonup & 0 \\ & & & \downarrow & \downarrow & & \downarrow \\ & & & C_1 & L_2 & & C_2 \end{array}$$

is correct because C_1 and C_2 are current-controlled, L_1 and L_2 are voltage-controlled, R_1 is voltage-controlled and R_2 is current-controlled. Thus causality shows the minimum dimensionality of the state space required for the description of a given system.

Causality is also important in stability considerations, especially if it is known that the element has a non-invertible constitutive relation in a certain region. For example, a tunnel diode having the familiar S-shaped bi-stable constitutive region shown in Figure 1 will produce a relaxation oscillation if the device is voltage-controlled

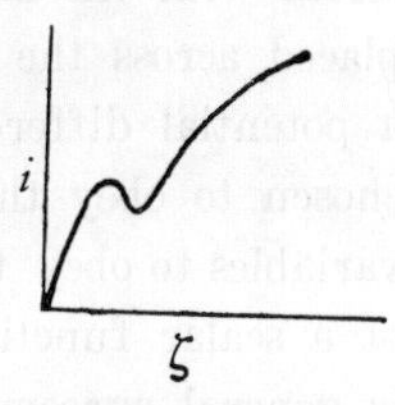

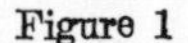
Figure 1

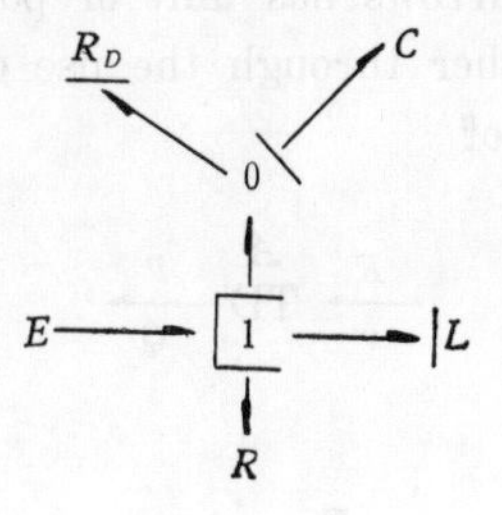

rather than current-controlled

$$
\begin{array}{c}
R_D \\
\uparrow \\
E \rightarrow \;|\,1\; \rightarrow |L. \\
\downarrow \\
R
\end{array}
$$

Isothermal Non-conservative Systems

Electrical networks are essentially isothermal. This implies that power dissipated in resistors is instantaneously transferred to the environment, leaving the system at constant temperature.

In the preceding sections, fluxes and forces are chosen from the time derivative of the Gibbs equation, i.e., based on extensive and intensive variables. However, in the derivation of the Onsager law (§16), it is indicated that fluxes are chosen from the time derivative of state variables and forces are chosen from the derivative of entropy with respect to the same state variable. Engineers prefer to use "across" and "through" variables. For across variables a measuring instrument is placed in series with the flow but for through variables the instrument is placed across the two points along the flow direction to measure a potential difference. In bond graphs flux-like variables are chosen to obey the flow of a conserved quantity and force-like variables to obey the generalized Kirchhoff's voltage law (gradient of a scalar function, for example). In isothermal systems there is general agreement among the various choices except in fluids.

In the schemes of Table 1, the product of the terms connected by the diagonal arrows has unit of power and they can be converted into each other through the use of a TD in the bond graphs. For example, of

$$\xrightarrow[v]{F} \overset{A}{\mathrm{TD}} \xrightarrow[Q]{p},$$

we have

$$F = Ap$$

and

$$v = \frac{Q}{A}.$$

Of

$$\xrightarrow[i]{\zeta} \overset{\alpha}{\mathrm{TD}} \xrightarrow[v]{F},$$

we have

$$\zeta = \alpha F$$

and

$$i = \frac{v}{\alpha}.$$

In thermodynamices $T\sigma$ has unit of power per unit volume. The products of fluxes and forces can also be converted into each other through a TD if we bear in mind that mechanical forces then have units of $N \cdot m^{-3}$. In chemical reactions the scheme is

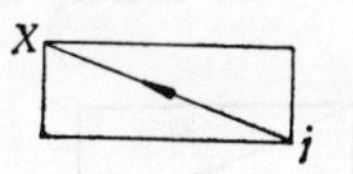

Of

$$\xrightarrow[v]{F} \overset{\rho/l}{\text{TD}} \xrightarrow[j]{X},$$

we have $F = (\rho/l)X$ and $v = (l/\rho)j$. Thus the rate of chemical reaction j has unit of $kg \cdot m^{-3} \cdot sec^{-1}$ and X has unit of $N \cdot m \cdot kg^{-1}$, which we assign as the affinity. In transfer of mass the scheme is

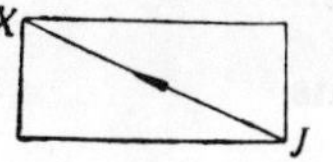

Of

$$\xrightarrow[v]{F} \overset{\rho}{\text{TD}} \xrightarrow[J]{X},$$

we have $F = \rho X$ and $v = J/\rho$. Thus the flux J has unit of $kg \cdot m^{-2} \cdot sec^{-1}$ and X has unit of $N \cdot kg^{-1}$, which we assign as the specific mechanical force or as the negative gradient of potential energy per unit mass. In transfer of charge the scheme is

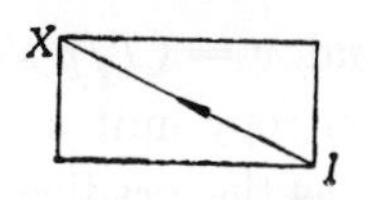

Of

$$\xrightarrow[v]{F} \overset{\rho Z}{\text{TD}} \xrightarrow[I]{X},$$

we have $F = \rho Z X$ and $v = (1/\rho Z)I$. Thus the current density (or the charge flux) I has unit of $\text{C}\cdot\text{sec}^{-1}\cdot\text{m}^{-2}$ and X has unit of $\text{N}\cdot\text{C}^{-1}$, which we assign as the electrical field or the negative gradient of scalar potential. In fluids the scheme is

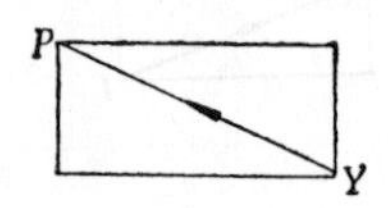

Of

$$\xrightarrow[v]{F} \overset{1/l}{\text{TD}} \xrightarrow[Y]{P} \text{GY} \xrightarrow[P]{Y},$$

we have $F = (1/l)P$ and $v = lY$. Thus the viscous pressure has unit of $\text{N}\cdot\text{m}^{-2}$ and Y has unit of sec^{-1}, which we assign as the gradient of velocity. In fluids, because of the use of a gyrator, the role of the flux and force in thermodynamics is reversed in hydrodynamics.

Non-conservative Systems

For non-conservative systems the scheme for the transfer of heat is

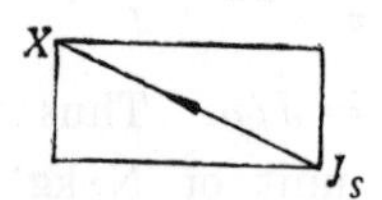

Of

$$\xrightarrow[v]{F} \overset{p/T}{\text{TD}} \xrightarrow[J_S]{X},$$

we have $F = (p/T)X$ and $v = (T/p)J_S$. Thus J_S has unit of $\text{m}\cdot\text{N}\cdot\text{sec}^{-1}\cdot\text{m}^{-2}\cdot{}^\circ\text{K}^{-1}$ = entropy unit$\cdot\text{sec}^{-1}\cdot\text{m}^{-2}$ and X has unit of ${}^\circ\text{K}\cdot\text{m}^{-1}$, which we assign as the gradient of temperature. Unlike mass (non-relativistic) or charge, heat (entropy) is not an absolute quantity, either in a reference sense or in the sense of separation from other forms of energy. Furthermore, it cannot be transferred without restrictions. For example, we can always write an ideal electrical transformer as

$$\xrightarrow[i_1]{\zeta_1} \mathrm{TF} \xrightarrow[i_2]{\zeta_2},$$

but an ideal thermal transformer as

$$\xrightarrow[(J_S)_1]{T_1} \mathrm{TF} \xrightarrow[(J_S)_2]{T_2} \quad T_1 - T_2 > 0,\ T_1 - T_3 > 0, \qquad \mathrm{TF} \searrow_{T_3}^{(J_S)_3}$$

where a full arrow indicates the uni-directional flow.

The second bond graph is a short form of

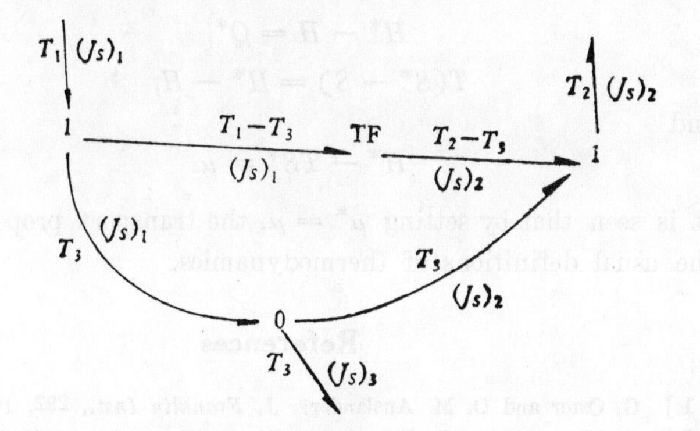

Thus part of the entropy must be disposed of even for an ideal thermal transformer.

For transfer of heat we define the heat of transfer as

$$\frac{\partial J_q^r}{\partial J} = Q^*, \tag{20}$$

the enthalpy of transfer as

$$\frac{\partial J_q}{\partial J} = H^*, \tag{21}$$

and the entropy of transfer as

$$\frac{\partial \boldsymbol{J}_S}{\partial \boldsymbol{J}} = S^*, \tag{22}$$

where* signifies transport properties.

From eq. (8, 7),

$$\boldsymbol{J}_q - \sum_k^n \mu_k \boldsymbol{J}_k = T\boldsymbol{J}_S. \tag{23}$$

From eq. (8, 12),

$$\boldsymbol{J}_q = \boldsymbol{J}_q^r + \sum_k^n H_k \boldsymbol{J}_k. \tag{24}$$

Thus we have

$$H^* - H = Q^*, \tag{25}$$

$$T(S^* - S) = H^* - H, \tag{26}$$

and

$$H^* - TS^* = \mu. \tag{27}$$

It is seen that by setting $\mu^* = \mu$, the transport properties follow the usual definitions of thermodynamics.

References

[1] G. Oster and D. M. Auslander: *J. Franklin Inst.*, **292**, 1(1971).

[2] D. Karnopp and R. Rosenburg: System Dynamics: A Unified Approach, Wiley-Interscience, New York (1975).

[3] D. Karnopp: *J. Franklin Inst.*, **291**, 211 (1971).

§23 *EXAMPLES OF POWER BALANCE*

In isothermal conservative systems, the differential equations obtained from bond graphs are equivalent to Lagrange's equations. In isothermal nonconservative systems, the entropy produced may be explicitly omitted from the bond graphs, with the understanding that heat so produced cannot be converted into work. In general cases, the entropy produced is so intricately woven with the heat conduction that the entropy flux is shown in the bond graphs. In this section we give examples of power balance.

Principal Modes and Transfer Functions of Vibration

Four different masses are connected by springs with a spring constant k (inverse of the capacitance C), while at one end a mass is fixed and at the other end a mass is subject to a mechanical force of $\boldsymbol{F}$.

The mechanical model is

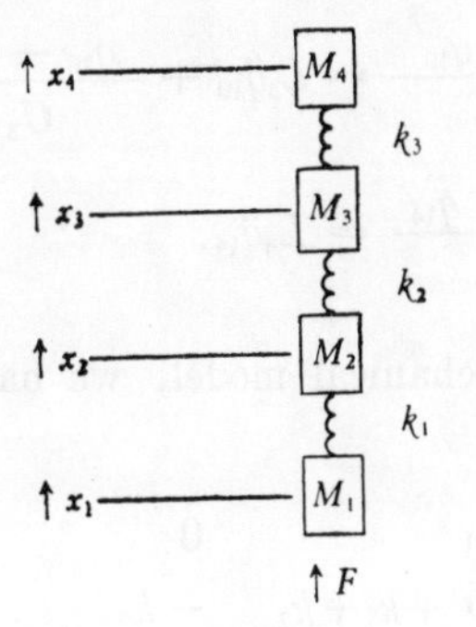

The electrical model is

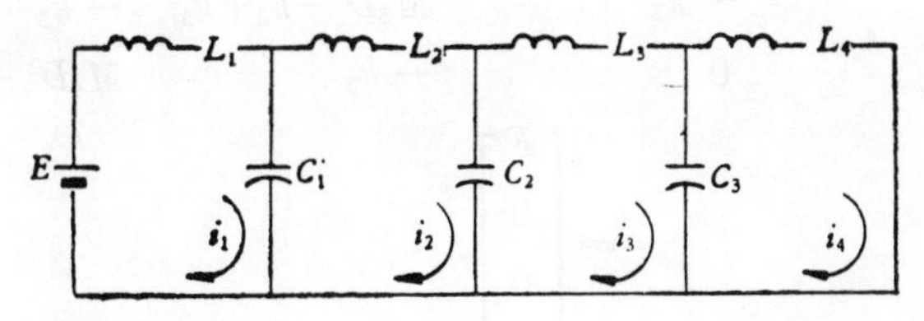

The bond graph is

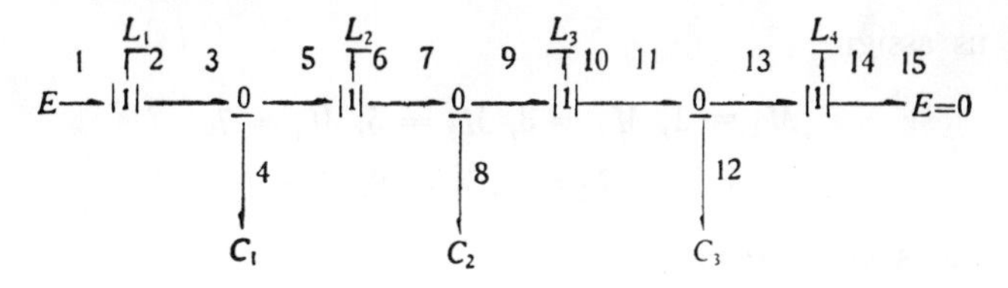

From the four 1-ports, we have

$$E = \zeta_2 + \zeta_3, \tag{1}$$

$$\zeta_3 = \zeta_6 + \zeta_7, \tag{2}$$

$$\zeta_7 = \zeta_{10} + \zeta_{11}, \tag{3}$$

$$\zeta_{13} = \zeta_{14}. \tag{4}$$

Substituting the constitutive relations,

$$E = L_1\ddot{q}_2 + \frac{q_2 - q_6}{C_1}, \tag{5}$$

$$\frac{q_2 - q_6}{C_1} = L_2\ddot{q}_6 + \frac{q_6 - q_{10}}{C_2}, \tag{6}$$

$$\frac{q_6 - q_{10}}{C_2} = L_3\ddot{q}_{10} + \frac{q_{10} - q_{14}}{C_3}, \tag{7}$$

$$\frac{q_{10} - q_{14}}{C_3} = L_4\ddot{q}_{14}. \tag{8}$$

Reverting to the mechanical model, we have the matrical equations

$$\begin{bmatrix} M_1D^2 + k_1 & -k_1 & 0 & 0 \\ -k_1 & M_2D^2+k_1+k_2 & -k_2 & 0 \\ 0 & -k_2 & M_3D^2+k_2+k_3 & -k_3 \\ 0 & 0 & -k_3 & M_4D^2 + k_3 \end{bmatrix} \begin{bmatrix} x_1 \\ x_2 \\ x_3 \\ x_4 \end{bmatrix}$$

$$= \begin{bmatrix} F \\ 0 \\ 0 \\ 0 \end{bmatrix}. \tag{9}$$

Let us assign

$$M_1 = 1,\ M_2 = 3,\ M_3 = 5,\ M_4 = 7,$$
$$k_1 = 6,\ k_2 = 4,\ k_3 = 2,\ F = 1.$$

Equations (9) become

$$\begin{bmatrix} D^2+6 & -6 & 0 & 0 \\ -6 & 3D^2+10 & -4 & 0 \\ 0 & -4 & 5D^2+6 & -2 \\ 0 & 0 & -2 & 7D^2+2 \end{bmatrix} \begin{bmatrix} x_1 \\ x_2 \\ x_3 \\ x_4 \end{bmatrix} = \begin{bmatrix} 1 \\ 0 \\ 0 \\ 0 \end{bmatrix}. \quad (10)$$

We solve eqs. (9) by the Laplace transformation[1].

The initial conditions are at $t=0$, $x_i = \dot{x}_i = 0$, $i = 1, 2, 3, 4$. Hence we have

$$\begin{bmatrix} S^2+6 & -6 & 0 & 0 \\ -6 & 3S^2+10 & -4 & 0 \\ 0 & -4 & 5S^2+6 & -2 \\ 0 & 0 & -2 & 7S^2+2 \end{bmatrix} \begin{bmatrix} \bar{x}_1 \\ \bar{x}_2 \\ \bar{x}_3 \\ \bar{x}_4 \end{bmatrix} = \begin{bmatrix} 1/S \\ 0 \\ 0 \\ 0 \end{bmatrix}. \quad (11)$$

By finding the inverse of the square matrix in eqs. (11) (We only need the first column of the inverse),

$$\begin{bmatrix} \bar{x}_1 \\ \bar{x}_2 \\ \bar{x}_3 \\ \bar{x}_4 \end{bmatrix} = \frac{1}{S\Delta} \begin{bmatrix} 105S^6 + 506S^4 + 432S^2 + 48 \\ 210S^4 + 312S^2 + 48 \\ 168S^2 + 48 \\ 48 \end{bmatrix}, \quad (12)$$

where Δ is the determinant of the square matrix in eqs. (11),

$$\begin{aligned} \Delta &= 105S^8 + 1136S^6 + 2208S^4 + 768S^2 \\ &= 105S^2(S^2 + 0.44591)(S^2 + 1.94660)(S^2 + 8.42654). \end{aligned}$$

Now the right side of each of the equations (12) is in the form of $f(S)$,

$$f(S) = \frac{P(S)}{Q(S)} = \frac{P(S)}{S\Delta}. \quad (13)$$

Because $Q(S)$ has 6 distinct pure imaginary roots,

$$\begin{aligned} \omega_1, \omega_4 &= \pm 0.66776\, i, \\ \omega_2, \omega_5 &= \pm 1.39521\, i, \\ \omega_3, \omega_6 &= \pm 2.90285\, i, \end{aligned}$$

and a triple root of 0, from the Heaviside expansion formula,

$$L^{-1}f(S) = \sum_{1}^{3} \left[\frac{P(\omega_k)}{\frac{dQ}{dS}(\omega_k)} \right] e^{\omega_k t}$$

$$+ \sum_{1}^{3} \left[\frac{P(-\omega_k)}{\frac{dQ}{dS}(-\omega_k)} \right] e^{-\omega_k t}$$

$$+ \frac{t^2}{2} [S^3 f(S)] |_{S=0} + t \frac{d}{dS} [S^3 f(S)] |_{S=0}$$

$$+ \frac{1}{2} \frac{d^2}{dS^2} [S^3 f(S)] |_{S=0}. \tag{14}$$

Thus

$$\begin{bmatrix} x_1 \\ x_2 \\ x_3 \\ x_4 \end{bmatrix} = \begin{bmatrix} 0.03125\,t^2 + 0.38281 & -0.21330 & -0.09044 & -0.07908 \\ 0.03125\,t^2 + 0.22656 & -0.19744 & -0.06110 & 0.03198 \\ 0.03125\,t^2 + 0.03906 & -0.10763 & 0.07212 & -0.00355 \\ 0.03125\,t^2 - 0.17969 & 0.19197 & -0.01241 & 0.00012 \end{bmatrix}$$

$$\begin{bmatrix} 1 \\ \cos \omega_1 t \\ \cos \omega_2 t \\ \cos \omega_3 t \end{bmatrix}. \tag{15}$$

We can check the results by noting that the sum of the equations in eqs. (10) is unity,

$$\ddot{x}_1 + 3\ddot{x}_2 + 5\ddot{x}_3 + 7\ddot{x}_4 = 1. \tag{16}$$

We find that the left side of eq. (16) is

$$\frac{1}{16} + \frac{3}{16} + \frac{5}{16} + \frac{7}{16} - 0.00002\omega_1^2 \cos \omega_1 t$$

$$+ 0.00001\omega_2^2 \cos \omega_2 t + 0.00005\omega_3^2 \cos \omega_3 t.$$

The last three terms of the above expression are not zero because of rounding errors.

In this example there are four degrees of freedom; hence there are four principal oscillations. So far as frequencies are concerned, one is degenerate and the other three are pure harmonic motions. So far as amplitudes are concerned, we notice from eqs. (15) that in the first mode there is one change of sign, in the second mode there are two, in the third mode there are three and in the fourth mode there are four.

In practical considerations it is the ratio of amplitudes that counts. Holzer[2] suggested a method of successive approximations to find directly the ratios. Duncon and Collar[3] also suggested a method of successive approximations to avoid the expansion of the determinantal equation and the solution of the resulting high-degree equation. With the wide use of computers, the labour in numerical calculations is not much a problem.

Holzer's idea of a transfer function does have the advantage of simplifying the representation by block diagrams. Two factors contribute to this fruitful idea. From the square matrix in eqs. (11), we notice that the entries are zero except in the diagonal and the second diagonal. We interpret from this that one end of the system does not feel the "influence" of the other end. Also, because of the transformation into Laplacian space, we can manipulate the D's as numbers.

From eqs. (9),

$$\frac{\bar{x}_4}{\bar{x}_3} = \frac{k_3}{M_4 S^2 + k_3}. \tag{17}$$

Now $M_4 S^2 k_3 / (M_4 S^2 + k_3)$ is the ratio of the product to the sum. It resembles the addition of parallel impedances Z. Let us call

$$Z_3 = \frac{M_4 S^2 k_3}{M_4 S^2 + k_3}. \tag{18}$$

Or

$$\frac{\bar{x}_4}{\bar{x}_3} = \frac{Z_3}{M_4 S^2}. \tag{19}$$

From eqs. (9) and (19),

$$\frac{\bar{x}_3}{\bar{x}_2} = \frac{k_2}{M_3S^2 + k_2 + k_3 - \dfrac{k_3Z_3}{M_4S^2}} = \frac{k_2}{M_3S^2 + k_2 + Z_3}$$

$$= \frac{k_2(M_3S^2 + Z_3)}{(M_3S^2 + Z_3)(k_2 + M_3S^2 + Z_3)}. \tag{20}$$

Let us call

$$Z_2 = \frac{k_2(M_3S^2 + Z_3)}{k_2 + M_3S^2 + Z_3}. \tag{21}$$

Or

$$\frac{\bar{x}_3}{\bar{x}_2} = \frac{Z_2}{M_3S^2 + Z_3}. \tag{22}$$

From eqs. (9) and (22),

$$\frac{\bar{x}_2}{\bar{x}_1} = \frac{k_1}{M_2S^2 + k_1 + Z_2}. \tag{23}$$

Let us call

$$Z_1 = \frac{k_1(M_2S^2 + Z_2)}{k_1 + M_2S^2 + Z_2}. \tag{24}$$

Or

$$\frac{\bar{x}_2}{\bar{x}_1} = \frac{Z_1}{M_2S^2 + Z_2}. \tag{25}$$

From eqs. (9) and (25),

$$\frac{\bar{x}_1}{\bar{F}} = \frac{1}{S(M_1S^2 + Z_1)}. \tag{26}$$

Thus

$$\frac{\bar{x}_2}{\bar{F}} = \frac{1}{S(M_1S^2 + Z_1)} \frac{Z_1}{M_2S^2 + Z_2}. \tag{27}$$

$$\frac{\bar{x}_3}{\bar{F}} = \frac{1}{S(M_1S^2 + Z_1)} \frac{Z_1}{M_2S^2 + Z_2} \frac{Z_2}{M_3S^2 + Z_3}. \tag{28}$$

$$\frac{\bar{x}_4}{\bar{F}} = \frac{1}{(SM_1S^2 + Z_1)} \frac{Z_1}{M_2S^2 + Z_2}$$
$$\cdot \frac{Z_2}{M_3S^2 + Z_3} \frac{Z_3}{M_4S^2} . \tag{29}$$

Equations (26)—(29) can be represented by block diagrams in Laplacian space:

$$\bar{F} \rightarrow \boxed{\frac{1/S}{M_1S^2 + Z_1}} \xrightarrow{\bar{x}_1} \boxed{\frac{Z_1}{M_2S^2 + Z_2}} \xrightarrow{\bar{x}_2} \boxed{\frac{Z_2}{M_3S^2 + Z_3}} \xrightarrow{\bar{x}_3} \boxed{\frac{Z_3}{M_4S^2}} \xrightarrow{\bar{x}_4}$$

Furthermore, for the electrical model with dissipation of power,

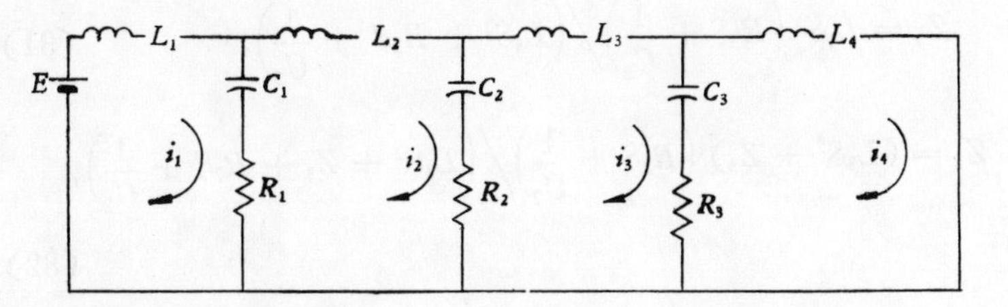

The bond graph is

$$
\begin{array}{ccccccccccccccccc}
 & & L_1 & & & & L_2 & & & & L_3 & & & & L_4 & & \\
 & & 2 & & & & 7 & & & & 12 & & & & 17 & & \\
E & \xrightarrow{1} & 1 & \xrightarrow{4} & 0 & \xrightarrow{6} & 1 & \xrightarrow{9} & 0 & \xrightarrow{11} & 1 & \xrightarrow{14} & 0 & \xrightarrow{16} & 1 & \xrightarrow{18} & E=0 \\
 & & 3 & & 5 & & 8 & & 10 & & 13 & & 15 & & & & \\
 & & R_1 & & C_1 & & R_2 & & C_2 & & R_3 & & C_3 & & & &
\end{array}
$$

The matrical equatious are

$$
\begin{bmatrix}
L_1D^2+R_1D+\dfrac{1}{C_1} & -\dfrac{1}{C_1} \\
-\dfrac{1}{C_1} & L_2D^2+R_2D+\dfrac{1}{C_1}+\dfrac{1}{C_2} \\
0 & -\dfrac{1}{C_2} \\
0 & 0
\end{bmatrix}
$$

$$\begin{matrix} 0 & 0 \\ -\dfrac{1}{C_2} & 0 \\ L_3D^2+R_3D+\dfrac{1}{C_2}+\dfrac{1}{C_3} & -\dfrac{1}{C_3} \\ -\dfrac{1}{C_3} & L_4D^2+\dfrac{1}{C_3} \end{matrix}\Bigg] \begin{bmatrix} q_1 \\ q_2 \\ q_3 \\ q_4 \end{bmatrix} = \begin{bmatrix} E \\ 0 \\ 0 \\ 0 \end{bmatrix}. \qquad (30)$$

The block diagrams are still valid for the electrical model with dissipation of power if we let

$$Z_3 = L_4S^2\left(R_3S + \frac{1}{C_3}\right)\Big/\left(L_4S^2 + R_3S + \frac{1}{C_3}\right), \qquad (31)$$

$$Z_2 = (L_3S^2 + Z_3)\left(R_2S + \frac{1}{C_2}\right)\Big/\left(L_3S^2 + Z_3 + R_2S + \frac{1}{C_2}\right), \qquad (32)$$

$$Z_1 = (L_2S^2 + Z_2)\left(R_1S + \frac{1}{C_1}\right)\Big/\left(L_2S^2 + Z_2 + R_1S + \frac{1}{C_1}\right). \qquad (33)$$

The entropy produced is conducted away as heat. Because the temperature is constant, we need not show the entropy flux in the bond graph.

Thermistor[4]

The electrical model is

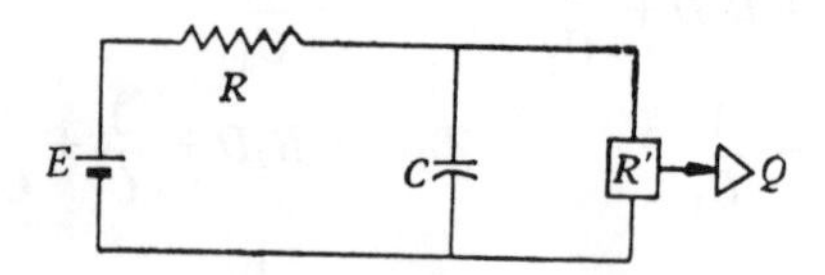

R' is a thermistor, a resistor with strong temperature dependence. The ambient temperature effects the resistance of the thermistor.

The bond graph is

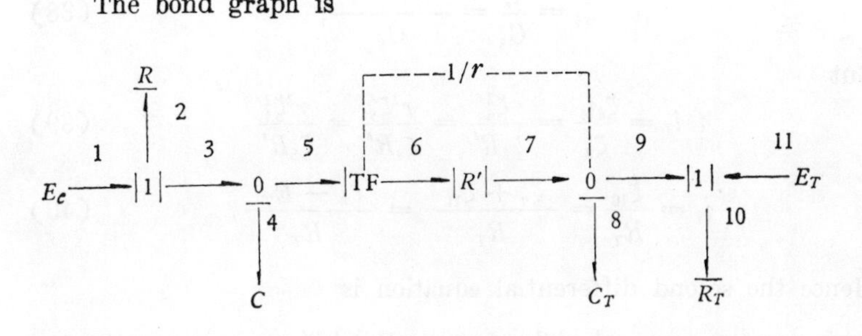

The thermistor could be shown in the bond graph as a transformer with a 2-terminal output. The input electrical power is split into electric power and thermal power. The electrical power is linked with potential and current of the electrical circuit and the thermal power is linked with temperature and entropy flux of the thermal circuit. As shown in the bond graph the thermal circuit is suppressed. By junction structures the output of a 2-terminal resistor consists of a capacitor C_T and a resistor R_T related to E_T. The double arrow in R' means uni-directional flow. The dashed line shows the no-power feedback to the transformer and the modulus $1/r$ being a variable.

Starting from the constitutive relation for the capacitor C,

$$\dot{\zeta}_4 = \frac{i_4}{C_4} = \frac{i_3 - i_5}{C_4}. \tag{34}$$

But

$$i_3 = i_2 = \frac{\zeta_1 - \zeta_3}{R} = \frac{E_e - \zeta_4}{R}, \tag{35}$$

$$i_5 = r i_6 = r\,\frac{\zeta_6}{R'} = r^2\,\frac{\zeta_5}{R'} = r^2\,\frac{\zeta_4}{R'}. \tag{36}$$

Hence the first differential equation is

$$\dot{\zeta}_4 = \frac{1}{C_4}\left[\frac{E_e - \zeta_4}{R} - \frac{r^2\zeta_4}{R'}\right] \tag{37}$$

Starting from the constitutive relation for the capacitor C_T,

$$\dot{\zeta}_8 = \frac{i_8}{C_8} = \frac{i_7 - i_9}{C_8}. \tag{38}$$

But

$$i_7 = \frac{\zeta_6 i_6}{\zeta_8} = \frac{\zeta_6^2}{\zeta_8 R'} = \frac{r^2\zeta_5^2}{\zeta_8 R'} = \frac{r^2\zeta_4^2}{\zeta_8 R'}. \tag{39}$$

$$i_9 = \frac{\zeta_{10}}{R_T} = \frac{\zeta_9 + \zeta_{11}}{R_T} = \frac{\zeta_8 - E_T}{R_T}. \tag{40}$$

Hence the second differential equation is

$$\dot{\zeta}_8 = \left[\frac{r^2\zeta_4^2}{\zeta_8 R'} - \frac{\zeta_8 - E_T}{R_T}\right] \Big/ C_8. \tag{41}$$

For nonlinear constitutive relations, we merely replace eq. (34) by $\zeta_4 = f_C^{-1}(i_3 - i_4)$, eq. (35) by $i_3 = f_R^{-1}(E_e - \zeta_4)$ etc. However, we have to investigate whether ζ_4, for example, is single valued.

Diffusion through a Membrane[5]

Consider 2 compartments I and II of a solution separated by a membrane m (Figure 1). The dimensions of the compartments are so large compared with those of the membrane that the flow Q (see Table 1 of §22) in each compartment is time independent. Representing the membrane by two resistors and one capacitor, the bond graph of the membrane is

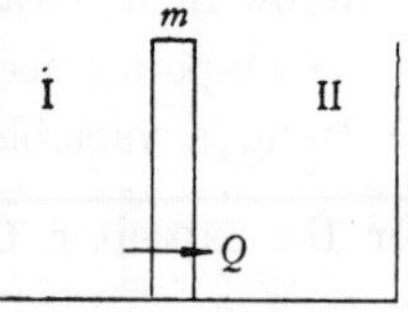

Figure 1

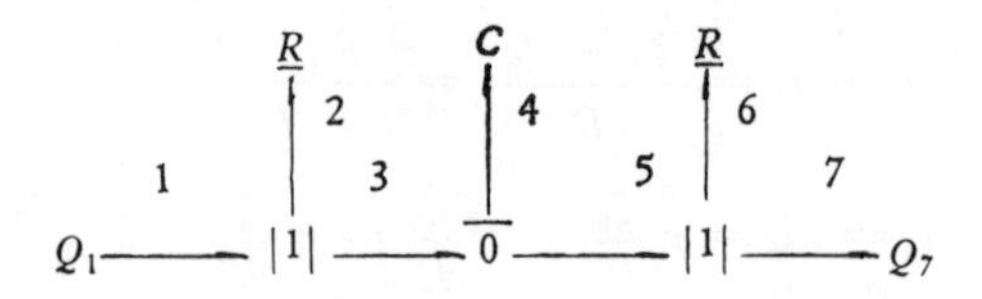

Assuming linear relations and a homogeneous membrane,

$$R_2 = R_6 = R_m = \frac{p_m}{Q_m}, \tag{42}$$

$$C_4 = C_m = \frac{v_m}{p_m}. \tag{43}$$

For capacitor 4, because of junction structures,

$$C_4 \dot{p}_4 = Q_4 = Q_3 - Q_5 = Q_2 - Q_6. \tag{44}$$

For resistors 2 and 6, because of junction structures,

$$C_4 \dot{p}_4 = \frac{p_2}{R_2} - \frac{p_6}{R_6} = \frac{p_1 - p_3}{R_2} - \frac{p_5 - p_7}{R_6}. \tag{45}$$

Or

$$\dot{p}_4 = \frac{2(p_{av} - p_m)}{C_m R_m}. \tag{46}$$

where $p_{av} = (p_1 + p_7)/2$ is the average pressure of the solvent in the solutions and $p_m = (p_3 + p_5)/2$ is the pressure within the membrane.

The relaxation time for current to decrease to $1/e$ of E/R in a circuit of $R - C$ in series is RC. Let us call

$$\Gamma_m = \frac{R_m C_m}{2}. \tag{47}$$

Reverting to the diffusion,

$$C_m = \frac{v_m}{p_m} = \frac{(A\Delta x)^2}{NRT}, \tag{48}$$

where N is the number of moles of the solute per unit volume, and A and Δx are the area and the thickness of the membrane respectively. Also

$$R_m = \frac{p_m}{Q_m} = \frac{NRT}{D_m A^2}, \tag{49}$$

where D_m is the diffusion coefficient within the membrane. Because $R_m A^2$ is equivalent to mobility, eq. (49) is Einstein's law of diffusion (eq. 44, 34). From eqs. (47—49),

$$D_m = \frac{(\Delta x)^2}{2\Gamma_m}. \tag{50}$$

In 1905, Einstein showed that the Brownian motion of microscopic particles is caused by the continued bombardment of the molecules of the surrounding medium. The displacement after a time t of a particle from its starting point is normally distributed, with mean 0 and variance $2Dt$. A model of the Brownian motion may be provided by a particle undergoing a random walk. Equation (50) may be interpreted as follows: the diffusion coeffient is half of the product of jump distance squared and jump frequency. Because the jump distance is in the order of the atomic distance, eq. (50) provides a rough method of estimating the diffusion coefficient if a reasonable value of the jump frequency can be assigned.

Resistance Membranes

Let a membrane m separate the two compartments of an aqueous solution (Figure 2). The phenomenological equations* are

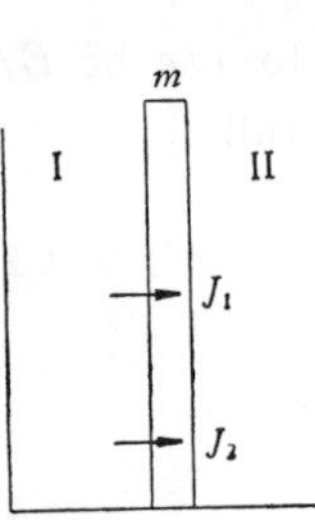

Figure 2

$$\begin{bmatrix} J_1 \\ J_2 \end{bmatrix} = \begin{bmatrix} L_{11} & L_{12} \\ L_{21} & L_{22} \end{bmatrix} \begin{bmatrix} X_1 \\ X_2 \end{bmatrix}, \tag{51}$$

where L_{11} is the permeability of the solvent, L_{22} is the permeability of the solute, and L_{12} and L_{21} are the cross coefficients.

Equations (51) can be interpreted in many ways. Suppose we have a 2-port "tee" circuit,

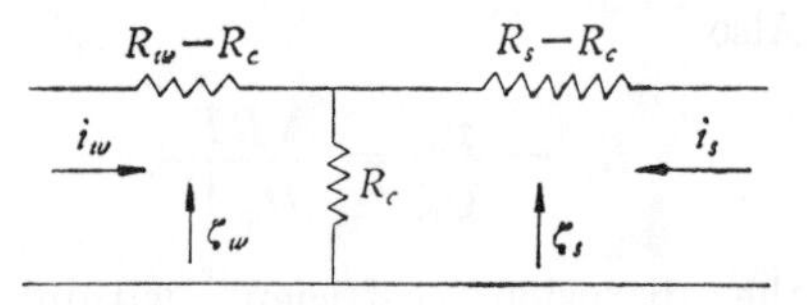

* Thermodynamically the right-side of eqs. (51) should be begun with a negative sign because we emphasize the positive rate of entropy-production. In electrical engineering, the negative sign will be only associated with the dissipated energy. Equations (51) are written without negative sign so that the other equations will not be cluttered with negative signs.

The loop equations in matrix form are

$$\begin{bmatrix} \zeta_w \\ \zeta_S \end{bmatrix} = \begin{bmatrix} R_w & R_c \\ R_c & R_S \end{bmatrix} \begin{bmatrix} i_w \\ i_S \end{bmatrix}. \tag{52}$$

Reverting to the permeability,

$$\begin{bmatrix} \Delta p \\ \Delta \mu \end{bmatrix} = \begin{bmatrix} R_w & R_c \\ R_c & R_S \end{bmatrix} \begin{bmatrix} Q_w \\ J_S \end{bmatrix}. \tag{53}$$

In view of eqs. (53), the matrix in eqs. (51) is the inverse of the matrix of frictional coefficients. The bond graph is

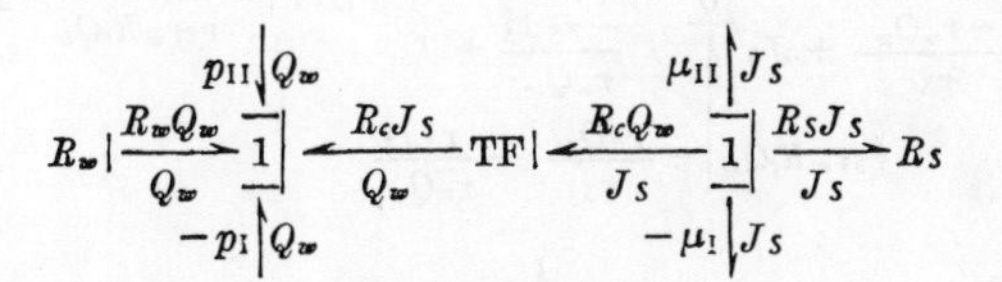

Suppose we have a 2-port "pi" circuit,

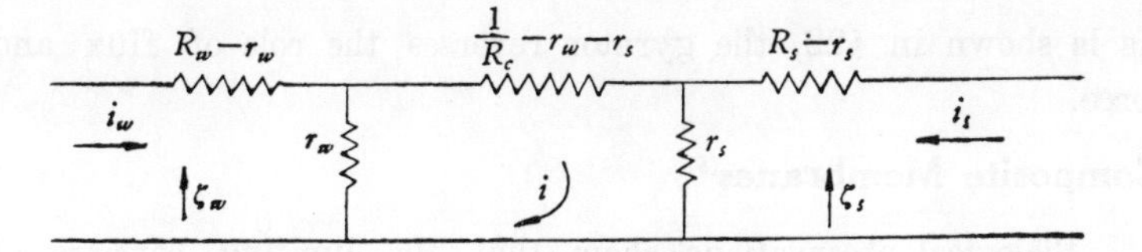

The loop equations are

$$\begin{aligned} -\zeta_w + R_w i_w - i r_w &= 0, \\ \frac{i}{R_c} - i_w r_w + i_S r_S &= 0, \\ -\zeta_S + R_S i_S + i r_S &= 0. \end{aligned} \tag{54}$$

Eliminating i,

$$\begin{bmatrix} \zeta_w \\ \zeta_S \end{bmatrix} = \begin{bmatrix} R_w - r_w^2 R_c & r_S r_w R_c \\ r_S r_w R_c & R_S - r_S^2 R_c \end{bmatrix} \begin{bmatrix} i_w \\ i_S \end{bmatrix}. \tag{55}$$

Reverting to the permeability,

$$\begin{bmatrix}\Delta p\\ \Delta\mu\end{bmatrix}=\begin{bmatrix}R_w - r_w^2R_c & r_S r_w R_c\\ r_S r_w R_c & R_S - r_S^2 R_c\end{bmatrix}\begin{bmatrix}Q_w\\ J_S\end{bmatrix}. \tag{56}$$

The bond graph is

$$R_w \xrightarrow[Q_w]{R_wQ_w} \underset{-p_{\mathrm{I}}\,\uparrow\, Q_w}{\overset{p_{\mathrm{II}}\,\downarrow\, Q_w}{1}} \xleftarrow[Q_w]{-r_w^2R_cQ_w + r_Sr_wR_cJ_S} |\mathrm{TF} \xleftarrow[r_Sr_wR_cQ_w]{\frac{-r_wQ_w}{r_S}+J_S} |\mathrm{GY}|$$

$$\xleftarrow[\frac{-r_wQ_w}{r_S}+J_S]{r_Sr_wR_cQ_w} 0 \xleftarrow[\frac{-r_SJ_S^2}{r_wQ_w}+J_S]{r_Sr_wR_cQ_w} |\mathrm{GY}| \xleftarrow[r_Sr_wR_cQ_w]{\frac{-r_SJ_S^2}{r_wQ_w}+J_S} \mathrm{TF}|$$

$$r_Sr_wR_cQ_w \,\Big|\, -\frac{r_wQ_w}{r_S}+\frac{r_SJ_S^2}{r_wQ_w}$$

$$\xleftarrow[J_S]{r_Sr_wR_cQ_w - r_S^2R_cJ_S} \underset{-\mu_1\,\downarrow\, J_S}{\overset{\mu_2\,\uparrow\, J_S}{1}} \xrightarrow[J_S]{R_SJ_S} R_S$$

As is shown in §22, the gyrator reverses the role of flux and force.

Composite Membranes[6]

Biological observations show that the limiting barriers of single cells and tissues are complex composite membranes of varying composition and varying permeability. For simplicity, we shall assume that no water flows through the charged membranes and the solute is a uni-uni salt. Also we consider two membranes arranged in a single layer (parallel array) and two membranes arranged in two layers (series array) only.

In the parallel array (Figure 3), the fractions of the total membrane area occupied by membrane 1 and 2 are Γ_1 and Γ_2 respectively. Because of the parallel array, the potential difference E and the osmotic-pressure difference $\Delta\pi$ are the same across the membranes. The phenomenological equations are

$$Q_i = (\omega_i, t_i)\begin{bmatrix}\Delta\pi \\ \dfrac{I_i}{\rho Z}\end{bmatrix}, \quad i = 1, 2, \tag{57}$$

Figure 3

$$I_i = (k_i t_i,\ k_i)\begin{bmatrix}\dfrac{\Delta\pi}{\rho Z} \\ E\end{bmatrix}, \tag{58}$$

where Q is the volume flow, I is the current, ω is the permeability, k is the electrical conductance, t is the transference number of anion or cation (for definiteness, in the following discussion t_1 means the transference number of positive ions in the membrane 1), and Z is the charge density. The network is represented by

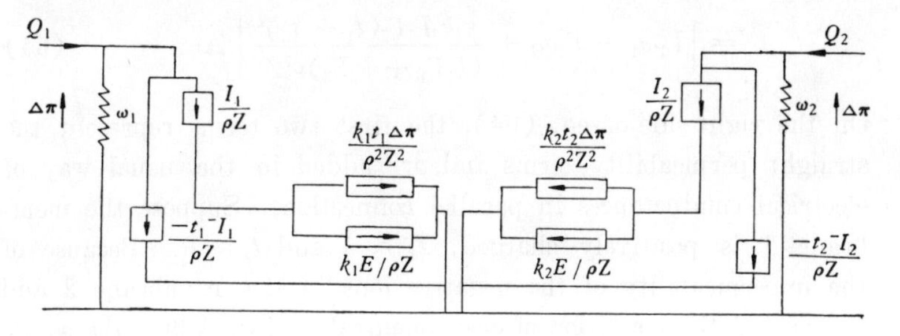

where ⇥ represents a constant-current-generator.

In the open-circuit (potentiometric) experiment, the electrical current is zero. Or

$$I = \Gamma_1 I_1 + \Gamma_2 I_2 = 0. \tag{59}$$

The node equations are

$$-\Gamma_1 I_1 + \frac{\Gamma_1 k_1 t_1 \Delta\pi}{\rho Z} + \Gamma_1 k_1 E = 0, \tag{60}$$

$$-\Gamma_2 I_2 + \frac{\Gamma_2 k_2 t_2 \Delta\pi}{\rho Z} + \Gamma_2 k_2 E = 0. \tag{61}$$

Solving eqs. (59—61),

$$E = -\frac{(\Gamma_1 k_1 t_1 + \Gamma_2 k_2 t_2)\Delta\pi}{(k_1\Gamma_1 + k_2\Gamma_2)\rho Z}. \tag{62}$$

$$\Gamma_1 I_1 = \frac{k_1 k_2 \Gamma_1 \Gamma_2 (t_1 - t_2)\Delta\pi}{(k_1\Gamma_1 + k_2\Gamma_2)\rho Z}. \tag{63}$$

$\Gamma_1 I_1$ is not zero unless $(t_1 - t_2)$ is zero. Then there is an electrical current through one of the membranes, compensated by an equal flow in the other membrane.

The circulation of electrical current by the increased permeability is shown by

$$\begin{aligned}\Gamma_1 Q_1 + \Gamma_2 Q_2 &= (\Gamma_1\omega_1 + \Gamma_2\omega_2)\Delta\pi + \frac{\Gamma_1 t_1 I_1 + \Gamma_2 t_2 I_2}{\rho Z} \\ &= (\Gamma_1\omega_1 + \Gamma_2\omega_2)\Delta\pi + \frac{k_1 k_2 \Gamma_1 \Gamma_2 (t_1 - t_2)^2 \Delta\pi}{(k_1\Gamma_1 + k_2\Gamma_2)\rho^2 Z^2} \\ &= \left[\Gamma_1\omega_1 + \Gamma_2\omega_2 + \frac{k_1 k_2 \Gamma_1 \Gamma_2 (t_1 - t_2)^2}{(k_1\Gamma_1 + k_2\Gamma_2)\rho^2 Z^2}\right]\Delta\pi.\end{aligned} \tag{64}$$

On the right-side of eq. (64), the first two terms represent the straight permeability terms and are added in the usual way of electrical conductances in parallel connection. Suppose the membrane 1 is positively charged, $t_1^- \approx 1$ and $t_1 \approx 0$. Because of the impermeability of the negative ions in the membrane 2 and because of the condition of electroneutrality (see §39), the solute is impermeable. However, because of the last term, positive ions carrying a positive current move from the higher to lower salt concentration, while negative ions carrying a positive current move from the lower to higher salt concentration. Both flows determine a closed circuit (For a specfic example, see Figure 4).

In biology[7] the coupling between metabolism——energy-production by the hydrolysis of ATP (adenosine-5′-triphosphate) to ADP (adenosine-5′-diphosphate)——and gradients of Na^+, K^+, Ca^{++} and Cl^- are well known. Both sodium and calcium pump can be reversed in synthesis experiments without the benefit of ATP. The normal mode of sodium pump is to extrude the sodium ion and to take the potassium ion at the same time. The ratio of Na^+ to K^+ is not always 1:1 (with a net outward current). Chloride pump extrudes Cl^- but takes protons.

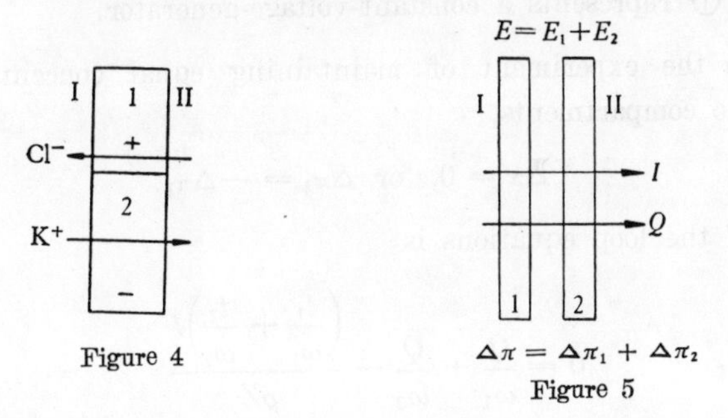

Figure 4

$\Delta\pi = \Delta\pi_1 + \Delta\pi_2$
Figure 5

In the series array, there is a thin layer of aqueous solution between the two membranes (Figure 5). We assume that salt in the intermediate space has the same chemical potential as that in the bordering surfaces. Then

$$E = E_1 + E_2, \tag{65}$$

and

$$\Delta\pi = \Delta\pi_1 + \Delta\pi_2. \tag{66}$$

Because Q or I is the same in either membrane, it is convenient to write the phenomenological equations (57) and (58) as

$$\Delta\pi_i = \left(\frac{1}{\omega_i}, -\frac{t_i}{\omega_i}\right)\begin{bmatrix} Q \\ \dfrac{I}{\rho Z} \end{bmatrix}, \; i = 1, 2, \tag{67}$$

$$E_i = \left(\frac{1}{k_i}, \ -t_i\right)\begin{bmatrix} I \\ \dfrac{\Delta\pi_i}{\rho Z} \end{bmatrix}. \tag{68}$$

The network is represented by

Q

$1/\omega_1$

$\Delta\pi_1$ $-t_1 I/\omega_1 \rho Z$

$1/\omega_2$

$\Delta\pi_2$ $-t_2 I/\omega_2 \rho Z$

I

$1/k_1$

E_1 $-t_1 \Delta\pi_1/\rho Z$

$1/k_2$

E_2 $-t_2 \Delta\pi_2/\rho Z$

where ⓣ represents a constant-voltage-generator.

In the experiment of maintaining equal concentration in the two compartments,

$$\Delta\pi = 0, \quad \text{or} \quad \Delta\pi_2 = -\Delta\pi_1. \tag{69}$$

One of the loop equations is

$$0 = \frac{Q}{\omega_1} + \frac{Q}{\omega_2} - \frac{\left(\dfrac{t_1}{\omega_1} + \dfrac{t_2}{\omega_2}\right)I}{\rho Z} \tag{70}$$

Therefore

$$Q = \frac{(\omega_2 t_1 + \omega_1 t_2)I}{(\omega_1 + \omega_2)\rho Z}, \tag{71}$$

and

$$\Delta\pi_1 = -\frac{(t_1 - t_2)I}{(\omega_1 + \omega_2)\rho Z}. \tag{72}$$

Equation (72) shows that the osmotic-pressure difference from compartment I to the intermediate space and the current have opposite signs. If the direction of current (not electrons) is positive (negative), there will be accumulation (depletion) of salt in the intermediate space. For specific example, see Figure 6.

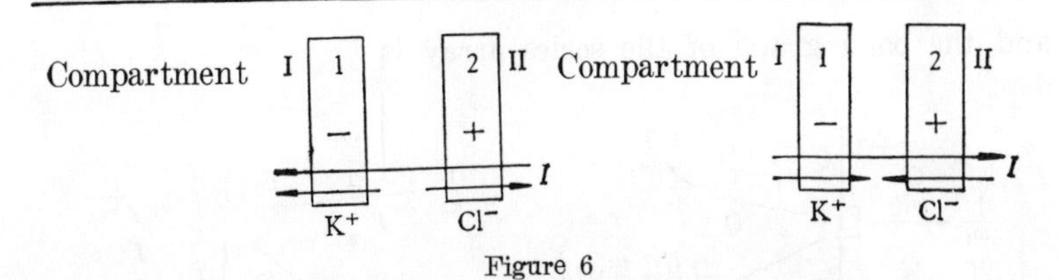

Figure 6

From the other loop equation and using eq. (72),

$$E = E_1 + E_2 = \left(\frac{1}{k_1} + \frac{1}{k_2}\right) I - \frac{(t_1 - t_2)\Delta\pi_1}{\rho Z}$$

$$= \left[\frac{1}{k_1} + \frac{1}{k_2} + \frac{(t_1 - t_2)^2}{(\omega_1 + \omega_2)\rho^2 Z^2}\right] I. \tag{73}$$

The first two terms on the right side of eq. (73) represent the straight electrical conductances and are added in the usual way of a series connection. The last term shows that, if there is accumulation of salt in the intermediate space, the effect is to decrease the overall resistance and the membrane system will become a better electrical conductor. When the accumulation is appreciable, the polarity term vanishes and the effect ceases. On the other hand, if there is depletion of salt in the intermediate space, the effect is to increase the overall resistance and the membrane system will become a poorer electrical conductor. When the depletion is almost complete, the passage of electricity is cut off. Thus the two charged membranes in series exhibit rectifying properties similar to a diode.

The bond graph of the parallel array is

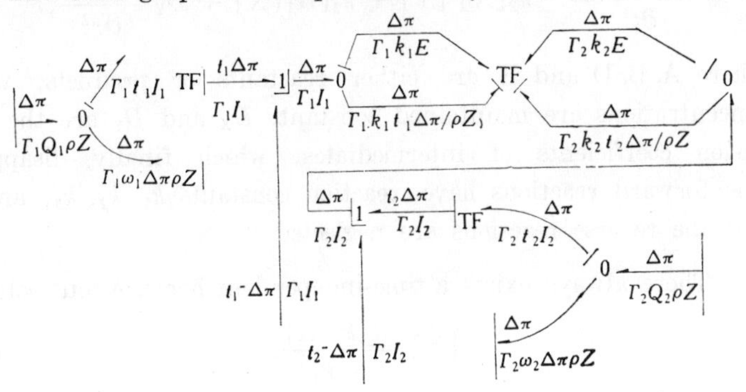

and the bond graph of the series array is

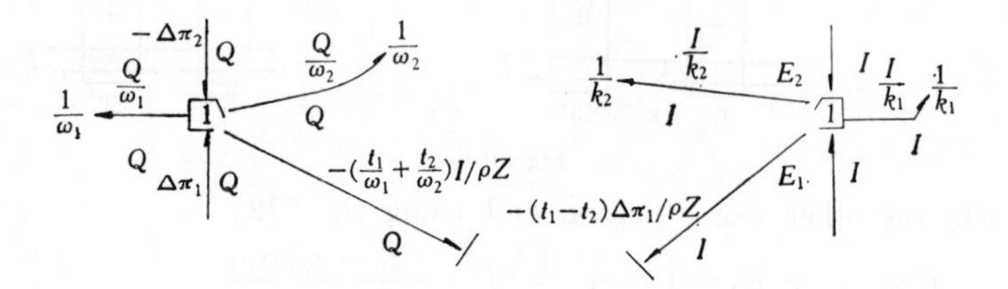

Kinetics of Chemical Reactions in Open Systems

Let us have a scheme of reactions[8] as in the following:

$$A \longrightarrow X, \tag{74}$$

$$2X + Y \longrightarrow 3X, \tag{75}$$

$$B + X \longrightarrow Y + D, \tag{76}$$

$$X \longrightarrow E. \tag{77}$$

The overall reaction is

$$A + B \longrightarrow D + E. \tag{78}$$

The kinetic equations are

$$\frac{\partial [X]}{\partial t} = k_1[A] + k_2[X]^2[Y] - k_3[B][X] - k_4[X] + D_X \frac{\partial^2 [X]}{\partial r^2}, \tag{79}$$

$$\frac{\partial [Y]}{\partial t} = -k_2[X]^2[Y] + k_3[B][X] + D_Y \frac{\partial^2 [Y]}{\partial r^2}, \tag{80}$$

where A, B, D and E are either reactants or products, whose concentrations are maintained constant. D_X and D_Y are the diffusion coefficients of intermediates, which finally disappear. The forward reactions have reaction constants k_1, k_2, k_3, and k_4 and the reverse reactions are neglected.

There always exists a time-independent homogeneous solution

$$[X]_0 = \frac{k_1[A]}{k_4}, \tag{81}$$

$$[Y]_0 = \left(\frac{k_3k_4}{k_1k_2}\right)\left(\frac{[B]}{[A]}\right), \tag{82}$$

which is the equilibrium solution. To investigate stability, we first investigate the dispersion equation within the framework of linear analysis. We consider a perturbed state of the form

$$[X] = [X]_0 + x \exp\left(\omega t + \frac{ir}{l}\right), \tag{83}$$

$$[Y] = [Y]_0 + y \exp\left(\omega t + \frac{ir}{l}\right), \tag{84}$$

with $\|x/[X]_0\| \ll 1$, and $\|y/[Y]_0\| \ll 1$.

Substituting eqs. (83)—(84) in eqs. (79)—(80), and cancelling equilibrium values and second-order small quantities,

$$\begin{bmatrix} \dfrac{\partial x}{\partial t} \\ \dfrac{\partial y}{\partial t} \end{bmatrix} = \begin{bmatrix} -k_3[B] - k_4 - \dfrac{D_X}{l^2} - \omega & \dfrac{k_1^2k_2[A]^2}{k_4^2} \\ k_3[B] & \dfrac{-k_1^2k_2[A]^2}{k_4^2} - \dfrac{D_Y}{l^2} - \omega \end{bmatrix} \begin{bmatrix} x \\ y \end{bmatrix} \tag{85}$$

We have shifted the origin from (0, 0) to (X_0, Y_0) and only need to examine the square matrix of eq. (85) after ω's are deleted. For simplicity, we put the reaction constants to unity. The characteristic roots are roots of the equation

$$S^2 - pS + q = 0, \tag{86}$$

where

$$p = -[B] - 1 - \frac{D_X}{l^2} - [A]^2 - \frac{D_Y}{l^2},$$

and

$$q = \frac{[B]D_Y}{l^2} + \frac{D_XD_Y}{l^4} + \frac{D_X[A]^2}{l^2} + \frac{D_Y}{l^2} + [A]^2.$$

Then $p = 0$ corresponds to the vanishing of real roots and $q<0$ corresponds to the unstable saddle points. For $q > 0$, the stable foci ($p > 0$) are separated from unstable ones ($p < 0$) by $p = 0$. Because the spatial part of exponents in eqs. (83—84) are represented by polynomials of sines and cosines, the new origin is really a point of bifurcation, separating asymptotical stable (returning to the old origin) cycle from the unstable one.

References

[1] J. C. Jager: An Introduction to the Laplace Transformation with Engineering Applications, Chapman, London (1966).
[2] H. Holzer: Die Brechung des Drechschwingen, Julius Springer, Berlin (1921).
[3] W. J. Duncon and A. R. Collar: *Phil. Mag.*, Ser. 7, **107**, 865 (1921).
[4] G. Oster and D. M. Auslander: *J. Franklin Inst.*, **292**, 1(1971).
[5] G. Oster, A. Perelson and A. Katchalsky: *Nature*, **234**, 393(1971).
[6] A. Katchalsky and P. F. Curran: Nonequilibrium Thermodynamics in Biophysics, Harvard Univ. Press, Cambridge, Mass. (1965).
[7] Energy Transformation in Biological Systems, Ciba Foundation Symposium 31, Elsevier, Associated Scientific Publishers, Amsterdam (1975).
[8] I. Prigogine and R. Lefever: *J. Chem. Phys.*, **48**, 1695 (1968).

§24 *TOPOLOGICAL THERMODYNAMICS*

Tellegen's Orthogonality Law

Tellegen's law states that $\sum i\zeta = 0$, where the summation is over all branches of the electrical network.

This can be proven by writing

$$\sum_{kj} i_{kj}\zeta_{kj} = \sum_{kj} i_{kj}(\zeta_k - \zeta_j) = \sum_{k} i_{kj}\zeta_k - \sum_{j} i_{jk}\zeta_j = 0. \quad (1)$$

Now i_{kj} is the current on the kth node and the jth mesh and ζ_{kj} is the potential on the branch connecting k and j. We can write $\zeta_{kj} = \zeta_k - \zeta_j$ because the voltage is unique from

Kirchhoff's voltage law. The summation $i_{kj}\ \zeta_k$ over k node is zero. The summation $i_{jk}\ \zeta_j$ over j node is also zero.

If the meshes are infinitesimal, we can prove the theorem from vector analysis. It can be shown that a vector $\boldsymbol{A}$ is uniquely determined by the specification of divergence and curl. The vector can be actually obtained from an arbitrary divergence but zero curl, and an arbitrary curl but zero divergence. The actual generality of these apparently special cases follows from the fact that any vector is the sum of a solenoidal vector (comparable with $\boldsymbol{i}$) and an irrotational vector (comparable with $\boldsymbol{\zeta}$). In other words, current and potential vectors are in different subspaces of a linear space; hence they are orthogonal to each other.

Tellegen's law is quite general. It does not specify whether the relationship between potential and current is linear or not. It is valid after integration or after differentiation.

Heaviside's Law of Rate of Entropy-production[1]

Let us consider a black box containing resistors, inductors and capacitors connected in some arbitrary manner. There are input terminals and output terminals. Suppose constant voltages, but which vary sinusoidally, are applied to the input terminals. Initially, there are transient effects which will die down. Eventually the output voltages will also be sinusoidal functions. The only difference between the two will be amplitudes and phases but the frequencies will be the same.

Let the frequencies be ω_a and ω_b. For positive time t, every n-terminal current is the sum of a term with exp $\omega_a t$, a term with exp $\omega_b t$ and a constant term. Because the n-terminal voltages are constant, in terms of the basis

$$\begin{bmatrix} e^{\omega_a t'} \\ e^{\omega_b t'} \\ 1 \end{bmatrix} [e^{\omega_a t}, e^{\omega_b t}, 1] = \begin{bmatrix} e^{\omega_a (t+t')} & e^{\omega_a t'+\omega_b t} & e^{\omega_a t'} \\ e^{\omega_a t+\omega_b t'} & e^{\omega_b (t+t')} & e^{\omega_b t'} \\ e^{\omega_a t} & e^{\omega_b t'} & 1 \end{bmatrix}, \tag{2}$$

the power delivered is

$$\begin{bmatrix} 0 & 0 & 0 \\ 0 & 0 & 0 \\ G_a & G_b & G_0 \end{bmatrix}. \tag{3}$$

The current and voltage of a resistance R_k can be written as

$$i_{R_k} = (A_{ak},\ A_{bk},\ A_{0k})(e^{\omega_a t},\ e^{\omega_b t},\ 1)^{\mathrm{T}} \tag{4}$$

$$\zeta_{R_k} = R_k(A_{ak}, A_{bk}, A_{0k})(e^{\omega_a t'},\ e^{\omega_b t'},\ 1)^{\mathrm{T}}. \tag{5}$$

The current and voltage of an inductance L_m can be written as

$$i_{L_m} = (B_{am}, B_{bm}, -B_{am}-B_{bm})(e^{\omega_a t},\ e^{\omega_b t},\ 1)^{\mathrm{T}} \tag{6}$$

$$\zeta_{L_m} = L_m(\omega_a B_{am}, \omega_b B_{bm}, 0)(e^{\omega_a t'},\ e^{\omega_b t'},\ 1)^{\mathrm{T}}, \tag{7}$$

because at $t = 0$, $i_{L_m} = 0$ and $\zeta_{L_m} = L_m di_{L_m}/dt$. The current and voltage of a capacitance C_p can be written as

$$i_{C_p} = C_p(\omega_a D_{ap}, \omega_b D_{bp}, 0)(e^{\omega_a t},\ e^{\omega_b t},\ 1)^{\mathrm{T}}, \tag{8}$$

$$\zeta_{C_p} = (D_{ap}, D_{bp}, -D_{ap}-D_{bp})(e^{\omega_a t'}, e^{\omega_b t'},\ 1)^{\mathrm{T}}, \tag{9}$$

because at $t' = 0$, $\zeta_{C_p} = 0$ and $i_{C_p} = C_p d\zeta_{C_p}/dt'$.

In terms of the basis (2), eqs. (3—9) yield

$$\begin{bmatrix} 0 & 0 & 0 \\ 0 & 0 & 0 \\ G_a & G_b & G_0 \end{bmatrix} = R_k \begin{bmatrix} A_{ak}^2 & A_{ak}A_{bk} & A_{ak}A_{0k} \\ A_{ak}A_{bk} & A_{bk}^2 & A_{bk}A_{0k} \\ A_{0k}A_{ak} & A_{0k}A_{bk} & A_{0k}^2 \end{bmatrix}$$

$$+ L_m \begin{bmatrix} \omega_a B_{am}^2 & \omega_a B_{am}B_{bm} & -\omega_a B_{am}^2 - \omega_a B_{am}B_{bm} \\ \omega_b B_{am}B_{bm} & \omega_b B_{bm}^2 & -\omega_b B_{bm}^2 - \omega_b B_{am}B_{bm} \\ 0 & 0 & 0 \end{bmatrix}$$

$$+ C_p \begin{bmatrix} \omega_a D_{ap}^2 & \omega_b D_{ap}D_{bp} & 0 \\ \omega_a D_{ap}D_{bp} & \omega_b D_{bp}^2 & 0 \\ -\omega_a D_{ap}^2 - \omega_a D_{ap}D_{bp} & -\omega_b D_{bp}^2 - \omega_b D_{ap}D_{bp} & 0 \end{bmatrix}. \tag{10}$$

In equalizing the 12, 21, 13, 23, 31 and 32 entries of eq. (10), we obtain

$$\sum_m L_m B_{am} B_{bm} = \sum_p C_p D_{ap} D_{bp}, \tag{11}$$

$$\sum_k R_k A_{0k} A_{ak} = \sum_m \omega_a L_m B_{am}^2 + \sum_p \omega_a C_p D_{ap} D_{bp}, \tag{12}$$

$$\sum_k R_k A_{0k} A_{bk} = \sum_m \omega_b L_m B_{bm}^2 + \sum_p \omega_b C_p D_{ap} D_{bp}, \tag{13}$$

$$G_a = \omega_a \Big(\sum_m L_m B_{am}^2 - \sum_p C_p D_{ap}^2 \Big), \tag{14}$$

$$G_b = \omega_b \Big(\sum_m L_m B_{bm}^2 - \sum_p C_p D_{bp}^2 \Big). \tag{15}$$

The magnetic (kinetic) energy is

$$T = \sum_m \frac{1}{2} L_m (B_{am}, B_{bm}, -B_{am} - B_{bm})^{\mathrm{T}} (B_{am}, B_{bm}, -B_{am} - B_{bm}). \tag{16}$$

The electrical (stored) energy is

$$\phi = \sum_p \frac{1}{2} C_p (D_{ap}, D_{bp}, -D_{ap} - D_{bp})^{\mathrm{T}} (D_{ap}, D_{bp}, -D_{ap} - D_{bp}). \tag{17}$$

Subtracting eq. (17) from eq. (16) and using eqs. (11), (14) and (15),

$$2(T - \phi) = \begin{bmatrix} \dfrac{G_a}{\omega_a} & 0 & \dfrac{-G_a}{\omega_a} \\ 0 & \dfrac{G_b}{\omega_b} & \dfrac{-G_b}{\omega_b} \\ -\dfrac{G_a}{\omega_a} & -\dfrac{G_b}{\omega_b} & \dfrac{G_a}{\omega_a} + \dfrac{G_b}{\omega_b} \end{bmatrix}. \tag{18}$$

Let $G_a/\omega_a = A$, $G_b/\omega_b = B$, $C^2 = A^2 - AB + B^2$,

$$A + C = s, \quad A - C = t, \quad B + C = m, \quad B - C = n.$$

The spectral decomposition of $2(T - \phi)$ is

$$2(T - \phi) = \lambda_1 E_1 + \lambda_2 E_2 + \lambda_3 E_3, \tag{19}$$

where

$$\lambda_1 = 0,\ \lambda_2 = A + B + C,\ \lambda_3 = A + B - C,$$

$$E_1 = \frac{1}{\Delta}\begin{bmatrix} -\frac{B}{s} + \frac{B}{t} & \frac{A}{m} - \frac{A}{n} & \frac{AB}{tm} - \frac{AB}{sn} \\ -\frac{B}{s} + \frac{B}{t} & \frac{A}{m} - \frac{A}{n} & \frac{AB}{tm} - \frac{AB}{sn} \\ -\frac{B}{s} + \frac{B}{t} & \frac{A}{m} - \frac{A}{n} & \frac{AB}{tm} - \frac{AB}{sn} \end{bmatrix},$$

$$E_2 = \frac{1}{\Delta}\begin{bmatrix} \frac{A}{m} + \frac{AB}{tm} & -\frac{A}{m} - \frac{A^2}{B^2 - C^2} & -\frac{AB}{tm} + \frac{A^2}{B^2 - C^2} \\ \frac{B}{s} + \frac{B^2}{A^2 - C^2} & -\frac{B}{s} - \frac{AB}{sn} & -\frac{B^2}{A^2 - C^2} + \frac{AB}{sn} \\ -1 - \frac{B}{t} & 1 + \frac{A}{n} & \frac{B}{t} - \frac{A}{n} \end{bmatrix},$$

$$E_3 = \frac{1}{\Delta}\begin{bmatrix} -\frac{A}{n} - \frac{AB}{sn} & \frac{A}{n} + \frac{A^2}{B^2 - C^2} & \frac{AB}{sn} - \frac{A^2}{B^2 - C^2} \\ -\frac{B}{t} - \frac{B^2}{A^2 - C^2} & \frac{B}{t} + \frac{AB}{tm} & \frac{B^2}{A^2 - C^2} - \frac{AB}{tm} \\ 1 + \frac{B}{s} & -1 - \frac{A}{m} & -\frac{B}{s} + \frac{A}{m} \end{bmatrix},$$

$\Delta = A/m - A/n + B/t - B/s - AB/sn + AB/tm$. λ's are characteristic roots of the square matrix in eq. (18),

$$P[2(T - \phi)]P^{-1} = \begin{bmatrix} \lambda_1 & & \\ & \lambda_2 & \\ & & \lambda_3 \end{bmatrix}, \qquad (20)$$

and E's are obtained by

$$P\begin{bmatrix} 1 & & \\ & 0 & \\ & & 0 \end{bmatrix}P^{-1} = E_1,\quad P\begin{bmatrix} 0 & & \\ & 1 & \\ & & 0 \end{bmatrix}P^{-1} = E_2,$$

$$P\begin{bmatrix} 0 & & \\ & 0 & \\ & & 1 \end{bmatrix}P^{-1} = E_3. \tag{21}$$

We see that

$$\sum_i^3 E_i = \overleftrightarrow{1}, i = 1, 2, 3, \tag{22}$$

and

$$\sum_{ij} E_i E_j = 0, \quad i \neq j. \tag{23}$$

Without going in details about linear transformation, the following heuristic remarks may be made on eq. (18):

(A) Intuitively a system cannot oscillate sinusoidally with arbitrary frequency, but only with certain discrete frequencies. The physical origin of this may be traced to the kind of boundary conditions.

(B) According to Tellegen's law, current and potential vectors belong to the null space having a dimension of 2. Because current may lead or lag behind potential, another invariant subspace has a dimension of 1. The eigenvector of this subspace is associated with zero frequency. The physical origin of this may be traced to the kind of initial conditions.

(C) The eigenvectors associated with zero frequency and the other two frequencies form an orthogonal space (a zero vector could assume any direction.). Equation (22) indicates that one subspace is supplementary to the other two subspaces. Equation (23) indicates that these three subspaces are mutually orthogonal. Although the admittance matrices and the initial conditions are arbitrary, the particular arrangement of the network to realize a certain $2(T - \phi)$ is irrelevant.

(D) The same cannot be said about T or ϕ separately.

(E) We can easily extend to nn-terminals by increasing the size of matrices in eq. (10).

(F) The application of a variable voltage can be conceived as the successive application of incremental constant voltages. For example, if the voltage ζ is zero for $t < 0$, $\zeta_0 t/T$ for $0 < t < T$ and ζ_0 for $t > T$ (Figure 1), it can be linearized as

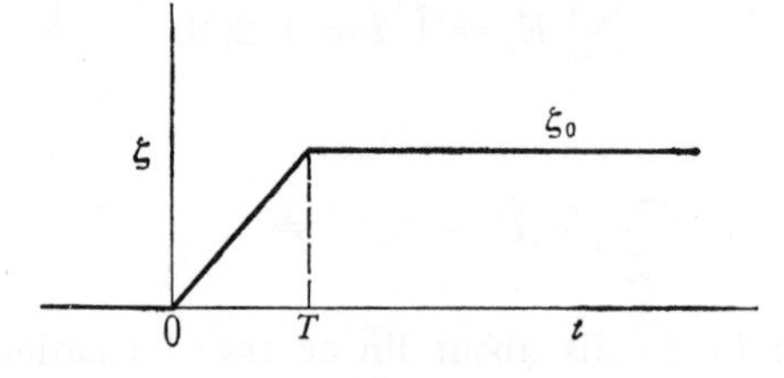

Figure 1

$$\frac{\zeta_0 t}{T} + \frac{\zeta_0 - \zeta_0 t}{T} u(t - T)$$
$$= \frac{\zeta_0}{T} [t - (t - T)u(t - T)], \tag{24}$$

where u is the Heaviside unit function.

Thus we have the theorem: If arbitrarily varying voltages are applied to a passive nn-terminal, the difference between the magnetic and the electrical energy will at any instant depend only on the admittance matrices and not on the particular network used for realizing it.

Because the energy delivered to the nn-terminal is $W + T + \phi$, where W is the dissipated energy, the above theorem not only holds for $T - \phi$ but also for $W + 2T$ and for $W + 2\phi$.

The above theorem states that at any instant the net power delivered is equal to $2(T - \phi)$. It is interesting to ask what is net power delivered at $t \to 0$ and at $t \to \infty$. Let the net power delivered at t be $P(t)$. According to the initial- and final-value theorems of the Laplacian transformation,

$$\lim_{t \to 0} P(t) = \lim_{s \to \infty} sp(s), \tag{25}$$

$$\lim_{t \to \infty} P(t) = \lim_{s \to 0} sp(s), \tag{26}$$

where $p(s)$ is the Laplace transform of $P(t)$. Because $p(s) = q(s)/r(s)$, where $q(s)$ has less degree than $r(s)$ usually, the net power delivered at $t \to 0$ may be zero and the net power delivered at $t \to \infty$ is certainly zero. Based on the power delivered at $t \to \infty$, the total useful energy delivered is equal to $-2(T - \psi)$. Hence we have a theorem due to Heaviside: If arbitrarily varying voltages are suddenly applied to a passive nn-terminal, when the final state has been reached, the total energy delivered to the nn-terminal exceeds the energy representing the loss by dissipation at the final rate, supposed to start at once, by twice the excess of the electrical over the magnetic energy. Because the system is isothermal, the law may be stated in terms of total entropy-production and rate of entropy-production instead of total energy and power.

Applications

Because the two laws mentioned above are general, the usefulness of their applications is limited. So far there is no general criterion to ascertain the existence of steady state in nonlinear systems; any limited information on steady state is helpful. However, because the two laws of topological origin and unrelated to trajectories, the applications along the trajectories could not bring out anything but nebulous conclusions.

Consider a simple linear network (Figure 2).

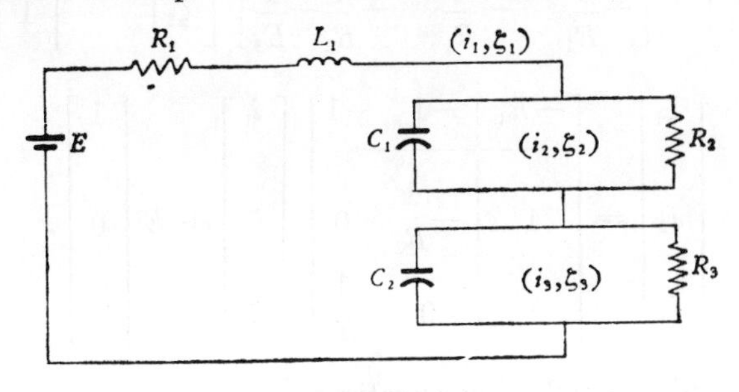

Figure 2

The bond graph is

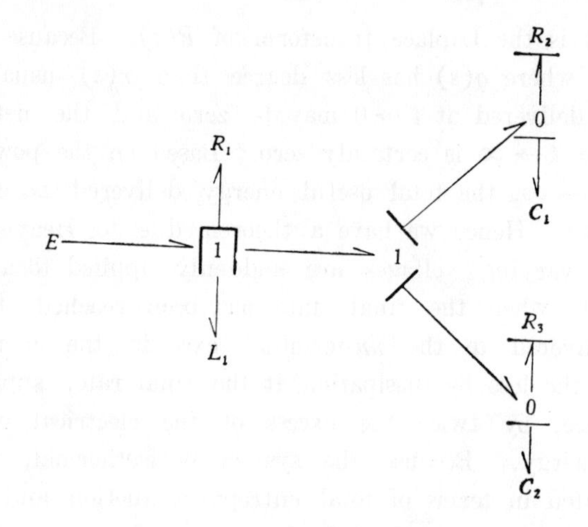

Flower and Evans[2] give the following equations:

$$\begin{bmatrix}\zeta_1\\ \zeta_2\\ \zeta_3\end{bmatrix}=\begin{bmatrix}-R_1-R_2-R_3 & R_2 & R_3\\ R_2 & -R_2 & 0\\ R_3 & 0 & -R_3\end{bmatrix}\begin{bmatrix}i_1\\ i_2\\ i_3\end{bmatrix}+E\begin{bmatrix}1\\ 0\\ 0\end{bmatrix}, \tag{27}$$

$$\begin{bmatrix}i_1\\ i_2\\ i_3\end{bmatrix}=\begin{bmatrix}-\dfrac{1}{R_1} & -\dfrac{1}{R_1} & -\dfrac{1}{R_1}\\ -\dfrac{1}{R_1} & -\dfrac{1}{R_1}-\dfrac{1}{R_2} & -\dfrac{1}{R_1}\\ -\dfrac{1}{R_1} & -\dfrac{1}{R_1} & -\dfrac{1}{R_1}-\dfrac{1}{R_3}\end{bmatrix}\begin{bmatrix}\zeta_1\\ \zeta_2\\ \zeta_3\end{bmatrix}+\frac{E}{R_1}\begin{bmatrix}1\\ 1\\ 1\end{bmatrix}, \tag{28}$$

$$\begin{bmatrix}\zeta_1\\ i_2\\ i_3\end{bmatrix}=\left[\begin{array}{c|cc}-R_1 & -1 & -1\\ \hline 1 & -\dfrac{1}{R_2} & 0\\ 1 & 0 & -\dfrac{1}{R_3}\end{array}\right]\begin{bmatrix}i_1\\ \zeta_2\\ \zeta_3\end{bmatrix}+E\begin{bmatrix}1\\ 0\\ 0\end{bmatrix}. \tag{29}$$

We may add the following equations:

$$\begin{bmatrix} i_1 \\ \zeta_2 \\ \zeta_3 \end{bmatrix} = \frac{1}{R_1+R_2+R_3}\left[\begin{array}{c|cc} -1 & R_2 & R_3 \\ \hline -R_2 & -R_1R_2-R_2R_3 & R_2R_3 \\ -R_3 & R_2R_3 & -R_1R_3-R_2R_3 \end{array}\right]\begin{bmatrix} \zeta_1 \\ i_2 \\ i_3 \end{bmatrix} + \frac{E}{R_1+R_2+R_3}\begin{bmatrix} 1 \\ R_2 \\ R_3 \end{bmatrix}. \tag{30}$$

Equations (27) are for independent current variables, eqs. (28) are for independent potential variables and eqs. (29) and (30) are for mixed independent variables. Equations (28—30) are equivalent to eq. (27). It is noticed that the Onsager relations are satisfied in eqs. (27) and (28). In eqs. (29) and (30), as the partition shows, a Casimir form of coefficient matrix arises. According to Meixner, quoted by Flower and Evans[2], the Casimir form appears because the independent variables are a mixed set of even and odd variables of time. If it is so, the square matrix of eqs. (29) and (30) will exhibit anti-symmetry. According to Karnopp and Rosenberg[3], if a gyrator is allowed in the bond graph, no definite statement about symmetry can be made. Otherwise it is tempting to conclude that Casimir form will appear in mixed varibles. This may be true in eqs. (29), but it is false in eqs. (30).

Tellegen's theorem states that the null space of current and potential is a subspace of dimension 2 (although each dimension may have many components). We can always find a 2×2 submatrix that is real and symmetric, because this is the necessary and sufficient condition for two orthogonal eigenvectors. It turns out that the symmetry of the submatrix in eqs. (29) is trivial but the symmetry of the submatrix in eqs. (30) is hard to explain in any other way.

Oster and Desoer[4] prove the following theorem: Given a system with energy-storage elements typified by capacitors and energy-dissipating elements typified by resistors. For (a) constant potentials on the boundaries, and (b) the stability conditions $f'_C(q) > 0$, where $\zeta = f_C(q)$ is the capacitive constitutive relation and

$$q(t) = q(0) + \int_0^t i \, dt.$$

Then

$$\sum \dot{\zeta}_k i_k \leqq 0. \tag{31}$$

Proof: From Tellegen's theorem,

$$\sum \dot{\overline{i^{\mathrm{T}}\zeta}} = 0.$$

Thus

$$\sum \dot{i}^{\mathrm{T}}\zeta = 0, \tag{32}$$

$$\sum i^{\mathrm{T}}\dot{\zeta} = 0. \tag{33}$$

Considering eq. (33),

$$\sum_R \dot{\zeta}_k i_k + \sum_C \dot{\zeta}_k i_k + \sum_{\text{ports}} \dot{\zeta}_k i_k = 0. \tag{34}$$

From conditions (a),

$$\sum_{\text{ports}} \dot{\zeta}_k i_k = 0. \tag{35}$$

From conditions (b),

$$\sum_C \dot{\zeta}_k i_k = f'_C(q) i_k^2 \geqq 0. \tag{36}$$

Hence

$$\sum_R \dot{\zeta}_k i_k \leqq 0. \tag{37}$$

According to Oster and Desoer, eq. (32) is not true for RC networks, because the sign of $\dot{i}_k = \ddot{q}_k$ is indeterminate in general. On the other hand, eq. (32) is true for RL networks, i.e., systems with inertia and dissipation. It is more straightforward to derive eq. (37), holding for constant power on the boundaries, as follows:

$$-[\boldsymbol{J}|\boldsymbol{X}] = [\boldsymbol{J}|\overline{\text{grad}\ \mu}] = [\boldsymbol{J}|\text{grad}\ \dot{\mu}] = \text{div}\ \boldsymbol{J}\dot{\mu} - \dot{\mu}\ \text{div}\ \boldsymbol{J}$$

$$= \left(\frac{\partial \mu}{\partial \rho}\right)\left(\frac{\partial \rho}{\partial t}\right)^2 \geqq 0, \tag{38}$$

provided that $\boldsymbol{X} = -\text{grad}\ \mu$ and $\text{div}\boldsymbol{J} = -\partial\rho/\partial t$. Because no constitutive relations are used, eq. (38) is quite general.

If we could define a potential function, known as thermo-kinetic potential, such that partial differentiation of this function with respect to any independent variable would yield the corresponding dependent variable, it would give us everything which we wished to know about the circuit. For example, instead of eq. (27), we may define

$$K = -\frac{1}{2}R_1 i_1^2 - \frac{1}{2}R_2(i_1 - i_2)^2$$

$$-\frac{1}{2}R_3(i_1 - i_3)^2 + E i_1, \tag{39}$$

then

$$\zeta_1 = \frac{\partial K}{\partial i_1},\ \zeta_2 = \frac{\partial K}{\partial i_2},\ \zeta_3 = \frac{\partial K}{\partial i_3}. \tag{40}$$

The first three terms on the right side of eq. (39) come from the inner product of the current and the potential and the last term is from the initial conditions. Similarly, instead of eq. (28), we may define

$$K = -\frac{1}{2}\frac{(\zeta_1 + \zeta_2 + \zeta_3)^2}{R_1} - \frac{1}{2}\frac{\zeta_2^2}{R_2} - \frac{1}{2}\frac{\zeta_3^2}{R_3}$$

$$+ \frac{E}{R_1}(\zeta_1 + \zeta_2 + \zeta_3). \tag{41}$$

But this is not possible for mixed independent variables when the Onsager law is not valid. Because

$$\sigma = \frac{dS_i}{dt} = -\frac{1}{T}[\boldsymbol{J}|\boldsymbol{X}] \geqq 0, \tag{42}$$

Prigogine[5] shows that, if the Onsager law is satisfied,

$$d\sigma = -\frac{1}{T}[d\boldsymbol{J}|\boldsymbol{X}] - \frac{1}{T}[\boldsymbol{J}|d\boldsymbol{X}] = \frac{1}{T^2}[Ld\boldsymbol{X}|\boldsymbol{X}]$$

$$-\frac{1}{T}[\boldsymbol{J}|d\boldsymbol{X}] = \frac{1}{T^2}[d\boldsymbol{X}|L\boldsymbol{X}] - \frac{1}{T}[\boldsymbol{J}|d\boldsymbol{X}]$$

$$-\frac{1}{T}[d\boldsymbol{X}|\boldsymbol{J}] = -\frac{1}{T}[\boldsymbol{J}|d\boldsymbol{X}] = -\frac{2}{T}[d\boldsymbol{X}|\boldsymbol{J}], \quad (43)$$

and

$$[d\boldsymbol{X}|\boldsymbol{J}] = -[\boldsymbol{J}|d\boldsymbol{X}] \geqq 0 \quad (44)$$

from eq. (38). But $d\sigma$ is not necessarily a perfect differential. Equations (38) and (44) cannot be put into practical use. Millar[6] defines the content G and its Legendre transform, the co-content G^* as follows:

$$G_k = \int_0^{i_k} \zeta_k di_k, \quad G = \sum_k G_k; \quad (45)$$

$$G_k^* = \int_0^{\zeta_k} i_k d\zeta_k, \quad G^* = \sum_k G_k^*. \quad (46)$$

From geometry, content and co-content become equal for linear constitutive relations.

The obvious application of the Heaviside law is that, if steady state exists, the total entropy-production is equal to twice the rate of entropy-production at steady state (compare the example at the end of §20).

References

[1] B. D. H. Tellegen: *Philips Res. Report*, **7**, 259(1952).
[2] J. O. Flower and F. G. Evans: *J. Franklin Inst.*, **291**, 121(1971).
[3] D. Karnopp and R. Rosenburg: System Dynamics: A Unified Approach, Wiley-Interscience, New York, p. 261 (1975).
[4] G. F. Oster and C. A. Desoer: *J. Theor. Biol.*, **32**, 219 (1971).
[5] I. Prigogine: Introduction to Thermodynamics of Irreversible Processes, Interscience, New York, Chaps. 6 and 7 (1961).
[6] W. Millar: *Phil. Mag.*, **42**, 1150(1951).

PART II

APPLICATIONS

CHAPTER VII

SCALAR PROCESSES

§25 *RATE OF CHEMICAL REACTIONS AND AFFINITIES*

We start from the defining equation (1, 6),

$$\sigma = -\frac{1}{T}\sum_i [j_i|X_i]. \tag{1}$$

For chemical reactions both fluxes and forces are scalars.

In general, for a chemical reaction i, the stoichiometric coefficients of the different components are different. Our choice of flux should be independent of the components. Let N be the total and N_k be the individual number of moles per unit volume and n_k be the molar concentration.

We define

$$\bar{j}_i = \frac{d\left(\dfrac{N_k}{\bar{\Lambda}_{ik}}\right)}{dt} = \frac{N}{\bar{\Lambda}_{ik}}\frac{dn_k}{dt} = N\frac{d\varepsilon_i}{dt}. \tag{2}$$

The flux $\bar{j}_i$ is the rate of chemical reaction i in moles per unit volume per unit time, where the bar denotes the molar basis. We agree that the atomic stoichiometric coefficient $\bar{\Lambda}_{ik}$ is positive or negative according to whether the component k is a reactant or a product in the reaction i respectively. Then,

$$\varepsilon_i = \frac{1}{N}\int_{t_0}^{t}\bar{j}_i dt = \frac{N_k - N_k^{\circ}}{N\bar{\Lambda}_{ik}} \tag{3}$$

is the degree of advancement first proposed by de Donder[1]. The name is derived from the observation that ε is zero if the components are at equilibrium and the absolute value of ε is 1 if at the elapse of time $t - t_0$ either reactants or products are absent.

de Donder also defined affinity $\bar{A}$ as

$$\bar{A}_i = \sum_k \bar{\Lambda}_{ik}\bar{\mu}_k = \Delta\bar{G}_i, \tag{4}$$

where $\bar{\mu}_k$ is the atomic chemical potential of the component k. The summation extends to all components of the chemical reaction i and $\Delta\bar{G}_i$ is the change of Gibbs free energy of the given reaction. Because $\Delta\bar{G}_i$ is necessarily negative for the reaction to proceed, the choice of affinity as a force is a good one.

Now eq. (1) can be written as

$$\sigma = -\frac{1}{T}\sum_i \bar{j}_i\bar{A}_i. \tag{5}$$

For irreversible processes σ is positive definite. For one chemica reaction, $\bar{j}$ and $\bar{A}$ have opposite signs. If $\bar{A}$ is negative, the reaction takes place in the forward direction, i.e., from left to right in the chemical reaction as written. If $\bar{A}$ is positive, the reaction takes place in the reverse direction. At equilibrium both $\bar{j}$ and $\bar{A}$ are zero. However, if $\bar{j} = 0$, it does not necessarily mean $\bar{A} = 0$. We could have labile equilibrium, for example, as in a mixture of hydrogen and oxygen gases at room temperature in the absence of a catalyst. The case of $\bar{j} \neq 0$ but $\bar{A} = 0$ is not possible. For more than one reaction, if the reactions are independent, $\bar{j}$ and $\bar{A}$ of each reaction must be opposite in sign. If they are not independent, it is possible for some reactions to have a negative rate of entropy-production so long as the total rate is positive.

Because the atomic stoichiometric coefficients are not "conservative", i.e., the sum of coefficients of the reactants is not equal to that of the products, we may prefer to use mass stoichiometric coefficients, or normalized stoichiometric coefficients:

$$\Lambda_{ik} = \frac{\bar{\Lambda}_{ik} M_k}{\sum_{k=1}^{r} \bar{\Lambda}_{ik} M_k} = \frac{\bar{\Lambda}_{ik} M_k}{-\sum_{k=r+1}^{r+p} \bar{\Lambda}_{ik} M_k} = \frac{N \bar{\Lambda}_{ik} M_k}{\rho}, \tag{6}$$

where M_k is the molecular weight of the component k. There are r reactants and p products in the chemical reaction i.

While the degree of advancement is absolute, the rate of chemical reaction on a mass basis, j_i, is different from that on an atomic basis, $\bar{j}_i$ j_i is similarly defined as

$$j_i = \frac{1}{\Lambda_{ik}} \frac{d\rho_k}{dt} = \frac{\rho}{\Lambda_{ik}} \frac{dc_k}{dt} = \rho \frac{d\varepsilon_i}{dt}, \tag{7}$$

where c_i is the mass concentration of the component i. Comparing eqs. (2) and (7),

$$j_i = \rho \frac{\bar{j}_i}{N}. \tag{8}$$

For chemical potentials,

$$\mu_k = \frac{\bar{\mu}_k}{M_k}. \tag{9}$$

For affinities,

$$A_i = \sum_k \Lambda_{ik} \mu_k = \frac{N}{\rho} \sum_k \bar{\Lambda}_{ik} \bar{\mu}_k = \frac{N \bar{A}_i}{\rho}. \tag{10}$$

If eq. (5) can be linealized,

$$\bar{j}_i = -\sum_k \frac{\bar{L}_{ik} \bar{A}_k}{T}, \tag{11}$$

the counterpart is

$$j_i = -\sum_k \frac{L_{ik} A_k}{T}. \tag{12}$$

Comparing eqs. (11) and (12) and using eqs. (8) and (10),

$$\bar{L}_{ik} = \frac{N^2}{\rho^2} L_{ik}. \tag{13}$$

The counterpart of eq. (5) is

$$T\sigma = -\sum_i [j_i | A_i], \tag{14}$$

which is the first term on the right of eq. (8, 11), the contribution to the rate of entropy-production due to chemical reactions.

Reference

[1] Th de Donder: *Bull. Acad. Roy. Belg.* (Classic Sciences), **24**, 15(1938).

§26 *KINETIC INTERPRETATIONS OF PHENOMENOLOGICAL COEFFICIENTS IN ONE CHEMICAL REACTION*

In this section we discuss the significance of phenomenological coefficients of chemical reactions.

Consider one chemical reaction only. There are r reactants and p products.

From the kinetic part of the law of mass action, admittedly a crude concept of linearity,

$$\bar{j}_i = k \prod_{i=1}^{r} N_i^{-\bar{\lambda}_i} - k' \prod_{i=r+1}^{r+p} N_i^{\bar{\lambda}_i}$$

$$= k' \prod_{i=r+1}^{r+p} N_i^{\bar{\lambda}_i} \left[\frac{k}{k'} \prod_{m=1}^{r+p} N_m^{-\bar{\lambda}_m} - 1 \right], \tag{1}$$

where N_i is the number of moles of the component i per unit volume, and k and k' are the velocity constants of the forward and backward reactions respectively. For ideal gases,

$$n_m = \frac{N_m}{N} = \exp \frac{\overline{\mu_m} - \overline{\mu_m^\circ}}{RT}, \tag{2}$$

where $\overline{\mu_m^\circ}$ is the standard chemical potential. In general $\overline{\mu_m^\circ}$ is a function of temperature and pressure. For ideal gases it is a function of temperature only. Substituting eq. (2) in eq. (1),

$$\bar{j}_i = k' \prod_{i=r+1}^{r+p} N^{\bar{\lambda}_i}_i \left[\frac{k}{k'} N^{\sum_{m=1}^{r+p} -\bar{\lambda}_m} \exp \left(\frac{-\bar{A}_i + \sum_{m=1}^{r+p} \overline{\mu_m^\circ} \bar{\Lambda}_m}{RT} \right) - 1 \right], \tag{3}$$

where eq. (25, 4) has been used. At equilibrium,

$$\bar{A}_i = \bar{j}_i = 0. \tag{4}$$

Substituting eq. (4) in eq. (3),

$$\frac{k}{k'} = N^{\sum_{m=1}^{r+p} \bar{\lambda}_m} \exp \left(- \sum_{m=1}^{r+p} \frac{\overline{\mu_m^\circ} \bar{\Lambda}_m}{RT} \right). \tag{5}$$

It is recalled that for any chemical reaction of ideal gases at a given temperature and pressure,

$$\Delta G - \Delta G^\circ = RT \ln Q_p - RT \ln Q_p^\circ, \tag{6}$$

where ΔG is the change of Gibbs free energy for the given reaction, Q_p is the proper quotient of partial pressures, and the superscript $\circ$ refers to the standard state. By proper quotient is meant the ratio of products of partial pressures, the numerator of which is the product of the products of the given reaction, each raised to a power equal to the absolute value of the stoichiometric coefficient, and the denominator of which is the corresponding product of the reactants. At equilibrium,

$$\Delta G = 0. \tag{7}$$

In the standard state,

$$\ln Q_p^\circ = 0. \tag{8}$$

Substituting eqs. (7) and (8) in eq. (6),

$$\Delta G^\circ = -RT \ln Q_p = -RT \ln K, \tag{9}$$

where K is the equilibrium constant. Although eq. (9) is derived from any reaction of ideal gases, it has general application for any substance in any state. But for ideal gases,

$$K = Q_p = \frac{k}{k'} N^{\sum_{m=1}^{r+p} -\bar{\Lambda}_m}. \tag{10}$$

From eqs. (5) and (10),

$$K = \exp\left(-\sum_{m=1}^{r+p} \frac{\overline{\mu_m^\circ \Lambda_m}}{RT}\right). \tag{11}$$

It is seen that eq. (11) is equivalent to eq. (9).

Substituting eq. (5) in eq. (3),

$$\bar{j}_i = k' \prod_{i=r+1}^{r+p} (N_i)_{\text{eq}}^{\bar{\Lambda}_i} \left[\exp\left(-\frac{\bar{A}_i}{RT}\right) - 1\right]. \tag{12}$$

Near equilibrium the absolute value of $\bar{A}_i$ is small, i. e., either $\bar{A}_i \to 0$ or $\|\bar{A}_i\| < RT$; the exponent in eq. (12) may be expanded and the first two terms retained,

$$\bar{j}_i = k' \prod_{i=r+1}^{r+p} \frac{(N_i)_{\text{eq}}^{\bar{\Lambda}_i} \ (-\bar{A}_i)}{RT}. \tag{13}$$

Comparing eq. (13) with eq. (25, 11),

$$\bar{L}_{ii} = k' \prod_{i=r+1}^{r+p} \frac{(N_i)_{\text{eq}}^{\bar{\Lambda}_i}}{R} = k \prod_{i=1}^{r} \frac{(N_i)_{\text{eq}}^{-\bar{\Lambda}_i}}{R}. \tag{14}$$

This is one of the few examples of phenomenological coefficients having kinetic interpretations. It must be noted that, while linearity is generally adequate for many types of irreversible processes, it is not satisfactory for chemical reactions.

However, when the affinity of a given reaction is large so that eq. (13) is not satisfied, sometimes the reaction may be split into a number of elementary reactions, each of which has an affinity sufficiently small to justify linearity. As an example, we consider a reaction of the form $z_1 \rightarrow z_n$, which proceeds in a number of steps, $z_1 \rightarrow z_2 \rightarrow \cdots \rightarrow z_n$, the intermediate compounds being unstable. As usual,

$$T\sigma = -\sum_{i=1}^{n} \bar{A}_i \bar{j}_i. \tag{15}$$

A state of steady equilibrium is established after a short time,

$$\bar{j}_i = \bar{j}_2 = \cdots = \bar{j}_n = \bar{j} \text{ and } \sum_{i=1}^{n} \bar{A}_i = A.$$

Then,

$$\sigma = \frac{1}{T} \|\bar{A}\| \bar{j} = \frac{1}{T} \sum_i \|\bar{A}_i\| \bar{j}_i.$$

Each reaction is still in the domain of linearity even if the total affinity $\|\bar{A}\| > RT$.

§27 *DETAILED BALANCING FROM A STATISTICAL VIEWPOINT*

In §15 we see that the Onsager law is a strong condition for detailed balancing, but we have not explained why detailed balancing is the favoured mechanism for tautomeric reactions nor have we given the stationary probability for detailed balancing.

Suppose a system has n possible states. It is a Markov chain if at any time the conditional probability of a transition from the present state to any other state is determined by the immediately preceding state but not by how the present state is reached. The matrix R of one-step transition probability determines the essential properties of the Markov chain and is called the matrix of the Markov chain.

For mechanism A of the tautomeric reactions in §15 we can write the transition probability as

$$R = \begin{bmatrix} 0 & 1 & 0 \\ 0 & 0 & 1 \\ 1 & 0 & 0 \end{bmatrix}. \tag{1}$$

Looking at the first row, for example, we see that there is no tendency for state 1 (z_1) to remain at 1, or to transform to state 3 (z_3), but there is definite tendency to transform to state 2 (z_2). At statistical equilibrium, the eventual behaviour of tautomeric reactions is reached; the Markov chain is ergodic,

$$\lim_{m\to\infty} (R_{ij})^m = \pi_j .$$

In other words, it is ergodic if as m tends to infinity, the m-step transition probability tends to a limit that depends on the final state and not on the initial state. For mechanism A, the limit does not exist, because the two-step transition probability is

$$R^2 = \begin{bmatrix} 0 & 0 & 1 \\ 1 & 0 & 0 \\ 0 & 1 & 0 \end{bmatrix}, \tag{2}$$

the three-step transition probability is

$$R^3 = \begin{bmatrix} 1 & 0 & 0 \\ 0 & 1 & 0 \\ 0 & 0 & 1 \end{bmatrix}, \tag{3}$$

and the four-step transition probability is

$$R^4 = \begin{bmatrix} 0 & 1 & 0 \\ 0 & 0 & 1 \\ 1 & 0 & 0 \end{bmatrix}. \tag{4}$$

The eventual behaviour is clearly oscillatory. As a matter of fact, if mechanism A is right, the concentration of three components will show oscillations during the approach to the equilibrium state. Studies of this kind have been made by Skrabel[1] and Bafi[2], but the results are not conclusive.

The mechanism B is not favoured from similar arguments as above. For mechanism C,

$$R = \begin{bmatrix} 0 & \dfrac{k_{12}N_1}{k_{12}N_1 + k_{13}N_1} & \dfrac{k_{13}N_1}{k_{12}N_1 + k_{13}N_1} \\ \dfrac{k_{21}N_2}{k_{21}N_2 + k_{23}N_2} & 0 & \dfrac{k_{23}N_2}{k_{21}N_2 + k_{23}N_2} \\ \dfrac{k_{31}N_3}{k_{31}N_3 + k_{32}N_3} & \dfrac{k_{32}N_3}{k_{31}N_3 + k_{32}N_3} & 0 \end{bmatrix}. \tag{5}$$

The sum of entries in each row of the matrix of the Markov chain must be equal to unity, because two mutually exclusive and exhaustive events are involved. To find the different powers of R is tedious unless R is diagonalized. To diagonalize R is also tedious. However, because

$$\lim_{m\to\infty} R^m = (\lim_{m\to\infty} R^m)R, \tag{6}$$

i.e., R times any column of $\lim R^m$ must be equal to this column, or each column is a characteristic vector of R with characteristic roots unity. Let

$$\lim_{m\to\infty} R^m = \begin{bmatrix} \pi_1 & 0 & 0 \\ 0 & \pi_2 & 0 \\ 0 & 0 & \pi_3 \end{bmatrix},$$

or

$$(\pi_1, \pi_2, \pi_3)^{\mathrm{T}} = R^{\mathrm{T}}(\pi_1, \pi_2, \pi_3)^{\mathrm{T}}. \tag{7}$$

Substituting eq. (5) in eq. (7), we have

$$\begin{bmatrix} -1 & \dfrac{k_{21}}{k_{21}+k_{23}} & \dfrac{k_{31}}{k_{31}+k_{32}} \\ \dfrac{k_{12}}{k_{12}+k_{13}} & -1 & \dfrac{k_{32}}{k_{31}+k_{32}} \\ \dfrac{k_{13}}{k_{12}+k_{13}} & \dfrac{k_{23}}{k_{21}+k_{23}} & -1 \end{bmatrix} \begin{bmatrix} \pi_1 \\ \pi_2 \\ \pi_3 \end{bmatrix} = 0. \tag{8}$$

The non-singular solution of eq. (8) is

$$\begin{bmatrix} \pi_1 \\ \pi_2 \\ \pi_3 \end{bmatrix} = \frac{1}{A+B+C} \begin{bmatrix} A \\ B \\ C \end{bmatrix}, \tag{9}$$

where

$$\left.\begin{aligned} A &= (k_{12}+k_{13})(k_{23}k_{31}+k_{21}k_{31}+k_{21}k_{32}), \\ B &= (k_{23}+k_{21})(k_{12}k_{31}+k_{12}k_{32}+k_{32}k_{13}), \\ C &= (k_{31}+k_{32})(k_{12}k_{23}+k_{23}k_{13}+k_{13}k_{21}), \end{aligned}\right\} \tag{10}$$

and the relation

$$\pi_1 + \pi_2 + \pi_3 = 1 \tag{11}$$

has been used.

Equation (9) is the stationary probability for detailed balancing. It is (1/3, 1/3, 1/3) if $k_{12}/k_{31} = k_{13}/k_{32}$ and if $k_{21}/k_{32} = k_{23}/k_{31}$, i.e., the sum of entries in each column of R is equal to unity. In this instance R is a doubly stochastic matrix.

References

[1] A Skrabel: Homogenkinetik, Steinkopf, Dresden (1941).
[2] T. A. Bafi: *Bull. Acad. Roy. Belg.* (Classic Sciences), **45**, 116 (1959).

§28 *DETAILED BALANCING VIEWED FROM THE RATE OF ENTROPY-PRODUCTION*

In this section, we treat the same problem as in the preceding section, except with the application of the Onsager law.

From eq. (8, 11),

$$\sigma = -\frac{1}{T}[\bar{j}|\bar{A}] = -\frac{1}{T}\sum_{i,k=1}^{3}\bar{j}_{ik}\bar{A}_{ki}, \tag{1}$$

if we consider a triangular reaction

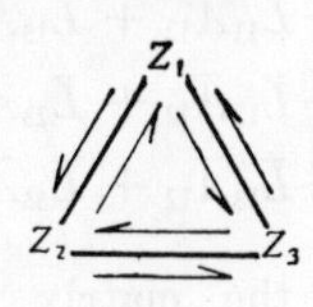

only. Clearly

$$\bar{j}_{ii} = \bar{A}_{kk} = 0;\ i, k = 1, 2, 3. \tag{2}$$

For a particular elementary reaction $Z_i \rightleftharpoons Z_k$, the stoichiometric coefficients for the forward and backward reactions have the same value, namely unity, but differ in sign.

$$\bar{A}_{ik} = -\bar{A}_{ki}, \quad k \neq i. \tag{3}$$

If we set

$$\bar{A}_{ik} = -\bar{\mu}_k + \bar{\mu}_i = -\bar{A}_{ki}, \tag{4}$$

$$\bar{j}_{ik} = \bar{j}_{ki}, \tag{5}$$

thus the fluxes and the forces are not independent,

$$\bar{j}_{23} = \bar{j}_{21} + \bar{j}_{13}, \tag{6}$$

$$\bar{A}_{12} - \bar{A}_{13} + \bar{A}_{23} = 0. \tag{7}$$

If we consider independent reactions,

$$\sigma = -\frac{1}{3T}\sum_{\substack{i,k=1\\ i\neq k}}^{3} \bar{j}_{ik}\bar{A}_{ki}; \tag{8}$$

but if we consider all possible reactions,

$$\sigma = -\frac{1}{2T}\sum_{\substack{i,k=1\\ i\neq k}}^{3} \bar{j}_{ik}\bar{A}_{ki}. \tag{9}$$

In the region of linearity,

$$-\begin{bmatrix} 0 & \bar{j}_{12} & \bar{j}_{13} \\ \bar{j}_{12} & 0 & \bar{j}_{23} \\ \bar{j}_{13} & \bar{j}_{23} & 0 \end{bmatrix} = \frac{1}{T}\begin{bmatrix} \bar{L}_{11} & \bar{L}_{12} & \bar{L}_{13} \\ \bar{L}_{12} & \bar{L}_{22} & \bar{L}_{23} \\ \bar{L}_{13} & \bar{L}_{23} & \bar{L}_{33} \end{bmatrix}\begin{bmatrix} 0 & -\bar{A}_{12} & \bar{A}_{13} \\ \bar{A}_{12} & 0 & -\bar{A}_{23} \\ -\bar{A}_{13} & \bar{A}_{23} & 0 \end{bmatrix} \tag{10}$$

$$= \frac{1}{T}\begin{bmatrix} \bar{L}_{12}\bar{A}_{12} - \bar{L}_{13}\bar{A}_{13} & -\bar{L}_{11}\bar{A}_{12} + \bar{L}_{13}\bar{A}_{23} & \bar{L}_{11}\bar{A}_{13} - \bar{L}_{12}\bar{A}_{23} \\ \bar{L}_{22}\bar{A}_{12} - \bar{L}_{23}\bar{A}_{13} & -\bar{L}_{12}\bar{A}_{12} + \bar{L}_{23}\bar{A}_{23} & \bar{L}_{12}\bar{A}_{13} - \bar{L}_{22}\bar{A}_{23} \\ \bar{L}_{23}\bar{A}_{12} - \bar{L}_{33}\bar{A}_{13} & -\bar{L}_{13}\bar{A}_{12} + \bar{L}_{33}\bar{A}_{23} & \bar{L}_{13}\bar{A}_{13} - \bar{L}_{23}\bar{A}_{23} \end{bmatrix}. \tag{11}$$

From Theorem 4 of §18, the matrix of $\bar{L}$ is not necessarily symmetric. The symmetry of the $\bar{L}$ matrix in eq. (11) is justified because the diagonal coefficients of the $\bar{j}$ matrix are zero and the determinant is positive definite. However, from the fraction of 1/3 in eq. (8) or from the fraction of 1/2 in eq. (9), only part of eq. (11) can be used at one time.

We may consider only two independent reactions. Then we retain only the equality of one column. For example, if we consider reactions 12 and 13 to be independent, we retain the first column:

$$-\begin{bmatrix} \bar{j}_{12} \\ \bar{j}_{13} \end{bmatrix} = \frac{1}{T}\begin{bmatrix} \bar{L}_{22} & \bar{L}_{23} \\ \bar{L}_{23} & \bar{L}_{33} \end{bmatrix}\begin{bmatrix} \bar{A}_{12} \\ -\bar{A}_{13} \end{bmatrix}. \tag{12}$$

There is no dependency but one coupling.

If we consider all possible reactions, we can retain either the upper or the lower diagonal.

For example, if we retain the lower diagonal,

$$-T\begin{bmatrix}\bar{j}_{12}\\ \bar{j}_{13}\\ \bar{j}_{23}\end{bmatrix}=\begin{bmatrix}\bar{L}_{22} & \bar{L}_{23} & 0\\ \bar{L}_{23} & \bar{L}_{33} & 0\\ -\bar{L}_{13} & 0 & \bar{L}_{33}\end{bmatrix}\begin{bmatrix}\bar{A}_{12}\\ -\bar{A}_{13}\\ \bar{A}_{23}\end{bmatrix}, \tag{13}$$

there is one dependency and one coupling. We can eliminate the dependency,

$$-T\begin{bmatrix}\bar{j}_{12}\\ \bar{j}_{13}\\ \bar{j}_{23}\end{bmatrix}=\begin{bmatrix}\bar{L}_{22}-\bar{L}_{23} & \bar{L}_{23}\\ \bar{L}_{23}-\bar{L}_{33} & \bar{L}_{33}\\ -\bar{L}_{13} & -\bar{L}_{33}\end{bmatrix}\begin{bmatrix}\bar{A}_{12}\\ -\bar{A}_{23}\end{bmatrix}. \tag{14}$$

If we set $\bar{j}_{13}=0$ (or $\bar{j}_{12}=0$), the coupling is eliminated,

$$-T\begin{bmatrix}\bar{j}_{12}\\ \bar{j}_{23}\end{bmatrix}=\begin{bmatrix}\bar{L}_{22}-\dfrac{\bar{L}_{23}^2}{\bar{L}_{33}} & 0\\ 0 & -\bar{L}_{33}+\dfrac{\bar{L}_{13}\bar{L}_{33}}{\bar{L}_{23}-\bar{L}_{33}}\end{bmatrix}\begin{bmatrix}\bar{A}_{12}\\ -\bar{A}_{23}\end{bmatrix}. \tag{15}$$

It is easy to extend to n-angular reactions. There are $\frac{n(n-1)}{2}$ possible reactions, but only $n-1$ of them are independent. Either we take $n-1$ independent reactions from any column of the matrix of fluxes, there are $\frac{n(n-1)}{2}-(n-1)=\frac{1}{2}(n-1)(n-2)$ couplings; or we take all possible reactions for the upper or the lower diagonal of the matrix of fluxes, there are $\frac{1}{2}(n-1)(n-2)$ dependencies and $\frac{1}{2}(n-1)(n-2)$ couplings, both of which can be eliminated.

§29 *ORDERING COEFFICIENTS AND RELAXATION TIMES*

In §3 we have defined ordering coefficients. In this section we show that the ordering coefficients are negative definite. Also, we introduce relaxation times.

From the Gibbs equation,

$$G = -SdT + Vdp + \sum_i \mu_i dc_i. \tag{1}$$

From eqs. (4, 52), (25, 7), and (25, 10), with chemical reactions only,

$$\sum_i \mu_i \frac{dc_i}{dt} = -\sum_k \frac{A_k j_k}{\rho} = -\sum_k A_k \frac{d\varepsilon_k}{dt}. \tag{2}$$

The negative sign in eq. (2) is due to our convention about the stoichiometric coefficients. Substituting eq. (2) in eq. (1),

$$\frac{dG}{dt} = -S\frac{dT}{dt} + V\frac{dp}{dt} - \sum_k A_k \frac{d\varepsilon_k}{dt}. \tag{3}$$

Thus

$$\left(\frac{\partial G}{\partial \varepsilon_k}\right)_{Tp\varepsilon_i, i\neq k} = -A_k. \tag{4}$$

Because chemical potential is not limited to partial Gibbs free energy at constant temperature and pressure, we have

$$-A_k = \left(\frac{\partial G}{\partial \varepsilon_k}\right)_{Tp} = \left(\frac{\partial F}{\partial \varepsilon_k}\right)_{TV} = \left(\frac{\partial U}{\partial \varepsilon_k}\right)_{SV} = \left(\frac{\partial H}{\partial \varepsilon_k}\right)_{Sp}. \tag{5}$$

If we limit ourselves to one chemical reaction not far from equilibrium (For more than one chemical reaction, see §31), we linearize,

$$A = A - A_{\mathrm{eq}} = \Delta A = \left(\frac{\partial A}{\partial \varepsilon}\right)\Delta\varepsilon. \tag{6}$$

From eqs. (6), (25, 7), and (25, 12),

$$\frac{\partial \varepsilon}{\partial t} = \frac{j}{\rho} = -\frac{L}{\rho T}\left(\frac{\partial A}{\partial \varepsilon}\right)_{Tp}\Delta\varepsilon. \tag{7}$$

If we define

$$\frac{L}{\rho T} = B, \tag{8}$$

and if we linearize

$$\frac{\partial \varepsilon}{\partial t} = \frac{\Delta \varepsilon}{\Gamma_{Tp}}, \tag{9}$$

then

$$\frac{\partial \varepsilon}{\partial t} = -B\left(\frac{\partial A}{\partial \varepsilon}\right)_{Tp} \Delta \varepsilon, \tag{10}$$

and

$$\left(\frac{\partial A}{\partial \varepsilon}\right)_{Tp} = -\frac{1}{B\Gamma_{Tp}}. \tag{11}$$

Γ is called the relaxation time, i.e., the time for the degree of advancement to decrease to e^{-1} of its original value. Because B and Γ are necessarily positive, the ordering coefficient, $(\partial A/\partial \varepsilon)_{Tp}$ is negative definite (Some authors use the opposite convention and the ordering coefficient is positive definite). We also have

$$\Gamma_{TV} = -\frac{1}{B}\left(\frac{\partial \varepsilon}{\partial A}\right)_{TV} > 0, \tag{12}$$

$$\Gamma_{Sp} = -\frac{1}{B}\left(\frac{\partial \varepsilon}{\partial A}\right)_{Sp} > 0, \tag{13}$$

and

$$\Gamma_{ST} = -\frac{1}{B}\left(\frac{\partial \varepsilon}{\partial A}\right)_{SV} > 0. \tag{14}$$

Besides four relaxation times, we have four specific heats: C_{pA}, $C_{p\varepsilon}$, C_{VA}, and $C_{V\varepsilon}$; and four compressibilities: κ_{TA}, $\kappa_{T\varepsilon}$, κ_{SA}, and $\kappa_{S\varepsilon}$. In each of the three categories, we can formulate four ratios of two coefficients with common subscript. These twelve ratios combine to form six relations:

$$\frac{\Gamma_{Tp}}{\Gamma_{TV}} = \frac{\kappa_{TA}}{\kappa_{T\varepsilon}}, \tag{15}$$

$$\frac{\Gamma_{Tp}}{\Gamma_{Sp}} = \frac{C_{pA}}{C_{p\varepsilon}}, \tag{16}$$

$$\frac{\Gamma_{TV}}{\Gamma_{SV}} = \frac{C_{VA}}{C_{V\varepsilon}}, \tag{17}$$

$$\frac{\Gamma_{Sp}}{\Gamma_{SV}} = \frac{\kappa_{SA}}{\kappa_{S\varepsilon}}, \tag{18}$$

$$\frac{\kappa_{T\varepsilon}}{\kappa_{S\varepsilon}} = \frac{C_{p\varepsilon}}{C_{V\varepsilon}}, \tag{19}$$

$$\frac{\kappa_{TA}}{\kappa_{SA}} = \frac{C_{pA}}{C_{VA}}. \tag{20}$$

We can prove

$$\Gamma_{Tp} > \Gamma_{TV} > \Gamma_{SV}, \tag{21}$$

and

$$\Gamma_{Tp} > \Gamma_{Sp} > \Gamma_{SV}. \tag{22}$$

For example,

$$\frac{\Gamma_{Tp}}{\Gamma_{TV}} = \frac{\left(\frac{\partial A}{\partial \varepsilon}\right)_{TV}}{\left(\frac{\partial A}{\partial \varepsilon}\right)_{Tp}} = \frac{\left(\frac{\partial A}{\partial \varepsilon}\right)_{Tp} + \left(\frac{\partial A}{\partial p}\right)_{T\varepsilon}\left(\frac{\partial p}{\partial \varepsilon}\right)_{TV}}{\left(\frac{\partial A}{\partial \varepsilon}\right)_{Tp}}$$

$$= 1 - \left(\frac{\partial \varepsilon}{\partial A}\right)_{Tp}\left(\frac{\partial V}{\partial \varepsilon}\right)_{Tp}\left(\frac{\partial p}{\partial \varepsilon}\right)_{TV}$$

$$= 1 + \left(\frac{\partial \varepsilon}{\partial A}\right)_{Tp}\left(\frac{\partial V}{\partial \varepsilon}\right)_{Tp}^{2}\left(\frac{\partial p}{\partial V}\right)_{T\varepsilon} > 1, \tag{23}$$

eq. (3, 19) having been used. We can also prove

$$C_{pA} > C_{VA}, \tag{24}$$

$$C_{p\varepsilon} > C_{V\varepsilon}, \tag{25}$$

$$C_{pA} > C_{p\varepsilon}, \tag{26}$$

$$C_{VA} > C_{V\varepsilon}. \tag{27}$$

For example,

$$C_{pA} - C_{VA} = T\left[\left(\frac{\partial S}{\partial T}\right)_{pA} - \left(\frac{\partial S}{\partial T}\right)_{VA}\right]$$
$$= T\left(\frac{\partial S}{\partial V}\right)_{TA}\left(\frac{\partial V}{\partial T}\right)_{pA} = T\left(\frac{\partial p}{\partial T}\right)_{VA}\left(\frac{\partial V}{\partial T}\right)_{pA}$$
$$= -T\left(\frac{\partial p}{\partial V}\right)_{TA}\left(\frac{\partial V}{\partial T}\right)_{pA}^{2} > 0. \tag{28}$$

From eqs. (15) and (18—20),

$$\kappa_{TA} > \kappa_{T\varepsilon}, \tag{29}$$
$$\kappa_{SA} > \kappa_{S\varepsilon}, \tag{30}$$
$$\kappa_{T\varepsilon} > \kappa_{S\varepsilon}, \tag{31}$$
$$\kappa_{TA} > \kappa_{SA}. \tag{32}$$

Equations (21), (22), (24)—(27), and (29)—(32) are reminiscent of

$$C_{pm} > C_{Vm} \tag{33}$$

and

$$\kappa_{Tm} > \kappa_{Sm}, \tag{34}$$

where m is the total mass [See eqs. (3, 7) and (3, 12)], and illustrative of the fact that the equilibrium coefficient is larger than the corresponding frozen coefficient.

§30 *SOUND PROPAGATION IN IDEAL FLUIDS*

In this section we treat the propagation of sound in fluids without any irreversibility.

The propagation of a periodic one-dimensional disturbance in a medium follows the relation

$$\frac{I}{I_0} = \exp i(\boldsymbol{K}\cdot\boldsymbol{x} - \omega t) \tag{1}$$

$$= \exp i[(\boldsymbol{k} - i\boldsymbol{a}) \cdot \boldsymbol{x} - \omega t], \tag{2}$$

where I stands for particle displacement, velocity, acceleration, pressure, temperature, density or affinity and I_0 is a corresponding amplitude. $\boldsymbol{K}$ is the complex propagation vector. The real part of $\boldsymbol{K}$ is $\boldsymbol{k}$, the phase constant; and the imaginary part of $\boldsymbol{K}$ is $\boldsymbol{a}$, the absorption coefficient.

The waves of eq. (1) are plane waves, because $\boldsymbol{K} \cdot \boldsymbol{x} - \omega t$ has a linear dependence on coordinates and time. In other words, the phase plane of eq. (1), or surfaces of equal phase, moves parallel to itself.

We have the familiar relation

$$k = \frac{2\pi}{l} = \frac{2\pi f}{v} \tag{3}$$

$$= \frac{\omega}{v}, \tag{4}$$

where $\boldsymbol{l}$ is the wave length, f is frequency, ω is angular velocity and $\boldsymbol{v}$ is velocity of propagation.

Although there is no common usage of terms like attenuation and absorption, the following distinction is useful. Attenuation is the decrease of amplitude (intensity) of a wave. Absorption is the transfer of wave energy into internal energy. However, attenuation in the broad sense includes absorption, scattering and decrease of intensity at a further distance. Attenuation in the narrow sense is synonymous with absorption in the loose sense, which includes both absorption and dissipation. Descriptive terms like thermal or viscous absorption may be used to emphasize the dissipative nature of absorption. While conductive absorption is always dissipative, radiation absorption is not if the frequency is low. Absorption is dissipative even if wave energy is not removed as heat but if the transfer into internal energy is not instantaneous. To emphasize this, relaxation or chemical absorption may be used. Except for real chemical reactions, these are not good descriptive terms. They merely indicate that we are ignorant of the true nature of irreversibility. In practice, the separation of different irreversible processes is not easy, especially if they are coupled.

Because $\boldsymbol{K}$ is a complex vector, we expect that attenuation is related to the length of the vector, dispersion (dependence of wave speed on frequency) is related to the real part and absorption to the imaginary part:

$$\zeta^2 = \frac{(k - ia)^2}{k_0^2}, \tag{5}$$

$$d = \left(\frac{v}{v_0}\right)^2 = \left(\frac{l}{l_0}\right)^2 = \left(\frac{k_0}{k}\right)^2 = \left(\frac{1}{\mathrm{Re}\,\zeta}\right)^2, \tag{6}$$

$$a = k_0 \| \mathrm{Im}\,\zeta \|. \tag{7}$$

In contrast to d, a is not dimensionless. We define

$$a_0 = l_0 a, \tag{8}$$

and

$$a' = la, \tag{9}$$

to make absorption dimensionless. We also define

$$\bar{a} = \frac{a'}{d}. \tag{10}$$

If

$$\zeta^2 = G \pm iH, \tag{11}$$

$$\zeta = (G \pm iH)^{\frac{1}{2}} = \sqrt{\frac{1}{2}}\,[\sqrt{\sqrt{G^2 + H^2} + G}$$

$$\pm\, i \sqrt{\sqrt{G^2 + H^2} - G}]. \tag{12}$$

The sign depends on the sign of H. The square root sign indicates the positive root but 1/2 power indicates the two roots. Once G and H in eq. (11) are known, eqs. (8—10) can be rewritten as follows:

$$a_0 = 2\pi \|\text{Im}\,\zeta\|, \tag{13}$$

$$a' = 2\pi \left\| \frac{\text{Im}\,\zeta}{\text{Re}\,\zeta} \right\|, \tag{14}$$

$$\bar{a} = 2\pi \|\text{Im}\,\zeta\,\text{Re}\,\zeta\|. \tag{15}$$

It is a general property of all actual media that, with entropy constant, the pressure increases with the increasing density or with decreasing volume. We can define a positive quantity C_0 with the dimension of speed,

$$C_0^2 = \left(\frac{\partial p}{\partial \rho}\right)_s = -V^2 \left(\frac{\partial p}{\partial V}\right)_s. \tag{16}$$

In the usual treatment of sound propagation in fluids, we ignore the difference between substantial and partial derivatives. From eq. (4, 53),

$$\rho \frac{\partial V}{\partial t} = \text{div}\,\boldsymbol{v}. \tag{17}$$

From eq. (5, 1) with no external forces and no viscosity,

$$\rho \frac{\partial \boldsymbol{v}}{\partial t} = -\text{grad}\,p. \tag{18}$$

Substituting eqs. (17) and (18) in eq. (1)

$$-\rho i\omega V_0 = i\boldsymbol{K} \cdot \boldsymbol{v}_0, \tag{19}$$

$$-\rho i\omega \boldsymbol{v}_0 = -i\boldsymbol{K} p_0. \tag{20}$$

Combining eqs. (19) and (20),

$$\frac{p_0}{V_0} = \left(\frac{\partial p}{\partial V}\right)_s = -\frac{\rho^2\omega^2}{\boldsymbol{K} \cdot \boldsymbol{K}}. \tag{21}$$

Because there is no irreversibility, $\boldsymbol{K}$ is real,

$$\boldsymbol{K} \cdot \boldsymbol{K} = k^2. \tag{22}$$

Or

$$v^2 = \frac{\omega^2}{k^2} = -V^2 \left(\frac{\partial p}{\partial V}\right)_s. \tag{23}$$

Comparing eqs. (16) and (23), we see that C_0 defined in eq. (16) is the reference speed of sound. From eqs. (3, 8) and (3, 11),

$$C_0^2 = -V^2 \frac{\kappa_T}{\kappa_S} \left(\frac{\partial p}{\partial V} \right)_T = -\gamma V^2 \left(\frac{\partial p}{\partial V} \right)_T = \gamma \left(\frac{\partial p}{\partial \rho} \right)_T, \quad (24)$$

where γ is the adiabatic exponent. In eqs. (16) and (24), $\partial p / \partial \rho$ are under adiabatic and isothermal conditions respectively (Strictly speaking, an isentropic process is different from an adiabatic process, which means no irreversibility due to heat conduction and radiation).

In the special example of ideal monoatomic gases, $\gamma = 5/3$ and $pV = RT/M$, where M is molecular weight, eq. (24) is reduced to

$$C_0 = \sqrt{\frac{5RT}{3M}}. \quad (25)$$

This is the result of Laplace-Poisson. In liquids such a formula does not hold at all. There is no general equation of state for liquids. Based on kinetic theory, it is commonly believed that $1 < \gamma \leqq 5/3$ for gases. For liquids no measured values greater than 3 has been reported. From eqs. (3, 11) and (3, 12),

$$\gamma - 1 = \frac{(C_p - C_V)}{C_V} = TV \frac{\alpha^2}{C_V \kappa_T} = \frac{TV\alpha^2}{C_p \kappa_S}$$

$$= -\frac{TV^2 \alpha^2}{C_p \left(\frac{\partial V}{\partial p} \right)_S} = \frac{T\alpha^2 C_0^2}{C_p} \geqq 0, \quad (26)$$

or $\gamma \geqq 1$ for any material.

We see that the ideal fluids cannot account for sound absorption.

§31 *SOUND PROPAGATION IN FLUIDS WITH CHEMICAL IRREVERSIBILITY*

In this section we treat the same problem of the preceding section except with "chemical irreversibility", i.e., the establishment of thermodynamical equilibrium may be delayed and dependent on frequency or time scale. One relaxation time is needed for each chemical reaction. For one chemical reaction only, it is easy to interpret the plot of $\ln\varepsilon$ against t, especially if the plot is linear.

With V and ε as independent variablies, we write

$$\begin{bmatrix} dp \\ dA \end{bmatrix} = \begin{bmatrix} \left(\frac{\partial p}{\partial V}\right)_{S\varepsilon} & \left(\frac{\partial p}{\partial \varepsilon}\right)_{SV} \\ \left(\frac{\partial A}{\partial V}\right)_{S\varepsilon} & \left(\frac{\partial A}{\partial \varepsilon}\right)_{SV} \end{bmatrix} \begin{bmatrix} dV \\ d\varepsilon \end{bmatrix}. \tag{1}$$

In the linearized region we can eliminate the second row of the matrix in eq. (1). From eq. (29, 11)

$$\left(\frac{\partial A}{\partial \varepsilon}\right)_{\alpha\beta} = -\frac{1}{B\Gamma_{\alpha\beta}}, \tag{2}$$

where α and β are fixed but unspecified state variables. From eqs. (25, 7), (25, 12), and (29, 8),

$$\frac{\partial \varepsilon}{\partial t} = -BA. \tag{3}$$

Substituting eq. (30, 1) in eq. (3),

$$-i\omega\varepsilon_0 = -BA_0. \tag{4}$$

Or

$$\frac{A_0}{\varepsilon_0} = \left(\frac{\partial A}{\partial \varepsilon}\right)_{\alpha\beta} = \frac{i\omega}{B}. \tag{5}$$

From eqs. (2) and (5),

$$\left(\frac{\partial A}{\partial V}\right)_{S\beta}=\left(\frac{\partial A}{\partial V}\right)_{S\varepsilon}+\left(\frac{\partial A}{\partial \varepsilon}\right)_{SV}\left(\frac{\partial \varepsilon}{\partial A}\right)_{S\beta}\left(\frac{\partial A}{\partial V}\right)_{S\beta}$$

$$=\left(\frac{\partial A}{\partial V}\right)_{S\varepsilon}+\left(-\frac{1}{B\Gamma_{SV}}\right)\left(\frac{B}{i\omega}\right)\left(\frac{\partial A}{\partial V}\right)_{S\beta}. \tag{6}$$

Or

$$\left(\frac{\partial A}{\partial V}\right)_{S\beta}=\frac{\left(\frac{\partial A}{\partial V}\right)_{S\varepsilon}(i\omega\Gamma_{SV})}{1+i\omega\Gamma_{SV}}. \tag{7}$$

From the first row of the matrix in eq. (1) and eqs. (2), (5) and (7),

$$\left(\frac{\partial p}{\partial V}\right)_{S\beta}=\left(\frac{\partial p}{\partial V}\right)_{S\varepsilon}+\left(\frac{\partial p}{\partial \varepsilon}\right)_{SV}\left(\frac{\partial \varepsilon}{\partial A}\right)_{S\beta}\left(\frac{\partial A}{\partial V}\right)_{S\beta}$$

$$=\left(\frac{\partial p}{\partial V}\right)_{S\varepsilon}+\frac{\left(\frac{\partial p}{\partial \varepsilon}\right)_{SV}\left(\frac{\partial \varepsilon}{\partial V}\right)_{SA}}{1+i\omega\Gamma_{SV}}$$

$$=\left(\frac{\partial p}{\partial V}\right)_{S\varepsilon}+\frac{\left[\left(\frac{\partial p}{\partial V}\right)_{SA}-\left(\frac{\partial p}{\partial V}\right)_{S\varepsilon}\right]}{1+i\omega\Gamma_{SV}}, \tag{8}$$

where β refers to a particular stage of development dependent on ω. Let us write

$$\chi(\omega)=\left(\frac{\partial p}{\partial V}\right)_{S\varepsilon}+\frac{\left[\left(\frac{\partial p}{\partial V}\right)_{SA}-\left(\frac{\partial p}{\partial V}\right)_{S\varepsilon}\right]}{1+i\omega\Gamma_{SV}}. \tag{9}$$

At zero frequency when the process at static equilibrium,

$$\chi(0)=\left(\frac{\partial p}{\partial V}\right)_{SA}=-\frac{1}{V\kappa_{SA}}<0. \tag{10}$$

At infinite frequency when the process is frozen,

$$\chi(\infty)=\left(\frac{\partial p}{\partial V}\right)_{S\varepsilon}=-\frac{1}{V\kappa_{S\varepsilon}}<0. \tag{11}$$

From eq. (29, 30),

$$\chi(0) - \chi(\infty) > 0. \tag{12}$$

Equation (9) becomes

$$\chi(\omega) = \chi(\infty) + \frac{\chi(0) - \chi(\infty)}{1 + i\omega\Gamma_{SV}}. \tag{13}$$

By separating $\chi(\omega)$ into real and imaginary parts,

$$\operatorname{Re}\chi(\omega) = -G = \frac{\chi(0) + \chi(\infty)\omega^2\Gamma_{SV}^2}{1 + \omega^2\Gamma_{SV}^2} < 0, \tag{14}$$

$$\operatorname{Im}\chi(\omega) = -H = \frac{[\chi(\infty) - \chi(0)]\omega\Gamma_{SV}}{1 + \omega^2\Gamma_{SV}^2} < 0. \tag{15}$$

From eqs. (30, 2), (30, 5) and (30, 21),

$$\chi(\omega) = -G - iH = -\frac{\rho^2\omega^2}{\boldsymbol{K}\cdot\boldsymbol{K}} = -\frac{\rho^2C_0^2k_0^2}{(k - ia)^2}$$
$$= -\frac{\rho^2C_0^2}{\zeta^2}. \tag{16}$$

Thus

$$\zeta = \frac{\rho C_0}{\sqrt{2}}\left[\sqrt{\frac{-\sqrt{G^2 + H^2} + G}{G^2 + H^2}}\right.$$
$$\left. - i\sqrt{\frac{-\sqrt{G^2 + H^2} - G}{G^2 + H^2}}\right]. \tag{17}$$

From eq. (30, 6),

$$d = \frac{-\left\{\frac{2}{C_0^2\rho^2}[\chi^2(0) + \chi^2(\infty)\omega^2\Gamma_{SV}^2]\right\}}{[\chi(0) + \chi(\infty)\omega^2\Gamma_{SV}^2 + \sqrt{1 + \omega^2\Gamma_{SV}^2}\sqrt{\chi^2(0) + \chi^2(\infty)\omega^2\Gamma_{SV}^2}]} \tag{18}$$

At $\omega = 0$,

$$d = -\frac{\chi(0)}{\rho^2C_0^2}. \tag{19}$$

At $\omega \to \infty$,

$$d = -\frac{\chi(\infty)}{\rho^2 C_0^2}. \tag{20}$$

Thus d is a monotonically increasing function of ω with increasing ω. From eq. (30, 15),

$$\bar{a} = \frac{\pi\rho^2 C_0^2 \omega \Gamma_{SV} [\chi(0) - \chi(\infty)]}{\chi^2(0) + \chi^2(\infty)\omega^2 \Gamma_{SV}^2}. \tag{21}$$

Thus $\bar{a}$ is zero at $\omega = 0$ and $\omega \to \infty$, and $\bar{a}$ is maximum at $\Gamma_{SV}^{-1}\chi(0)/\chi(\infty)$.

Instead of treating sound propagation in a single fluid as chemical irreversibility, we could treat it as viscosity-irreversibility. From eqs. (4, 53), (25, 7), (25, 12), and (29, 8),

$$\sigma \text{ with chemical irrevsibility only} = -\frac{\rho A}{T}\left(\frac{\partial \varepsilon}{\partial t}\right)_{Tp}$$

$$= \left(\frac{\partial \varepsilon}{\partial t}\right)_{Tp}^2 \frac{\rho}{BT} = \frac{\rho}{BT}\left(\frac{\partial \varepsilon}{\partial V}\right)_{Tp}^2 \left(\frac{\partial V}{\partial t}\right)_{Tp}^2$$

$$= \frac{1}{\rho BT}\left(\frac{\partial \varepsilon}{\partial V}\right)_{Tp}^2 (\operatorname{div} \boldsymbol{v})^2. \tag{22}$$

From eq. (19, 1),

σ with (volume) viscosity-irreversibility alone

$$= \frac{\eta_V}{T} (\operatorname{div} \boldsymbol{v})^2. \tag{23}$$

Comparing eqs. (21) and (22),

$$\eta_V = \frac{1}{\rho B}\left(\frac{\partial \varepsilon}{\partial V}\right)_{Tp}^2. \tag{24}$$

The essential feature of chemical irreversibility is that the state variables could have two values, an asymptotic one for slow variations and another for rapid changes. Einstein[1] treated sound propagation in a fluid mixture of two reacting components. Liebermann[2] treated it in an ionic solution. The chemical reactions are obvious. For diatomic gases the rotational degrees of freedom may not follow quick variations. Here the connection with chemistry is based on analogy. If the fluids are not ideal, i.e., viscosity and heat exchanges are taken into consideration, even for perfect monatomic gases, the situation is complex. For example, in comparing the experimental maximum absorption of real fluids with the theoretical absorption of ideal fluids, we could adjust relaxation time to fit with the frequency to give the maximum absorption, but the magnitude may be not large enough and more than one relaxation time may be needed.

To generalize to multiple relaxation times, we write

$$\begin{bmatrix} \dot{z} \\ \dot{A} \end{bmatrix} = \begin{bmatrix} C & \lambda \\ \lambda^{\tau} & \dfrac{\partial A}{\partial \varepsilon} \end{bmatrix} \begin{bmatrix} \dot{z} \\ \dot{\varepsilon} \end{bmatrix}, \tag{25}$$

where Z is a scalar intensive property, z is a corresponding extensive property, C is a frozen coefficient and the matrix is symmetric. $\dot{z}$ and $\dot{A}$ are connected to $\dot{z}$ and $\dot{\varepsilon}$ through the dynamic matrix. In the region of linearity, the dynamic matrix is the same as the matrix of differential coefficients. From eqs. (2) and (5), eq. (25) can be simplified,

$$\begin{bmatrix} \dot{z} \\ 0 \end{bmatrix} = \begin{bmatrix} C & \lambda \\ \lambda^{\mathrm{T}} & -\dfrac{1 + i\omega\Gamma}{B\Gamma} \end{bmatrix} \begin{bmatrix} \dot{z} \\ \dot{\varepsilon} \end{bmatrix}. \tag{26}$$

Hence

$$\begin{aligned} \dot{z} &= C\dot{z} + \lambda \left[\frac{1 + i\omega\Gamma}{B\Gamma} \right]^{-1} \lambda^{\mathrm{T}} \dot{z} \\ &= \left[C + \lambda \left(\frac{1}{B\Gamma} + \frac{i\omega}{B} \right)^{-1} \lambda^{\mathrm{T}} \right] \dot{z}. \end{aligned} \tag{27}$$

Now $1/B\Gamma$ can be diagonalized to a unit matrix and $1/B$ can be diagonalized to relaxation times Γ_1, Γ_2, etc. The simultaneous diagonalization is possible because $1/B\Gamma$ and $1/B$ commute.

Then

$$\dot{z}_i = \sum_k C_{ik}(\omega)\dot{z}_k, \tag{28}$$

where

$$C_{ik}(\omega) = C_{ik} + \sum_j \frac{\lambda_{ij}\lambda_{kj}}{1 + i\omega\Gamma_j}. \tag{29}$$

At static equilibrium

$$C_{ik}(0) = C_{ik} + \sum_j \lambda_{ij}\lambda_{kj}, \tag{30}$$

and at infinite frequency

$$C_{ik}(\infty) = C_{ik}. \tag{31}$$

Thus

$$C_{ik}(\omega) = C_{ik}(\infty) + \sum_j \frac{C_{ik}(0) - C_{ik}(\infty)}{1 + i\omega\Gamma_j}. \tag{32}$$

With one single relaxation time, the kinetic equation is one of exponential decay. For example, the solution of eq. (29, 9), with $\varepsilon_0 - \varepsilon = \Delta\varepsilon$, is

$$\varepsilon(t) = \exp\left(-\frac{t}{\Gamma}\right) + \varepsilon_0. \tag{33}$$

The quantity of $d \ln \varepsilon/dt$ is expected to correspond to an activation energy representing a barrier to the flow. When a spectrum of relaxation times is proposed, the order of differential equation is increased to n, where n is the number of relaxation times. The kinetic equation is no longer exponential and the interpretation is not simple. For example, the general linear solid may be represented by

$$\alpha S + \beta \dot{S} = \gamma e + \delta \dot{e}, \tag{34}$$

in which S is stress and e is strain.

If $\dot{S} = \dot{e} = 0$, we have an elastic element or spring. If $\dot{S} = e = 0$, we have a viscous element or a dashpot. If $e = 0$, we have an elastic element and a viscous element in series. It may be called an elasticoviscous element. If $\dot{S} = 0$, we have an elastic element and a viscous element in parallel. It may be called a fermicoviscous element (The nomenclature is not standardized. The elasticoviscous element is also called a Maxwell element. The fermicoviscous element is also called a Voigt, or a Kelvin, or a Voigt-Kelvin element).

The full equation (34) is represented by a spring element in series with another spring element and a viscous element in parallel. In the case of fluids, stresses may be considered as negative pressures and strains may be considered as volume changes. It may be possible to explore the propagation of sound in fluids from this equation alone[3]. However, whether the flow is isothermal or adiabatic must be specified.

References

[1] A. Einstein: *Sitzber. Akad. Wissen.*, Berlin, Math-Phys., **Kl**, 380(1920)
[2] L. Liebermann: *Phys. Rev.*, **76**, 1520(1949).
[3] J. Frenkel and J. Obraztsor; *J. Phys.* (U. S. S. R.), **3**, 131(1940).

§32 *SOUND PROPAGATION IN HEAT-EXCHANGING AND VISCOUS FLUIDS* (I)

In this section we treat the problem of §30, including irreversibility due to heat conduction, radiation and viscosity. As usual in the approximate treatment, we ignore the difference between substantial and partial derivatives. With $\boldsymbol{v}$, ρ, and T as independent variables, the appropriate relations are conservation of mass, balance of momentum and rate of entropy-production.

The equation for conservation of mass is eq. (4, 46),

$$\frac{\partial \rho}{\partial t} + \rho \operatorname{div} \boldsymbol{v} = 0. \tag{1}$$

The equation for balance of momentum is eq. (5, 23). With no external force and with the assumption curl $\boldsymbol{v} = 0$, eq. (5, 23) is reduced to

$$\rho \frac{\partial \boldsymbol{v}}{\partial t} + \operatorname{grad} p - \left(\eta_V + \frac{4\eta}{3}\right) \operatorname{Div} \operatorname{Grad} \boldsymbol{v} = 0. \tag{2}$$

Through the following manipulations p is separated into functions of ρ and T,

$$\begin{aligned}\operatorname{grad} p &= \left(\frac{\partial p}{\partial \rho}\right)_T \operatorname{grad} \rho + \left(\frac{\partial p}{\partial T}\right)_\rho \operatorname{grad} T \\ &= \left(\frac{\partial p}{\partial \rho}\right)_T \left[\operatorname{grad} \rho - \left(\frac{\partial \rho}{\partial T}\right)_p \operatorname{grad} T\right] \\ &= -V^2 \left(\frac{\partial p}{\partial V}\right)_T \left[\operatorname{grad} \rho + \frac{1}{V^2}\left(\frac{\partial V}{\partial T}\right)_p \operatorname{grad} T\right] \\ &= \frac{C_0^2}{\gamma}(\operatorname{grad} \rho + \rho\alpha \operatorname{grad} T),\end{aligned} \tag{3}$$

eq. (30, 24) having been used. Substituting eq. (3) in eq. (2),

$$\begin{aligned}\rho \frac{\partial \boldsymbol{v}}{\partial t} &+ \frac{C_0^2}{\gamma}(\operatorname{grad} \rho + \rho\alpha \operatorname{grad} T) \\ &- \left(\eta_V + \frac{4\eta}{3}\right) \operatorname{Div} \operatorname{Grad} \boldsymbol{v} = 0.\end{aligned} \tag{4}$$

Instead of using the transport equation of entropy, it is convenient to use the transport equation of temperature, eq. (9, 11). With no chemical reactions and no mass fluxes, eq. (9, 11) is reduced to

$$\begin{aligned}\rho C_V \frac{\partial T}{\partial t} &+ (\Pi - p\overleftrightarrow{1}) : \operatorname{Grad} v + \frac{\alpha}{\kappa} T \operatorname{div} \boldsymbol{v} \\ &+ \operatorname{div} \boldsymbol{J}_q^r = 0.\end{aligned} \tag{5}$$

From eq. (19, 1) the second term in eq. (5) is

$$-\left(\eta_V - \frac{2\eta}{3}\right)(\operatorname{div}\boldsymbol{v})^2 - 2\eta\,(\operatorname{Grad}\boldsymbol{v})^2,$$

which is considered as a second-order effect and neglected. In the third term the juxtaposition of T and v is avoided by using eq. (3, 7),

$$\frac{\alpha T}{\kappa} = \frac{C_p - C_V}{V\alpha} = \frac{\rho C_V}{\alpha}(\gamma - 1). \tag{6}$$

The heat exchange is approximately separated into conduction and radiation. The part of the reduced heat flux due to conduction is

$$\boldsymbol{J}_q^r = -\frac{L \operatorname{grad} T}{T^2} = -\theta \operatorname{grad} T, \tag{7}$$

where θ is the coefficient of thermal conductivity; the part of the reduced heat flux due to radiation is approximated, for small temperature differences, by Newton's law of cooling,

$$\operatorname{div}\boldsymbol{J}_q^r \approx C_V q \rho T, \tag{8}$$

where q is the radiation coefficient in $\sec^{-1}$. With eqs. (6—8), eq. (5) is reduced to

$$\frac{\partial T}{\partial t} + \frac{1}{\alpha}(\gamma - 1)(\operatorname{div}\boldsymbol{v}) - \frac{\theta}{\rho C_V}\operatorname{div}\operatorname{grad} T$$

$$+ qT = 0. \tag{9}$$

If the assumed solution of (30, 2) is substituted in eqs. (1), (4), and (9), the system of equations are

$$\left[\begin{array}{cc} i\rho(k - ia) & -i\omega \\ -i\omega\rho + \left(\eta_V + \frac{4\eta}{3}\right)(k - ia)^2 & i\,\frac{C_0^2}{\gamma}(k - ia) \\ \frac{i}{\alpha}(\gamma - 1)(k - ia) & 0 \end{array}\right.$$

$$\begin{matrix} 0 \\ i\,\dfrac{\rho\alpha C_0^2}{\gamma}(k-ia) \\ q - i\omega + \dfrac{\theta}{\rho C_V}(k-ia)^2 \end{matrix} \Bigg] \begin{bmatrix} \boldsymbol{v}_0 \\ \rho_0 \\ T_0 \end{bmatrix} = 0. \quad (10)$$

In order that eq. (10) has non-trivial solutions, the matrix must be singular,

$$-i\omega^3\rho + \omega^2\rho q + (k-ia)^2\left[i\omega\rho C_0^2 + i\omega\left(\eta_V + \frac{4\eta}{3}\right)q + \frac{\omega^2\theta}{C_V} - \frac{1}{\gamma}C_0^2\rho q + \omega^2\left(\eta_V + \frac{4\eta}{3}\right)\right] + (k-ia)^4\left[i\omega\theta\left(\eta_V + \frac{4\eta}{3}\right)\Big/\rho C_V - \frac{C_0^2\theta}{C_V\gamma}\right] = 0. \quad (11)$$

Multiplying eq. (11) by $i/\omega^3\rho$, and using eq. (30, 5),

$$1 + \frac{iq}{\omega} + \zeta^2\left[-1 - \frac{q\left(\eta_V + \frac{4\eta}{3}\right)}{\rho C_0^2} + \frac{i\omega\left(\eta_V + \frac{4\eta}{3}\right)}{\rho C_0^2} + \frac{i\omega\theta}{\rho C_0^2 C_V} - \frac{iqC_V}{\omega C_p}\right] - \zeta^4\left[\frac{\theta\omega^2\left(\eta_V + \frac{4\eta}{3}\right)}{\rho^2 C_0^4 C_V} + \frac{i\omega\theta}{\rho C_0^2 C_p}\right] = 0. \quad (12)$$

The coefficients of eq. (12) contain 9 parameters: C_p, C_V, η_V, η, ω, C_0, ρ, q, and θ. According to the Buckingham pi theorem of dimensional analysis, besides four basic dimensional units (mass, length, temperature and time), there are just five dimensional ratios:

$$\gamma = \frac{C_p}{C_V},\ N = \frac{1}{\eta}\left(\eta_V + \frac{4\eta}{3}\right),\ P = \frac{\eta C_p}{\theta},$$

$$\mathscr{S}' = \frac{\eta\omega}{\rho C_0^2}, \quad \mathscr{S} = \frac{\omega}{q},$$

where γ is the adiabatic exponent, N is the viscous number, P is the Prandtl number, and $\mathscr{S}'$ and $\mathscr{S}$ are Stokes' numbers. However, it is convenient to use γ, N, $\mathscr{S}$, X, and Y, where

$$X = N\mathscr{S}' = \omega \frac{\eta_V + \dfrac{4\eta}{3}}{\rho C_0^2}, \quad Y = \frac{1}{NP} = \frac{\theta}{\left(\eta_V + \dfrac{4\eta}{3}\right) C_p}.$$

Equation (12) is abbreviated as

$$1 + \frac{i}{\mathscr{S}} + \zeta^2 \left[-1 - \frac{X}{\mathscr{S}} + i\left(X + \gamma XY - \frac{1}{\mathscr{S}\gamma}\right)\right] - \zeta^4 XY(i + \gamma X) = 0. \tag{13}$$

It is seen that N does not appear in eq. (13). From eq. (13) ζ^2 can be found by the standard quadratic formula and ζ by the extraction of the square root. It is physically evident that the pair of complex roots with nonpositive imaginary part corresponding to real absorption is to be retained. Furthermore, because X, Y, γ, and $\mathscr{S}$ occur in one or more combinations as products, it is not certain whether the effects of viscosity and heat exchange are still linearly additive.

Before discussing various approximations, we mention briefly some terminlogy commonly used. Stokes[1] found that viscosity decreases the dispersion of free waves. The effect increases sharply as frequencies increase. For frequencies high enough, the motion may be overdamped and the disturbance may not propagated at all. Therefore, we speak of "anomalous" dispersion if dispersion increases with frequencies, in spite of the fact that we are dealing with forced waves and not with viscosity alone. In absorption the usual approximation $\eta_V = 0$ is known as Stokes' relation, although Stokes himself did not use it. The linear increase of absorption coefficient per unit wave length with frequency, together with Stokes' relation is called "classic".

One simplification is to consider effects of radiation alone[2]. Mathematically let X and XY approach zero but keep $\mathscr{S}$ finite. Equation (13) is reduced to

$$1 + \frac{i}{\mathscr{S}} - \zeta^2\left(1 + \frac{i}{\mathscr{S}\gamma}\right) = 0. \tag{14}$$

$$\zeta = \frac{\sqrt{\gamma}}{\sqrt{2}\sqrt{1+\gamma^2\mathscr{S}^2}}\left[\sqrt{\sqrt{(1+\gamma^2\mathscr{S}^2)(1+\mathscr{S}^2)} + (1+\gamma\mathscr{S}^2)}\right.$$
$$\left. + i\sqrt{\sqrt{(1+\gamma^2\mathscr{S}^2)(1+\mathscr{S}^2)} - (1+\gamma\mathscr{S}^2)}\right]. \tag{15}$$

From eqs. (30, 6), (30, 13), (30, 14), and (30, 15),

$$d = \frac{\dfrac{2(1+\gamma^2\mathscr{S}^2)}{\gamma}}{\sqrt{(1+\gamma^2\mathscr{S}^2)(1+\mathscr{S}^2)} + (1+\gamma\mathscr{S}^2)}, \tag{16}$$

$$a_0^2 = \frac{2\pi^2\gamma\left[\sqrt{(1+\gamma^2\mathscr{S}^2)(1+\mathscr{S}^2)} - (1+\gamma\mathscr{S}^2)\right]}{1+\gamma^2\mathscr{S}^2}, \tag{17}$$

$$a' = \frac{2\pi\mathscr{S}(\gamma-1)}{\left[\sqrt{(1+\gamma^2\mathscr{S}^2)(1+\mathscr{S}^2)} + (1+\gamma\mathscr{S}^2)\right]}, \tag{18}$$

$$\bar{a} = \frac{\pi\gamma\,\mathscr{S}\sqrt{\gamma-1}}{1+\gamma^2\mathscr{S}^2}. \tag{19}$$

Thus the dispersion d is a monotonically increasing function of ω, with a single point of inflection. The square of the speed C increases steadily from C_0^2/γ to C_0^2. All three absorption coefficients rise from zero to single maxima and fall off thereafter to zero. The values of $\mathscr{S}$ exhibiting maxima are

$$\mathscr{S}_1 = \frac{1}{\gamma}, \quad \bar{a}_{\max} = \frac{\pi\sqrt{\gamma-1}}{2}, \tag{20}$$

$$\mathscr{S}_2 = \frac{\sqrt{3\gamma+1}}{\gamma\sqrt{\gamma+3}}, \quad a_{0,\max} = \frac{\pi(\gamma-1)}{\sqrt{2}\sqrt{\gamma+1}}, \tag{21}$$

$$\mathscr{S}_3 = \frac{1}{\sqrt{\gamma}}, \quad a'_{\max} = \frac{2\pi(\sqrt{\gamma}-1)}{\sqrt{\gamma}+1}, \tag{22}$$

$$\mathscr{S}_1 < \mathscr{S}_2 < \mathscr{S}_3.$$

Another simplification is to ignore the effects of radiation and to deal with piezotropic fluids only. Mathematically let $\mathscr{S}$ approach infinity and γ be unity (From kinetic theory it is the lowest value of γ possible). Equation (13) is reduced to

$$(i - XY\zeta^2)(-i + X\zeta^2 + i\zeta^2) = 0. \tag{23}$$

We see that in this limiting example the effects of heat conduction and viscosity are separable.

For $\zeta^2 = i/XY$,

$$\zeta = \frac{1}{\sqrt{2}}\left(\frac{1}{\sqrt{XY}} + \frac{i}{\sqrt{XY}}\right). \tag{24}$$

The wave may be called a thermal wave. The result given above is essentially the same as that of Rayleigh[2].

For

$$\zeta^2 = \frac{1+iX}{1+X^2}, \tag{25}$$

$$\zeta = \frac{1}{\sqrt{2}}\left[\frac{\sqrt{\sqrt{1+X^2}+1}}{\sqrt{1+X^2}} + i\frac{\sqrt{\sqrt{1+X^2}-1}}{\sqrt{1+X^2}}\right], \tag{26}$$

The wave may be called a pressure wave. The result given above is the same as that of Stefan[3].

In the more general case, we only let $\mathscr{S}$ approach infinity and eq. (13) is modified to

$$\frac{1}{\zeta^4} + \frac{1}{\zeta^2}[-1 + iX(1+\gamma Y)] - (i + \gamma X)XY = 0. \tag{27}$$

The equation (26) is the Kirchhoff-Langevin equation. While Kirchhoff's treatment[4] is restricted to the case of perfect gases, Langevin, quoted by Biquard[5], shows that it remains valid for pure fluids obeying an arbitrary equation of state. Then

$$2\left(\frac{1}{\zeta^2}\right) = 1 - iX(\gamma Y + 1) \pm \sqrt{E - iF} = G - iH, \quad (28)$$

where

$$E = 1 - X^2(1 - \gamma Y)^2, \quad (29)$$

$$F = (1 - 2Y + \gamma Y)(2X), \quad (30)$$

$$G = 1 \pm \frac{1}{\sqrt{2}} \sqrt{\sqrt{E^2 + F^2} + E}\ , \quad (31)$$

$$H = X(\gamma Y + 1) \pm \frac{1}{\sqrt{2}} \sqrt{\sqrt{E^2 + F^2} - E}\,. \quad (32)$$

Truesdell[6] interpreted the sign as two different kinds of fluids and two different kinds of waves.

References

[1] G. G. Stokes: *Trans. Cambr. Phil. Soc.*, **8** (1884—1849), 287 (1845).
[2] Lord Rayleigh (J. W. Strutt): Theory of Sound, 2nd Ed., Denver Publications, New York (1945).
[3] J. Stefan: *Sitzber. Akad. Wiss.*, Wien, **53**, 529 (1866).
[4] G. Kirchhoff: *Ann. der Phys.*, **134**, 177 (1868).
[5] P. Biquard: *Ann. der Phys.*, **6**, 195 (1836).
[6] C. Truesdell: *J. Rational Mech. and Analysis*, **2**, 643 (1953).

§33 *SOUND PROPAGATION IN HEAT-EXCHANGING AND VISCOUS FLUIDS* (II)

In this section we treat the same problem as in the previous section, except that the hypothesis of incomplete reaction is introduced. The introduction of frequency is natural and some insight is gained in the interaction between heat conductivity and viscosity if linearity prevails.

The equation of conservation of mass is

$$\rho \frac{\partial V}{\partial t} = \operatorname{div} \boldsymbol{v}. \tag{1}$$

Substituting eq. (30, 1) in eq. (1) yields

$$-\rho\omega V_0 = \boldsymbol{K} \cdot \boldsymbol{v}_0. \tag{2}$$

The equation of balance of momentum is

$$\rho \frac{\partial \boldsymbol{v}}{\partial t} = -\operatorname{grad} p + \left(\eta_V + \frac{4\eta}{3}\right) \operatorname{Div} \operatorname{Grad} \boldsymbol{v}. \tag{3}$$

Substituting eq. (30, 1) in eq. (3) yields

$$-\rho i\omega \boldsymbol{v}_0 = -i\boldsymbol{K} p_0 - \left(\eta_V + \frac{4\eta}{3}\right) \boldsymbol{K} \cdot \boldsymbol{K} \boldsymbol{v}_0. \tag{4}$$

Combining eqs. (2) and (4),

$$\rho^2\omega^2 V_0 = -\boldsymbol{K} \cdot \boldsymbol{K} p_0 - i\left(\eta_V + \frac{4\eta}{3}\right) \rho\omega \boldsymbol{K} \cdot \boldsymbol{K} V_0. \tag{5}$$

Thus

$$\frac{p_0}{V_0} = \frac{\rho^2\omega^2}{\boldsymbol{K} \cdot \boldsymbol{K}} - i\left(\eta_V + \frac{4\eta}{3}\right) \rho\omega. \tag{6}$$

The equation of change of entropy, due to conduction of heat alone, (8, 9), is

$$\rho \frac{\partial S}{\partial t} = -\operatorname{div} \boldsymbol{J}_S - \frac{\boldsymbol{J}_S \cdot \operatorname{grad} T}{T}. \tag{7}$$

Or

$$\rho T \frac{\partial S}{\partial t} = -T \operatorname{div} \boldsymbol{J}_S - \boldsymbol{J}_S \cdot \operatorname{grad} T = -\operatorname{div} T\boldsymbol{J}_S$$

$$= -\operatorname{div} \boldsymbol{J}_q^r = \theta \operatorname{div} \operatorname{grad} T. \tag{8}$$

Substituting eq. (30, 1) in eq. (8) yields

$$-i\rho\omega T S_0 = -\theta \boldsymbol{K} \cdot \boldsymbol{K} T_0. \tag{9}$$

Or

$$\frac{S_0}{T_0} = - i\theta \frac{\boldsymbol{K} \cdot \boldsymbol{K}}{\rho \omega T}. \tag{10}$$

For the Jacobian matrix,

$$\begin{bmatrix} dp \\ dT \\ dA \end{bmatrix} = \begin{bmatrix} \left(\frac{\partial p}{\partial V}\right)_{\varepsilon S} & \left(\frac{\partial p}{\partial \varepsilon}\right)_{VS} & \left(\frac{\partial p}{\partial S}\right)_{V\varepsilon} \\ \left(\frac{\partial T}{\partial V}\right)_{\varepsilon S} & \left(\frac{\partial T}{\partial \varepsilon}\right)_{VS} & \left(\frac{\partial T}{\partial S}\right)_{V\varepsilon} \\ \left(\frac{\partial A}{\partial V}\right)_{\varepsilon S} & \left(\frac{\partial A}{\partial \varepsilon}\right)_{VS} & \left(\frac{\partial A}{\partial S}\right)_{V\varepsilon} \end{bmatrix} \begin{bmatrix} dV \\ d\varepsilon \\ dS \end{bmatrix}. \tag{11}$$

From the last row of the matrix in eqs. (11) and from eqs. (31, 2) and (31, 5),

$$dA = \left(\frac{\partial A}{\partial V}\right)_{\varepsilon S} dV + \left(\frac{\partial A}{\partial S}\right)_{V\varepsilon} dS - \frac{1}{B\Gamma_{VS}} \left(\frac{B}{i\omega}\right) dA. \tag{12}$$

Or

$$dA = \frac{\left[\left(\frac{\partial A}{\partial V}\right)_{\varepsilon S} dV + \left(\frac{\partial A}{\partial S}\right)_{V\varepsilon} dS\right] i\omega\Gamma_{VS}}{1 + i\omega\Gamma_{VS}}. \tag{13}$$

From the first row of the matrix in eqs. (11) and from eq. (13),

$$dp = \left[\left(\frac{\partial p}{\partial V}\right)_{\varepsilon S} - \frac{\left(\frac{\partial p}{\partial \varepsilon}\right)_{VS}\left(\frac{\partial \varepsilon}{\partial A}\right)_{VS}\left(\frac{\partial A}{\partial V}\right)_{\varepsilon S}}{1 + i\omega\Gamma_{VS}}\right] dV + \left[\left(\frac{\partial p}{\partial S}\right)_{\varepsilon V} - \frac{\left(\frac{\partial p}{\partial \varepsilon}\right)_{VS}\left(\frac{\partial \varepsilon}{\partial A}\right)_{VS}\left(\frac{\partial A}{\partial S}\right)_{V\varepsilon}}{1 + i\omega\Gamma_{VS}}\right] dS. \tag{14}$$

Now the first square bracket in eq. (14) is from eq. (31, 8),

$$\left(\frac{\partial p}{\partial V}\right)_{\varepsilon S} + \frac{\left(\frac{\partial p}{\partial \varepsilon}\right)_{VS}\left(\frac{\partial \varepsilon}{\partial V}\right)_{AS}}{(1 + i\omega\Gamma_{VS})} = \left(\frac{\partial p}{\partial V}\right)_{\varepsilon S} + \frac{\left[\left(\frac{\partial p}{\partial V}\right)_{AS} - \left(\frac{\partial p}{\partial V}\right)_{\varepsilon S}\right]}{(1 + i\omega\Gamma_{VS})}$$

$$= \left(\frac{\partial p}{\partial V}\right)_{\alpha S} = \chi(\omega). \tag{15}$$

The second square bracket in eq. (14) is, from eqs. (3, 16), (3, 20) and (3, 21),

$$\left(\frac{\partial p}{\partial S}\right)_{\varepsilon V} - \frac{\left(\frac{\partial p}{\partial A}\right)_{VS}\left(\frac{\partial A}{\partial S}\right)_{V\varepsilon}}{(1 + i\omega\Gamma_{VS})} = -\left(\frac{\partial T}{\partial V}\right)_{\varepsilon S} - \frac{\left(\frac{\partial \varepsilon}{\partial V}\right)_{AS}\left(\frac{\partial T}{\partial \varepsilon}\right)_{VS}}{(1 + i\omega\Gamma_{VS})}$$

$$= -\left\{\left(\frac{\partial T}{\partial V}\right)_{\varepsilon S} + \frac{\left[\left(\frac{\partial T}{\partial V}\right)_{AS} - \left(\frac{\partial T}{\partial V}\right)_{\varepsilon S}\right]}{1 + i\omega\Gamma_{VS}}\right\}$$

$$= -\left(\frac{\partial T}{\partial V}\right)_{\alpha S}$$

$$= -\chi_T(\omega). \tag{16}$$

Substituting eqs. (6), (10), (15) and (16) in eq. (14),

$$\frac{\rho^2\omega^2}{\boldsymbol{K}\cdot\boldsymbol{K}} + i\left(\eta_V + \frac{4\eta}{3}\right)\rho\omega + \left(\frac{\partial p}{\partial V}\right)_{\alpha S}$$

$$+ \frac{i\theta\left(\frac{\partial T}{\partial V}\right)^2_{\alpha S}\boldsymbol{K}\cdot\boldsymbol{K}}{\rho\omega T} = 0. \tag{17}$$

Now from eq. (31, 16),

$$\boldsymbol{K}\cdot\boldsymbol{K} = \frac{\zeta^2\omega^2}{C_0^2}. \tag{18}$$

Also

$$\chi(\omega) = \left(\frac{\partial p}{\partial V}\right)_{\alpha S} = \left(\frac{\partial p}{\partial V}\right)_{AS}(1 - A), \tag{19}$$

$$\chi_T^2(\omega) = \left(\frac{\partial T}{\partial V}\right)_{aS}^2 = \left(\frac{\partial T}{\partial V}\right)_{AS}^2 (-1 + B), \tag{20}$$

where A and B are dimensional ratios, whose values are as yet undetermined. We know that A and B must be pure imaginary functions of ω, because $(\partial p/\partial V)_{AS}$ and $(\partial T/\partial V)_{AS}^2$ are real. Furthermore, we may consider A and B as interactions. Because $(\partial p/\partial V)_{AS}$ is related to viscosity from eq. (6), A is related to thermal conductivity and because $(\partial T/\partial V)_{AS}^2$ is related to conductivity from eq. (10), B is related to viscosity. Or

$$\frac{A}{B} = \frac{\theta}{\left(\eta_V + \dfrac{4\eta}{3}\right) C_p} = Y = \frac{i\gamma XY}{i\gamma X}. \tag{21}$$

Thus

$$\left(\frac{\partial p}{\partial V}\right)_{aS} = -\rho^2 C_0^2 (1 - i\gamma XY), \tag{22}$$

$$\left(\frac{\partial T}{\partial V}\right)_{aS}^2 = \left(-\frac{\rho\alpha T C_0^2}{C_p}\right)^2 (-1 + i\gamma X). \tag{23}$$

Substituting eqs. (18), (22) and (23) in eq. (17), dividing by $\rho^2 C_0^2 \zeta^2$, and replacing $(\eta_V + 4\eta/3)\omega/\rho C_0^2$ by X and $\omega\theta/\rho C_0^2\ C_p$ by XY,

$$\frac{1}{\zeta^4} + \frac{iX}{\zeta^2} - \frac{1 - i\gamma XY}{\zeta^2} + i(-1 + i\gamma X)\left(\frac{\alpha^2 T C_0^2}{C_p}\right) XY = 0. \tag{24}$$

From eqs. (24) and (30, 26)

$$\frac{1}{\zeta^4} - \frac{1}{\zeta^2}(1 - iX - i\gamma XY) + i(-1 + i\gamma X)(\gamma - 1)XY = 0. \tag{25}$$

This is the same as eq. (32, 27) except that there is an extra factor of $(\gamma - 1)$ in the last term on the left of eq. (25).

§34 *ELASTIC WAVES IN SOLIDS*

The propagation of disturbances in an elastic solid is non-dispersive and non-dissipative.

Let us denote $\boldsymbol{u}$ as the vector of the elastic displacement. We denote elements of the strain tensor by e so that

$$e_{ij} = \frac{1}{2}\left(\frac{\partial u_i}{\partial x_j} + \frac{\partial u_j}{\partial x_i}\right) = e_{ji}, \quad i,\ j = 1, 2, 3. \tag{1}$$

Thus the strain tensor is symmetric by definition. Physically, we assume that there are no components of rotation about the displacement of a point within the medium. The rotation can occur bodily. The elements of the stress tensor are

$$S_{ij} = S_{ji}. \tag{2}$$

The symmetry of the stress tensor is due to the invariance of the strain energy $1/2\ S{:}e$. The elements of the stress tensor are connected to those of the strain tensor by the equations of the generalized Hooke's law

$$S_{ij} = \sum_{k,l=1}^{3} \mathscr{C}_{ijkl} e_{kl}. \tag{3}$$

The elastic coefficients $\mathscr{C}_{ijkl}$ satisfy the symmetry relations:

$$\mathscr{C}_{ijkl} = \mathscr{C}_{klij} = \mathscr{C}_{jilk}. \tag{4}$$

There are not more than 21 independent coefficients in an anisotropic solid.

If the solid is isotropic and rotational equilibrium is achieved, there are two independent elastic coefficients:

$$\mathscr{C}_{iiii} = \lambda + 2G, \quad \mathscr{C}_{iikk} = \lambda, \quad \mathscr{C}_{ikik} = G, \tag{5}$$

where λ and G are Lamé's constants [According to § 13, an isotropic tensor of the fourth rank has three independent components in three dimensions. However, elastic coefficients are the ratio of symmetric stresses to symmetric strains. There is an additional restraint $\mathscr{C}_{1212} = \mathscr{C}_{1221}$. The restraint of (13, 9) becomes $\mathscr{C}_{1111} = \mathscr{C}_{1122} + \mathscr{C}_{1212} + \mathscr{C}_{1221} = \mathscr{C}_{1122} + 2\mathscr{C}_{1212}]$. In matrix form,

$$\begin{bmatrix} S_{11} \\ S_{22} \\ S_{33} \\ S_{23} \\ S_{31} \\ S_{12} \end{bmatrix} = \begin{bmatrix} \lambda + 2G & \lambda & \lambda & 0 & 0 & 0 \\ \lambda & \lambda + 2G & \lambda & 0 & 0 & 0 \\ \lambda & \lambda & \lambda + 2G & 0 & 0 & 0 \\ 0 & 0 & 0 & 2G & 0 & 0 \\ 0 & 0 & 0 & 0 & 2G & 0 \\ 0 & 0 & 0 & 0 & 0 & 2G \end{bmatrix} \begin{bmatrix} e_{11} \\ e_{22} \\ e_{33} \\ e_{23} \\ e_{31} \\ e_{12} \end{bmatrix}. \quad (6)$$

Equations (6) can be inverted,

$$\begin{bmatrix} e_{11} \\ e_{22} \\ e_{33} \\ e_{23} \\ e_{31} \\ e_{12} \end{bmatrix} = \begin{bmatrix} A & -B & -B & 0 & 0 & 0 \\ -B & A & -B & 0 & 0 & 0 \\ -B & -B & A & 0 & 0 & 0 \\ 0 & 0 & 0 & \frac{1}{2G} & 0 & 0 \\ 0 & 0 & 0 & 0 & \frac{1}{2G} & 0 \\ 0 & 0 & 0 & 0 & 0 & \frac{1}{2G} \end{bmatrix} \begin{bmatrix} S_{11} \\ S_{22} \\ S_{33} \\ S_{23} \\ S_{31} \\ S_{12} \end{bmatrix}, \quad (7)$$

where

$$A = \frac{\lambda + G}{G(3\lambda + 2G)}, \qquad B = \frac{\lambda}{2G(3\lambda + 2G)}.$$

Lamé's constants can be expressed in familiar physical quantities. From eqs. (6) we see that G is just the modulus of rigidity, the ratio of shear stress to shear strain and that λ and G are related to K_V, the bulk modulus, which is defined as the ratio of hydrostatic pressure to the dilation it produces. Now $S_{11} = S_{22} = S_{33} = -p$, and the dilation is div $\boldsymbol{u}$,

$$K_V = -\frac{p}{\operatorname{div} \boldsymbol{u}} = \frac{S_{11} + S_{22} + S_{33}}{3(e_{11} + e_{22} + e_{33})} = \lambda + \frac{2}{3}G. \tag{8}$$

The coefficient of compressibility (Compare eq. (3), (5) is the reciprocal of the bulk modulus. Young's modulus E is defined as the ratio of tensile stress to the associated strain for an elastic solid rod, which is under axial tension and unrestricted laterally. Poisson's ratio p is defined as the ratio of lateral contractions to longitudinal extension.

By setting $S_{22} = S_{33} = 0$ in eqs. (7), we have

$$E = \frac{S_{11}}{e_{11}} = \frac{1}{A} = \frac{G(3\lambda + 2G)}{\lambda + G}, \tag{9}$$

$$p = -\frac{e_{22}}{e_{11}} = \frac{B}{A} = \frac{\lambda}{2(\lambda + G)}. \tag{10}$$

From eq. (5, 1), with no hydrostatic pressure and external force,

$$\rho \frac{\partial^2 u_i}{\partial t^2} = \frac{\partial S_{ij}}{\partial x_j}. \tag{11}$$

Substituting eq. (3) in eq. (11),

$$\rho \frac{\partial^2 u_i}{\partial t^2} = \mathscr{C}_{ijlk} \frac{\partial^2 u_k}{\partial x_j \, \partial x_l} = \mathscr{C}_{ijkl} \frac{\partial^2 u_l}{\partial x_j \, \partial x_k}. \tag{12}$$

Substituting eq. (30, 1) in eq. (12),

$$-\rho\omega^2 u_i = -\mathscr{C}_{ijkl} K_j K_k u_l, \tag{13}$$

the relation

$$\frac{\partial(\exp iK_l x_l)}{\partial x_j} = iK_j \exp i(\boldsymbol{K} \cdot \boldsymbol{x}) \tag{14}$$

having been used.

Or

$$\left(\mathscr{C}_{ijkl} \frac{K_j K_k}{\rho} - \omega^2 \delta_{il}\right) u_i = 0, \tag{15}$$

Or

$$\left(\mathscr{C}_{ijkl}\frac{n_j n_k}{\rho} - v^2\delta_{il}\right)u_i = 0, \tag{16}$$

where $\boldsymbol{n}$ is the wave normal and

$$\boldsymbol{K} = k\boldsymbol{n}. \tag{17}$$

In condensed form

$$(\Lambda - v^2\overleftrightarrow{1}) \cdot \boldsymbol{u} = 0, \tag{18}$$

where Λ is the reduced elastic coefficient tensor ($\mathscr{C}$ divided by ρ) contracted along two directions of the wave normal. Equation (18) is Christoffel's equation. It shows that $\boldsymbol{u}$ is an eigenvector of the tensor Λ and v^2 is an eigenvalue. Solving for a given Λ, we can find the directions of a displacement vector (magnitude uncertain with a constant factor), and the corresponding phase velocities of the wave that propagates in the solid.

To arrange Λ in a matrix, we first change the matrix of elastic coefficients from three-dimensional to six-dimensional subscripts, i.e.,

$$\begin{gathered}(11) \rightleftarrows (1),\ (22) \rightleftarrows (2),\ (33) \rightleftarrows (3),\\ (23) = (32) \rightleftarrows (4),\ (31) = (13) \rightleftarrows (5),\\ (12) = (21) \rightleftarrows (6).\end{gathered} \tag{19}$$

The contraction is affected by

$$\begin{aligned}\Lambda_{il} = {} & \mathscr{C}_{i11l}\frac{n_1^2}{\rho} + \mathscr{C}_{i22l}\frac{n_2^2}{\rho} + \mathscr{C}_{i33l}\frac{n_3^2}{\rho} + \frac{1}{\rho}(\mathscr{C}_{i23l}n_2 n_3 \\ & + \mathscr{C}_{i32l}n_3 n_2) + \frac{1}{\rho}(\mathscr{C}_{i31l}n_3 n_1 + \mathscr{C}_{i13l}n_1 n_3) \\ & + \frac{1}{\rho}(\mathscr{C}_{i12l}n_1 n_2 + \mathscr{C}_{i21l}n_2 n_1) \\ & \text{for } i, l = 11, 22, 33, 23, 31, 12.\end{aligned} \tag{20}$$

Specifically for an isotropic solid, the corresponding equations for eqs. (6) are

$$\begin{bmatrix} S_1 \\ S_2 \\ S_3 \\ S_4 \\ S_5 \\ S_6 \end{bmatrix} = \begin{bmatrix} \lambda+2G & \lambda & \lambda & 0 & 0 & 0 \\ \lambda & \lambda+2G & \lambda & 0 & 0 & 0 \\ \lambda & \lambda & \lambda+2G & 0 & 0 & 0 \\ 0 & 0 & 0 & G & 0 & 0 \\ 0 & 0 & 0 & 0 & G & 0 \\ 0 & 0 & 0 & 0 & 0 & G \end{bmatrix} \begin{bmatrix} e_1 \\ e_2 \\ e_3 \\ e_4 \\ e_5 \\ e_6 \end{bmatrix}, \tag{21}$$

and the Λ matrix in eqs. (20) is

$$\frac{1}{\rho}\begin{bmatrix} \lambda+4G & \lambda+G & \lambda+G \\ \lambda+G & \lambda+4G & \lambda+G \\ \lambda+G & \lambda+G & \lambda+4G \end{bmatrix}. \tag{22}$$

Then

$$\frac{1}{3}\begin{bmatrix} 0 & \frac{1}{\sqrt{2}} & -\frac{1}{\sqrt{2}} \\ -\frac{2}{\sqrt{6}} & \frac{1}{\sqrt{6}} & \frac{1}{\sqrt{6}} \\ \frac{1}{\sqrt{3}} & \frac{1}{\sqrt{3}} & \frac{1}{\sqrt{3}} \end{bmatrix} (\Lambda) \begin{bmatrix} 0 & -\frac{2}{\sqrt{6}} & \frac{1}{\sqrt{3}} \\ \frac{1}{\sqrt{2}} & \frac{1}{\sqrt{6}} & \frac{1}{\sqrt{3}} \\ -\frac{1}{\sqrt{2}} & \frac{1}{\sqrt{6}} & \frac{1}{\sqrt{3}} \end{bmatrix}$$

$$= \frac{1}{\rho}\begin{bmatrix} G & 0 & 0 \\ 0 & G & 0 \\ 0 & 0 & \lambda+2G \end{bmatrix}. \tag{23}$$

We see that two of eigenvalues of Λ are equal, G/ρ, and the third eigenvalue is $(\lambda+2G)/\rho$. Thus for an isotropic solid any direction along the wave normal is longitudinal. Any two directions in the plane perpendicular to the longitudinal direction are transverse. The longitudinal displacement is associated with the faster speed $\sqrt{(\lambda+2G)/\rho}$. The transverse displacements are associated with the slower speed $\sqrt{G/\rho}$. The longitudinal waves are pressure or primary waves and transverse waves are shear or secondary waves. Because $\boldsymbol{K}$ is real and positive, the phase velocities are independent of frequencies. There is no dispersion nor dissipation.

Fedorov[1] has treated the propagation of sound in anisotropic solids.

Reference

[1] A. F. I. Fedorov (translated by J. E. S. Bradley): Theory of Elastic Waves in Crystals, Plenum Press, New York (1968).

§35 *PLASMAS*

A plasma is a nearly neutral fluid (Compared with fluids, electrons in solids cannot move great distances without making collisions) composed of free electrons, charged ions and molecules. At an elementary level, plasmas are familiar to everyone. The glowing vapor of burning fuels is a plasma; however, the temperature limit of combustion is about 5000°K. The gas in the gap between the electrodes of an electric arc and "high-energy" gas achieved by magnetohydrodynamic effects are also plasmas.

Although the difference is not sharp, it is useful to distinguish magnetodynamics and plasma physics. At low frequencies the fluid moves as a whole. A one-fluid model is adequate to describe magnetodynamics. At high frequencies there is charge separation and two-fluid model is necessary to describe plasma physics.

Magnetodynamics

Consider a neutral, conducting, compressible and non-viscous fluid in the presence of a uni-directional magnetic field. For non-separating charges,

$$\rho Z = 0. \tag{1}$$

This is equivalent to

$$\frac{\partial \boldsymbol{D}}{\partial t} = 0 \tag{2}$$

and

$$\operatorname{div} \boldsymbol{I} = 0. \tag{3}$$

The Maxwell equations are reduced to

$$\operatorname{curl} \boldsymbol{E} = -\frac{\partial \boldsymbol{B}}{\partial t} \tag{4}$$

and

$$\frac{1}{\mu} \operatorname{curl} \boldsymbol{B} = \boldsymbol{I}. \tag{5}$$

Because the fluid is compressible,

$$\frac{\partial \rho}{\partial t} + \rho \operatorname{div} \boldsymbol{v} = 0. \tag{6}$$

Because the fluid is non-viscous, from eq. (5, 1),

$$\rho \frac{\partial \boldsymbol{v}}{\partial t} + \operatorname{grad} p + \boldsymbol{B} \times \boldsymbol{I} = 0. \tag{7}$$

In writing eqs. (6) and (7) we have neglected the difference between the substantial and partial derivatives. Substituting eq. (5) in eq. (7),

$$\rho \frac{\partial \boldsymbol{v}}{\partial t} + C_0^2 \operatorname{grad} \rho + \frac{1}{\mu} \boldsymbol{B} \times (\operatorname{curl} \boldsymbol{B}) = 0, \tag{8}$$

where

$$C_0^2 = \frac{dp}{d\rho}$$

is the square of velocity of sound. Although, because of eq. (1), the charge flux $\boldsymbol{I}$ on the frame at rest is the same as that on the moving frame, $\boldsymbol{E}'$ on the moving frame is different from $\boldsymbol{E}$ on the frame at rest. From eq. (7, 43),

$$\boldsymbol{E} = \boldsymbol{E}' - \boldsymbol{v} \times \boldsymbol{B} = r\boldsymbol{I} - \boldsymbol{v} \times \boldsymbol{B}, \tag{9}$$

where r is the resistivity. Because the fluid is conducting, r

is small and may be neglected. Equation (4) becomes

$$\frac{\partial \boldsymbol{B}}{\partial t} - \operatorname{curl}(\boldsymbol{v} \times \boldsymbol{B}) = 0. \tag{10}$$

Equations (6), (8) and (10) can be combined as

$$\rho \frac{\partial^2 \boldsymbol{v}}{\partial t^2} - \rho C_0^2 \operatorname{grad} \operatorname{div} \boldsymbol{v} + \frac{1}{\mu} \boldsymbol{B} \times \operatorname{curl} \operatorname{curl}(\boldsymbol{v} \times \boldsymbol{B}) = 0. \tag{11}$$

Similar to C_0^2, we introduce a Alfrén velocity vector,

$$\boldsymbol{v}_A = \frac{\boldsymbol{B}}{\sqrt{\rho\mu}}. \tag{12}$$

Equation (11) becomes

$$\frac{\partial^2 \boldsymbol{v}}{\partial t^2} - C_0^2 \operatorname{grad} \operatorname{div} \boldsymbol{v} + \boldsymbol{v}_A \times \operatorname{curl} \operatorname{curl}(\boldsymbol{v} \times \boldsymbol{v}_A) = 0. \tag{13}$$

Substituting eq. (30, 1) in eq. (13), and noting

$$\begin{aligned} \boldsymbol{v}_A \times \operatorname{curl} \operatorname{curl}(\boldsymbol{v} \times \boldsymbol{v}_A) &= \operatorname{grad}[(\boldsymbol{v}_A \cdot \operatorname{grad})(\boldsymbol{v} \cdot \boldsymbol{v}_A)] \\ &- \operatorname{grad}[v_A^2(\operatorname{div} \boldsymbol{v})] - (\boldsymbol{v}_A \cdot \operatorname{grad})[(\boldsymbol{v}_A \cdot \operatorname{grad})\boldsymbol{v}] \\ &+ (\boldsymbol{v}_A \cdot \operatorname{grad})\, \boldsymbol{v}_A(\operatorname{div} \boldsymbol{v}), \end{aligned}$$

we have

$$\begin{aligned} -\omega^2 \boldsymbol{v} + (C_0^2 + v_A^2)\boldsymbol{K}(\boldsymbol{K} \cdot \boldsymbol{v}) + (\boldsymbol{v}_A \cdot \boldsymbol{K})[(\boldsymbol{v}_A \cdot \boldsymbol{K})\boldsymbol{v} \\ - (\boldsymbol{v} \cdot \boldsymbol{v}_A)\boldsymbol{K} - (\boldsymbol{K} \cdot \boldsymbol{v})\boldsymbol{v}_A] = 0. \end{aligned} \tag{14}$$

If $\boldsymbol{v}_A \cdot \boldsymbol{K} = 0$,

$$\frac{\omega^2}{K^2} = C_0^2 + v_A^2. \tag{15}$$

This is the magnetosonic wave with phase velocity $\sqrt{C_0^2 + v_A^2}$.

If $\boldsymbol{K} \| \boldsymbol{v}_A \| \boldsymbol{v}$,

$$\frac{\omega^2}{K^2} = C_0^2. \tag{16}$$

This is the ordinary longitudinal sonic wave with phase velocity C_0.

If $\boldsymbol{K} \| \boldsymbol{v}_A$, and if $\boldsymbol{v}_A \cdot \boldsymbol{v} = 0$,

$$\frac{\omega^2}{K^2} = v_A^2. \tag{17}$$

This is the transverse Alfrén wave with phase velocity $\boldsymbol{v}_A$.

So far we have elucidated the phase velocities of the three waves parallel and perpendicular to the magnetic field. In general, let θ be the inclination of any direction with respect to the magnetic field. For the Alfrén wave,

$$v_3^2 = v_A^2 \cos^2\theta. \tag{18}$$

For the fast or the slow wave,

$$v_1^2 \text{ or } v_2^2 = \frac{1}{2}(C_0^2 + v_A^2) \pm \frac{1}{2} \cdot \sqrt{(C_0^2 + v_A^2)^2 - 4C_0^2 v_A^2 \cos^2\theta}. \tag{19}$$

Thus in any direction the velocity of the fast wave always exceeds that of the others. In laboratory, the velocity of the Alfrén wave is usually less than that of the slow wave; but in astronomical bodies the reverse might be true. We could represent eqs. (18) and (19) with three surfaces such that the distance from the origin is equal to the velocity of the wave in that direction. The fast wave is represented by a surface like an ellipse and the others are like lemniscates.

Because we have neglected viscosity and resistivity, the three waves are dispersive but not dissipative. Otherwise they are both dispersive and dissipative. Take Alfrén waves for example, from eq. (32, 25),

$$\zeta^2 = \frac{1 + iX}{1 + X^2} \approx 1 + iX = 1 + \frac{i4\eta\omega}{3\rho C_0^2} + \frac{i\omega r}{C_0^2 \mu}. \tag{20}$$

Thus the attenuation increases with increasing frequency and also increases with increasing viscosity and resistivity.

At higher frequencies charge separations occur. These separations produce coulombian forces and oscillations of an electrostatic nature are set up. Because of the charge separations,

$$\rho Z \neq 0.$$

From one of Maxwell's equations,

$$\varepsilon \operatorname{div} \boldsymbol{E} - \rho Z = 0. \tag{21}$$

From eq. (4, 46),

$$\frac{\partial \rho}{\partial t} + \rho \operatorname{div} \boldsymbol{v} = 0. \tag{22}$$

From eq. (5, 1),

$$\rho \frac{\partial \boldsymbol{v}}{\partial t} + \operatorname{grad} p - \rho Z \boldsymbol{E} - \boldsymbol{I} \times \boldsymbol{B} = 0. \tag{23}$$

The separation of charges means that charge flux and electric field are out of phase. Conductivity or resistivity in the ordinary sense breaks down. Also, so long as magnetic field is time-invariant or fluctuations small, the last term on the left side of eq. (23) can be neglected. Or

$$\frac{\partial \boldsymbol{v}}{\partial t} + \frac{1}{\rho}\left(\frac{\partial p}{\partial \rho}\right)_s \operatorname{grad} \rho - Z\boldsymbol{E} = 0. \tag{24}$$

For positive ions,

$$\left(\frac{\partial p}{\partial \rho}\right)_s = C_0^2,$$

where C_0 is the velocity of sound. For a degenerate electron gas,

$$\left(\frac{\partial p}{\partial \rho}\right)_s = 3C'^2,$$

where C' is the root-mean-square velocity. We may say that the adiabatic exponent γ is 3 instead of 5/3 for monoatomic gases. However, γ is no longer the ratio of specific heats, but is interpreted in terms of energy according to Fermi statistics.

So far with the assumed solutions of $\exp i(\boldsymbol{K}\cdot\boldsymbol{x}-\omega t)$, we have considered $\boldsymbol{K}$ to be complex. There is a spatial attenuation. In this example it is convenient to assume that $\boldsymbol{K}$ is real but ω is complex, i.e.,

$$\exp i[\boldsymbol{K}\cdot\boldsymbol{x}-(\omega_r - ia)t].$$

The imaginary part of ω corresponds to absorption, or the attenuation in amplitude. This may be called temporal attenuation. Or

$$\boldsymbol{K}\cdot\boldsymbol{K} = k^2, \tag{25}$$

and

$$\zeta^2 = \frac{(\omega_r - ia)^2}{\omega_0^2}. \tag{26}$$

With these assumed solutions, eqs. (21), (22), and (24) become

$$\begin{bmatrix} -i\rho\boldsymbol{K} & 0 & (\omega_r - ia)i \\ 0 & -i\varepsilon\boldsymbol{K} & Z \\ (\omega_r - ia)i & Z & -\dfrac{1}{\rho}\left(\dfrac{\partial p}{\partial \rho}\right)_s i\boldsymbol{K} \end{bmatrix} \begin{bmatrix} \boldsymbol{v}_0 \\ \boldsymbol{E}_0 \\ \rho_0 \end{bmatrix} = 0. \tag{27}$$

For non-trivial solutions, the matrix must be singular,

$$\varepsilon\left(\frac{\partial p}{\partial \rho}\right)_s k^2 - \varepsilon(\omega_r - ia)^2 + \rho Z^2 = 0. \tag{28}$$

For $\left(\dfrac{\partial p}{\partial \rho}\right)_s = 0$,

$$\omega_0^2 = \frac{\rho Z^2}{\varepsilon}. \tag{29}$$

Thus pressure, velocity and electric field oscillate with the same frequency. This means that there is no effect of magnetic field which influences the pressure.

For $\left(\dfrac{\partial p}{\partial \rho}\right)_s = C_0^2$,

$$(\omega_r - ia)^2 = \omega_0^2 + C_0^2 k^2. \tag{30}$$

Or

$$\zeta^2 = 1 + \frac{C_0^2 k^2}{\omega_0^2}. \tag{31}$$

These are the transverse waves (two).

For $\dfrac{\partial p}{\partial \rho} = 3C'^2$,

$$(\omega_r - ia)^2 = 3C'^2 k^2 + \omega_0^2, \tag{32}$$

Or

$$\zeta^2 = 1 + \frac{3C'^2 k^2}{\omega_0^2}. \tag{33}$$

This is the longitudinal wave. We could add the complex term $i4\eta\omega/3\rho C_0^2$ to ζ^2 to account for viscosity.

§36 *RAYLEIGH WAVES*

The body waves in an elastic solid are non-attenuated and non-dissipative and the surface waves in a viscoelastic solid are dispersive and dissipative.

As is shown in §34, in an infinite homogeneous (isotropic) elastic medium, the primary and the secondary waves are the only waves which appear. In the semi-infinite elastic medium, the linear combination of the independent solutions to satisfy the boundary condition of zero stress gives surface waves known as Rayleigh waves. In simple Rayleigh waves, the waves propagate in one direction but the amplitude diminishes exponentially in a plane perpendicular to this direction. There are also simple Rayleigh waves in an inhomogeneous (isotropic) viscoelastic medium; for example, transmitted Rayleigh waves in the earth's crust from epicentrum in the interior of the earth.

For simple Rayleigh wave, the solution is assumed to be

$$u_i = u_i \exp i[(kx - iaz) - \omega t]. \tag{1}$$

In words, the x-axis is taken as the direction of propagation and the x-axis is taken along a normal to the surface so that $Z = 0$ is the surface and $Z < 0$ is the isotropic medium; a is the absorption coefficient. In contrast to eq. (30, 1), $\boldsymbol{K}$ is a constant complex wave propagation vector with

$$K_1 = k, K_2 = 0 \text{ and } K_3 = -ia. \tag{2}$$

(If K_2 and K_3 are complex instead of zero and pure imaginary, we are dealing with generalized Rayleigh waves.)

From eq. (34, 20), eq. (34, 15) is written in full as

$$\left.\begin{aligned} -\rho\omega^2 u_1 + (\mathscr{C}_{1111}K_1^2 + \mathscr{C}_{1331}K_3^2)u_1 + (\mathscr{C}_{1313} + \mathscr{C}_{1133})K_1K_3u_3 = 0,\\ -\rho\omega^2 u_2 + (\mathscr{C}_{2112}K_1^2 + \mathscr{C}_{2332}K_3^2)u_2 = 0,\\ -\rho\omega^2 u_3 + (\mathscr{C}_{3311} + \mathscr{C}_{3131})K_1K_3u_1 + (\mathscr{C}_{3113}K_1^2 + \mathscr{C}_{3333}K_3^2)u_3 = 0. \end{aligned}\right\} \tag{3}$$

Substituting values of $\mathscr{C}$ from eqs. (34, 21) and values of K from eq. (2) in eq. (3), eq. (3) in matrix form is

$$(\Lambda - \omega^2 \overleftrightarrow{1}) \cdot \boldsymbol{u} = 0, \tag{4}$$

where

$$\Lambda = \frac{1}{\rho}\begin{bmatrix} (\lambda+2G)k^2 - Ga^2 & 0 & -iak(\lambda+G) \\ 0 & G(k^2-a^2) & 0 \\ -iak(\lambda+G) & 0 & -(\lambda+2G)a^2 + Gk^2 \end{bmatrix}. \tag{5}$$

Because Λ's eigenvalues are

$$\frac{G}{\rho}(k^2 - a^2)\,(\text{double}),$$

and

$$\left[\frac{(\lambda+2G)}{\rho}\right](k^2 - a^2),$$

$$\omega^2 = \left[\frac{(\lambda+2G)}{\rho}\right](k^2 - a_\alpha^2) = \alpha^2(k^2 - a_\alpha^2), \tag{6}$$

or

$$\omega^2 = \frac{G}{\rho}(k^2 - a_\beta^2) = \beta^2(k^2 - a_\beta^2), \tag{7}$$

where α and β are phase velocities of the primary and the secondary waves respectively. If $\gamma = \omega/k$ is the phase velocity of the Rayleigh waves,

$$\gamma^2 = \alpha^2\left(1 - \frac{a_\alpha^2}{k^2}\right), \tag{8}$$

or

$$\gamma^2 = \beta^2\left(1 - \frac{a_\beta^2}{k^2}\right). \tag{9}$$

The solutions of eq. (4), after using eqs. (6) and (7), are

$$\frac{u_1}{u_3} = \frac{ia_\beta}{k} \text{ or } \frac{ik}{a_\alpha}. \tag{10}$$

The general solution is

$$u_1 = ikf_1T_\alpha + ia_\beta f_2T_\beta,$$
$$u_3 = a_\alpha f_1T_\alpha + kf_2T_\beta, \tag{11}$$

where

$$T_\alpha = \exp(a_\alpha z + ikx - i\omega t)$$
$$T_\beta = \exp(a_\beta z + ikx - i\omega t)$$

and f_1 and f_2 are constants to be determined by the boundary conditions, namely $Z = 0$ is a stress-free surface. Or

$$S_3 = S_5 = 0. \tag{12}$$

By using eqs. (34, 21) and (11), the conditions of eq. (12) are

$$\begin{bmatrix} 2 - \dfrac{\gamma^2}{\beta^2} & 2\sqrt{1 - \dfrac{\gamma^2}{\beta^2}} \\ 2\sqrt{1 - \dfrac{\gamma^2}{\alpha^2}} & 2 - \dfrac{\gamma^2}{\beta^2} \end{bmatrix} \begin{bmatrix} f_1 \\ f_2 \end{bmatrix} = 0. \tag{13}$$

Or, for non-trivial solutions,

$$\left(2 - \frac{\gamma^2}{\beta^2}\right)^2 = 4\sqrt{1 - \frac{\gamma^2}{\alpha^2}}\sqrt{1 - \frac{\gamma^2}{\beta^2}}. \tag{14}$$

In changing from a non-attenuated to an attenuated medium, we change $v^2 = \frac{\mathscr{C}}{\rho}$ into

$$v^2 = \frac{\mathscr{C}}{\rho}\left(1 - \frac{a^2}{k^2}\right) \tag{15}$$

from eqs. (8) and (9). Thus we see that the wave energy of Rayleigh waves comes from that of the primary and secondary waves. From eq. (11) we also see that they are so proportioned that no energy is stored at the free surface. Equation (14) is such a criterion.

In a non-attenuated medium the phase velocity is the same as the energy velocity, i.e., energy density times phase velocity is equal to energy flux. In an attenuated medium, energy velocity is no longer equal to phase velocity, but is equal to group velocity. Group velocity was originally proposed for the superposition of numerous waves with similar frequency.

The velocity at the direction of propagation, u_i, after eq. (1), is $-i\omega u_i$. The strain $e_{ij} = \frac{1}{2}\left(\frac{\partial u_i}{\partial x_j} + \frac{\partial u_j}{\partial x_i}\right)$ is $ik_j u_i$ or $ik_i u_j$. The kinetic energy $\frac{1}{2}\rho(\dot{u}_i)^2$ is $-\frac{1}{2}\rho\omega^2 u_i^2$. The stored potential energy

$$\phi = \frac{1}{2} S:e = \frac{1}{2}\mathscr{C}_{ijkl} e_{lk} e_{ij} \text{ is } -\frac{1}{2}\mathscr{C}_{ijkl} k_j u_i k_k u_l.$$

From eq. (34, 13),

$$\rho\omega^2 u_i = \mathscr{C}_{ijkl} k_j k_k u_l. \tag{16}$$

Thus the kinetic energy is equal to the potential energy and the total energy density is

$$E = -\rho\omega^2 u_i^2 = -\rho\omega^2. \tag{17}$$

The energy flux is $S_{ij}\dot{u}_j = \mathscr{C}_{ijkl} e_{kl} \dot{u}_j$; after using eq. (1), it is

$$\omega\mathscr{C}_{ijkl} k_k u_l u_j.$$

The ratio of energy flux to energy density in absolute value is

$$\frac{\mathscr{C}_{ijkl}k_k u_l u_j}{\rho\omega} = \frac{\mathscr{C}_{jikl}n_k u_l u_j}{\rho}\frac{k}{\omega}$$

$$= \frac{v^2}{n_i}\frac{1}{v} = \frac{v}{n_i} = g_i, \tag{18}$$

where $\boldsymbol{n}$ is the wave normal and $\boldsymbol{g}$ is the group velocity. In condensed form,

$$\boldsymbol{g}\cdot\boldsymbol{n} = v. \tag{19}$$

According to eq. (18), group velocity is the ratio of the energy flux normal to the direction of propagation and the average energy density, which is what we mean by energy velocity in a non-attenuated medium. Also we see that the group velocity cannot be less than the phase velocity. They are equal if the two are in the same direction.

Group velocity can be defined in another way.

From eq. (34, 13),

$$\rho\omega^2 = \mathscr{C}_{ijkl}k_j k_k u_l u_i.$$

Differentiating with respect to one component of $\boldsymbol{K}$,

$$2\rho\omega\frac{\partial\omega}{\partial k_k} = 2\mathscr{C}_{ijkl}k_j u_l u_i.$$

Or

$$\frac{\partial\omega}{\partial k_k} = \frac{\mathscr{C}_{ijkl}n_j u_l u_i}{\rho\dfrac{\omega}{k}} = \frac{v}{n_k}.$$

In condensed form,

$$\frac{\partial\omega}{\partial\boldsymbol{K}} = \boldsymbol{g}. \tag{20}$$

Because we have assumed a small attenuation, we only use the weak version of eq. (20), i.e., the linear relationship between

ω and k.

In changing from an elastic to a viscoelastic medium, we have to decide the cause of the dissipation. If the dissipation is from viscosity (internal friction) alone,
from eq. (32, 25),

$$\zeta^2 = \frac{1 + iX}{1 + X^2}. \tag{21}$$

If the dissipation is small and if we neglect volume viscosity,

$$\zeta^2 \approx 1 + iX = 1 + \frac{i\omega \frac{4\eta}{3}}{\rho C_0^2} = 1 + \frac{1}{\mathscr{C}} i\omega \frac{4\eta}{3} = 1 + \frac{i}{Q}, \tag{22}$$

where

$$Q = \frac{3\mathscr{C}}{4\eta\omega} \tag{23}$$

is called the dissipation coefficient. Because

$$\zeta = \frac{k - ia}{k_0} \approx 1 + \frac{i}{2Q}, \tag{24}$$

equalizing the imaginary parts of eq. (24) yields

$$\|Q\| = \frac{\omega}{2a}\left(\frac{\mathscr{C}}{\rho}\right)^{\frac{1}{2}}. \tag{25}$$

We have superposed dissipation with dispersion. This would not be possible if we had not assumed small dispersion and dissipation and their linear relationship to frequency. From eq. (25) the dissipation coefficient Q[1,2], and therefore η in a viscoelastic medium, can be found from the phase velocity $(\mathscr{C}/\rho)^{1/2}$ and absorption coefficient a.

Because there are only two absorption coefficients in an elastic medium, the dissipation coefficient of the Rayleigh waves is to be obtained from the additional relation,

$$4\sqrt{\left[1-\frac{v_\gamma^2}{\alpha^2\left(1+\dfrac{i}{Q_\alpha}\right)}\right]\left[1-\frac{v_\gamma^2}{\beta^2\left(1+\dfrac{i}{Q_\beta}\right)}\right]}=\left[2-\frac{v_\gamma^2}{\beta^2\left(1+\dfrac{i}{Q_\beta}\right)}\right]^2, \tag{26}$$

where

$$v_\gamma=\gamma\left(1+\frac{i}{Q_\gamma}\right).$$

Equation (26) corresponds to eq. (14) in the elastic medium· In changing from a non-dissipative to a dissipative medium, we change real v to complex v. The result[1] is

$$a_\gamma=a_\beta\frac{B}{\gamma}\frac{\beta}{A+B}+a_\alpha\frac{\alpha}{\gamma}\frac{A}{(A+B)}, \tag{27}$$

where

$$A=4\left(1-\frac{\gamma^2}{\beta^2}\right)\frac{\beta^2}{\alpha^2},$$

$$B=4\left(1-\frac{\gamma^2}{\alpha^2}\right)-\left(2-\frac{\gamma^2}{\beta^2}\right)^3.$$

References

[1] F. Press and J. Healy: *J. Applied Phys.*, **28**, 1323 (1957).
[2] J. Lamb and J. Richter: *Proc. Royal Soc.* (London), A **293**, 479 (1966).

§37 *SHOCK WAVES*

We all know that shocks are due to explosions or motions of supersonic bodies, but the conditions of a shock wave are more stringent. There is an abrupt change of pressure or stress in a material medium. The abrupt change occurs at the shock front and the velocity of propagation of the shock is greater than that of sound in the material. At any moment immediately ahead of the front, the material is undisturbed. At the front the additional velocity acquired at the shock, added to the velocity of propagation behind the front permits the rarefaction phase to overtake the front. The shock front is the compressive phase of the shock wave. Behind the front is the rarefaction phase, where the pressure drops rapidly from the peak to its pre-shock ambient value. As time evolves, the entire wave simultaneously lengthens and decreases in amplitude, which causes its final decay.

Mathematically, the shock wave is characterized by nonlinearity and discontinuity. Because the wave equation is no longer linear, due to its nonlinear coefficients, the velocity of propagation of wave is no longer constant. Furthermore, the coefficients are functions of the initial conditions. The smooth continuous conditions may lead to shock discontinuity. This is possible because the particular solution for the wave equation cannot be found by superposition of the general solutions.

To simplify the situation, we assume that, at any moment, the irreversibility is concentrated in a narrow region perpendicular to the direction of the propagation.

From eqs. (8, 9) and (8, 21), in the irreversible region,

$$\rho \frac{dS}{dt} = -\operatorname{div} \boldsymbol{J}_S - \frac{1}{T}(\boldsymbol{J}_S \cdot \operatorname{grad} T) - \frac{1}{T}(P{:}\operatorname{Grad} \boldsymbol{v}). \quad (1)$$

Outside the region equation (4, 12) reads

$$\frac{\partial \rho}{\partial t} = -\operatorname{div} \rho \boldsymbol{v}. \quad (2)$$

Equation (5, 1) reads

$$\frac{\partial \rho \boldsymbol{v}}{\partial t} = -\operatorname{Div} \rho \boldsymbol{v}\boldsymbol{v} - \operatorname{Div} p\overleftrightarrow{1}. \quad (3)$$

Equation (6, 20) reads

$$\frac{\partial \rho U}{\partial t} = -\operatorname{div} \rho U \boldsymbol{v} - p \operatorname{div} \boldsymbol{v}. \tag{4}$$

Equation (6, 4) reads

$$\frac{\partial \frac{1}{2} \rho v^2}{\partial t} = -\operatorname{div} \frac{1}{2} \rho v^2 \boldsymbol{v} - \operatorname{div} p\boldsymbol{v} + p \operatorname{div} \boldsymbol{v}. \tag{5}$$

We attach a moving coordinate system to the shock front. Then the shock front is stationary. If the subscripts 0 and 1 refer to the two sides of the front, the integration of eq. (2) yields

$$\rho_0 \boldsymbol{v}_0 = \rho_1 \boldsymbol{v}_1, \tag{6}$$

the integration of eq. (3) yields

$$p_0 + \rho_0 v_0^2 = p_1 + \rho_1 v_1^2, \tag{7}$$

and the integration of the sum of eqs. (4) and (5) yields

$$U_0 + \frac{1}{2} v_0^2 + p_0 V_0 = U_1 + \frac{1}{2} v_1^2 + p_1 V_1. \tag{8}$$

Equation (1) becomes

$$\operatorname{div} \boldsymbol{J}_s = -\frac{(\boldsymbol{J}_s \cdot \operatorname{grad} T)}{T} - \frac{(P : \operatorname{Grad} \boldsymbol{v})}{T}. \tag{9}$$

Equations (6) and (7) can be combined as

$$\frac{p_1 - p_0}{\boldsymbol{v}_0 - \boldsymbol{v}_1} = \rho_0 \boldsymbol{v}_0 = \rho_1 \boldsymbol{v}_1, \tag{10}$$

or

$$\frac{p_1 - p_0}{V_0 - V_1} = \rho_0^2 v_0^2 = \rho_1^2 v_1^2, \tag{11}$$

or

$$\frac{p_1 - p_0}{\rho_1 - \rho_0} = \boldsymbol{v}_0 \cdot \boldsymbol{v}_1. \tag{12}$$

Equations (6—8) can be combined as

$$U_1 - U_0 = \frac{1}{2} (p_0 + p_1)(V_0 - V_1), \tag{13}$$

which is the Rankine-Hugoniot relation. It is significant that this relation contains no thermodynamical variables involving irreversibility.

The shock strength is defined as $(p_1 - p_0)/p_0$. As $p_1 \to p_0$, from eq. (10), $\boldsymbol{v}_0 \to \boldsymbol{v}_1$, and from eqs. (12) and (30, 16),

$$(p_1 - p_0)/(\rho_1 - \rho_0) \to C_0^2,$$

where C_0 is the sonic speed. Thus a sound wave can be interpreted as an infinitely weak shock wave. Also

$$\boldsymbol{v}_0 = \boldsymbol{u}, \tag{14}$$

$$\boldsymbol{v}_1 = \boldsymbol{D} + \boldsymbol{u}, \tag{15}$$

where $\boldsymbol{D}$ is the additional velocity acquired at the front and $\boldsymbol{u}$ is the velocity behind the front. Thus only relative velocities are involved according to the principle of the Galilean relativity.

If the Rankine-Hugoniot relation is combined with the equation of state of the medium, unique relations of p, V and T are obtained. For example, in polytropic gases, we assume that the ideal gas law is valid and there is a linear relation between internal energy and temperature. Thus

$$Tp^{\frac{1-\gamma}{\gamma}} = \text{constant}, \tag{16}$$

and

$$U = C_V T = \frac{pV}{\gamma - 1}. \tag{17}$$

Using the Rankine-Hugoniot relation,

$$\mathscr{R}(pV) = U_1 - U_0 - \frac{1}{2}(p_0 + p_1)(v_0 - v_1)$$

and eq. (17), we have

$$\frac{2(\gamma - 1)\mathscr{R}}{\gamma + 1} = \left[V_1 - \frac{(\gamma - 1)V_0}{\gamma + 1}\right] p_1$$

$$-\left[V_0 - \frac{(\gamma-1)V_1}{\gamma+1}\right]p_0; \tag{18}$$

with $\mathscr{R}=0$, we have the Rankine-Hugoniot curve, a rectangular hyperbola with the center (p_0, V_0).

The Mach number is defined as $\|v_0\|/C_0$.

For polytropic gases,

$$M_0 = \frac{\|v_0\|}{\sqrt{\gamma p_0 V_0}}. \tag{19}$$

From eq. (18) with $\mathscr{R}=0$, and eq. (19),

$$\frac{p_1 - p_0}{p_0} = \frac{2\gamma(M_0^2 - 1)}{\gamma+1}. \tag{20}$$

Also

$$\frac{T_1}{T_0} = \frac{p_1 V_1}{p_0 V_0}. \tag{21}$$

In the p–V diagram, the Rankine-Hugoniot curve is just like an isothermal or an adiabatic curve. It is the path through which parameters like p, V, ρ, etc. can be reached by shock compression. The Rankine-Hugoniot curve differs from the isothermal curve in that the transition region is viscous and heat-conducting, and it differs from the adiabatic curve in that heat moves out of the transition region. Starting from the same initial state, the Rankine-Hugoniot and the adiabatic curve have the same slope and curvature, but at higher pressures the former lies above the latter. Provided that the shock strength is not too great, the average temperature and the average pressure can be approximated by $\frac{1}{2}(T_1 + T_0)$ and $\frac{1}{2}(p_1 + p_0)$ respectively. The input of entropy to compensate for the entropy-production in the transition region is expressed by eq. (9). The entropy-production can be actually found by dividing the dissipated energy by the average temperature. The dissipated energy can be found approximately by a graphical method. Draw a straight line connecting (p_1, V_1) and (p_0, V_0) in the p–V diagram. The difference between the area under the straight line and the area under the Rankine-Hugoniot curve is the dissipated energy. The evaluation is approximate because some stresses are left behind in the material even if the shock pressure has been relieved. For solids there is no equation of state; we can only make a crude estimation of the dissipated energy.

CHAPTER VIII

VECTORIAL PROCESSES

§38 *ISOTHERMAL DIFFUSION*

In considering a non-reactive, non-viscous and isothermal mixture with no external forces, eq. (8, 17) is reduced to

$$\sigma = -\frac{1}{T}\sum_{i=1}^{n}[\boldsymbol{J}_i|\operatorname{grad}_T \mu_i]. \tag{1}$$

Because diffusion is a slow process, mechanical equilibrium is reached first. According to (10,1), the mixture is also isobaric,

$$\sigma = -\frac{1}{T}\sum_{i=1}^{n}[\boldsymbol{J}_i|\operatorname{grad}_{Tp} \mu_i]. \tag{2}$$

From eq. (4,41),

$$\sum_{i=1}^{n}\boldsymbol{J}_i = 0. \tag{3}$$

From the Gibbs-Duhem relation at constant temperature and pressure,

$$\sum_{i=1}^{n} c_i \operatorname{grad}_{Tp} \mu_i = 0. \tag{4}$$

Thus both fluxes and forces are dependent. According to Theorem (4) of §18, the Onsager law is not necessarily held. By eliminating dimensions from n to $n-1$, the dependency is eliminated.

At mechanical equilibrium, according to the Prigogine theorem, §10, the reference velocity is immaterial. Thus there are different ways of expressing diffusion fluxes.

In the region of linearity, diffusion fluxes are proportional to the gradients of chemical potential. Experimentally it is easier to determine the gradients of concentration than the gradients of chemical potential. Then the proportional constants are called diffusion coefficients (Fick's first law). There are different ways of expressing concentrations. Thus there are different forms of diffusion equations. Some are simpler than others.

The diffusion fluxes can be expressed in partial mass or molar densities. The concentrations can also be expressed in mass units or moles. Frequently it is desirable to convert from one basis to the other.

Because the Onsager law does not hold for diffusion coefficients, it is necessary to convert diffusion coefficients into phenomenological coefficients to test the Onsager law. There is only one diffusion coefficient in binary diffusion. The Onsager law is not involved. The choice of one kind of diffusion coefficient is so convenient that it has been elevated to the position of a standard coefficient. The test of the Onsager law begins with ternary diffusion. The choice of certain kinds of diffusion coefficients is also convenient in simplifying the diffusion equations.

Conversion between Fluxes

In §4 we have defined four reference velocities. If we call one reference velocity $\boldsymbol{v}^a$ and another $\boldsymbol{v}^b$,

$$\begin{aligned} \boldsymbol{J}_i^a &= \rho_i(\boldsymbol{v}_i - \boldsymbol{v}^b) - \rho_i(\boldsymbol{v}^a - \boldsymbol{v}^b) \\ &= \boldsymbol{J}_i^b - \rho_i \sum_k^n \frac{a_k}{\rho_k} \rho_k(\boldsymbol{v}_k - \boldsymbol{v}^b) \\ &= \boldsymbol{J}_i^b - \rho_i \sum_k^n \frac{a_k}{\rho_k} \boldsymbol{J}_k^b, \end{aligned} \tag{5}$$

where eqs. (4,25) and (4,26) have been used.

Although the equation (5) is simple to use, the matrix $\delta_{ik} - \frac{c_i a_k}{c_k}$, $i, k = 1, 2, \cdots, n$, is singular; it is more convenient to use

$$J_i^a = B_{ik}^{ab} J_k^b, \tag{6}$$

where

$$B_{ik}^{ab} = \delta_{ik} + \frac{c_i}{c_k} \sum_k^{n-1} \left(\frac{b_k}{b_n} a_n - a_k \right). \tag{7}$$

The elegant formula (6) is due to de Groot and his co-workers[1] and it can be proved as follows:

$$\left[\delta_{ik} + \frac{c_i}{c_k} \sum_k^{n-1} \left(\frac{b_k}{b_n} a_n - a_k \right) \right] \rho_k (\boldsymbol{v}_k - \boldsymbol{v}^b)$$

$$= \rho_i (\boldsymbol{v}_i - \boldsymbol{v}^b) + \rho_i \sum_k^{n-1} \left(\frac{a_n}{b_n} b_k - a_k \right) \boldsymbol{v}_k$$

$$- \rho_i \sum_k^{n-1} \left(\frac{a_n}{b_n} b_k - a_k \right) \boldsymbol{v}^b = \rho_i \Big[(\boldsymbol{v}_i - \boldsymbol{v}^b)$$

$$+ \frac{a_n}{b_n} (\boldsymbol{v}^b - b_n \boldsymbol{v}_n) - (\boldsymbol{v}^a - a_n \boldsymbol{v}_n) - \boldsymbol{v}^b \frac{a_n}{b_n} (1 - b_n)$$

$$+ \boldsymbol{v}^b (1 - a_n) \Big] = \rho_i (\boldsymbol{v}_i - \boldsymbol{v}^a) = J_i^a.$$

The matrix in eq. (7) is not singular. As a matter of fact the reciprocal matrix is

$$(B_{ik}^{ab})^{-1} = \delta_{ik} + \frac{c_i}{c_k} \sum_k^{n-1} \left(\frac{a_k}{a_n} b_n - b_k \right), \; i, k = 1, 2, \cdots, (n-1). \tag{8}$$

Or

$$(B_{ik}^{ab})^{-1} = B_{ik}^{ba}. \tag{9}$$

Hereby we give five examples to illustrate the use of eq. (6). Example 1:

$$J_i^r = J_i^c - \frac{c_i}{c_n} J_n^c.$$

Example 2:

$$\boldsymbol{J}_i^0 = \boldsymbol{J}_i^c - \rho_i \sum_k^n V_k \boldsymbol{J}_k^c .$$

Example 3:

$$\boldsymbol{J}_i^0 = \boldsymbol{J}_i^r - \rho_i \sum_k^{n-1} V_k \boldsymbol{J}_k^r .$$

Example 4:

$$\boldsymbol{J}_i^n = \boldsymbol{J}_i^c - \frac{\rho_i}{N} \sum_k^n \frac{\boldsymbol{J}_k^c}{M_k} .$$

Example 5:

$$\boldsymbol{J}_i^r = \boldsymbol{J}_i^0 - \frac{\rho_i}{\rho_n} \boldsymbol{J}_n^0 .$$

Reduction of Dimensions from n to $(n-1)$

By substituting eqs. (3) and (4) in eq. (2) and using (4,28),

$$\begin{aligned} T\sigma &= -\sum_i^{n-1} [\boldsymbol{J}_i^a | \mathrm{grad}_{Tp}\, \mu_i] - [\boldsymbol{J}_n^a | \mathrm{grad}_{Tp}\, \mu_n] \\ &= -\sum_i^{n-1} [\boldsymbol{J}_i^a | \mathrm{grad}_{Tp}\, \mu_i] - \left[\sum_i^{n-1} \frac{a_i c_n}{a_n c_i} \boldsymbol{J}_i^a \Big| \sum_k^{n-1} \frac{c_k}{c_n} \mathrm{grad}_{Tp} \mu_k \right] \\ &= -\sum_i^{n-1} \left[\boldsymbol{J}_i^a \Big| \sum_k^{n-1} \left(\delta_{ik} + \frac{a_i c_k}{a_n c_i} \right) \mathrm{grad}_{Tp}\, \mu_k \right] \\ &= -\sum_{i,k}^{n-1} [\boldsymbol{J}_i^a | A_{ik}^a \, \mathrm{grad}_{Tp}\, \mu_k], \end{aligned} \tag{10}$$

where the "A" matrix is

$$A_{ik}^a = \left(\delta_{ik} + \frac{c_k a_i}{c_i a_n} \right), \quad i, k = 1, 2, \cdots, (n-1). \tag{11}$$

The reciprocal matrix is

$$(A_{ik}^a)^{-1} = \left(\delta_{ik} - \frac{c_k a_i}{c_i} \right). \tag{12}$$

The matrix A is also due to de Groot and his co-workers[1]. It is useful to decrease the dimensions from n to $(n-1)$. It is noted that A^r is the unit matrix.

Conversion among Concentrations

Ordinarily for concentrations we use mass fractions c_i, molar fractions n_i, partial density ρ_i, or partial molar density N_i with restrictions:

$$\sum_i^n c_i = 1, \quad \sum_i^n n_i = 1, \quad \sum_i^n \rho_i = \rho, \quad \sum_i^n N_i = N. \tag{13}$$

(For concentrations of aqueous solutions expressed in molalities, see §41.) The following partial differential coefficients of concentrations in $(n-1)$ dimensions can be derived:

$$\frac{\partial \rho_i}{\partial N_j} = M_i \frac{\partial N_i}{\partial N_j} = M_i \delta_{ij}, \tag{14}$$

$$\begin{aligned} \frac{\partial \rho_i}{\partial c_j} &= \frac{\partial \rho c_i}{\partial c_j} = \rho \delta_{ij} + c_i \frac{\partial \rho}{\partial c_j} \\ &= \rho \delta_{ij} - \frac{c_i}{V^2}\left(\frac{\partial V}{\partial c_j}\right) = \rho[\delta_{ij} + \rho_i(V_n - V_j)] \\ &= \rho B_{ij}^{0c}, \end{aligned} \tag{15}$$

$$\begin{aligned} \frac{\partial c_i}{\partial n_j} &= \frac{\partial \left(\dfrac{n_i M_i}{\sum_k^n n_k M_k} \right)}{\partial n_j} \\ &= \frac{c_i}{n_i} \frac{\partial n_i}{\partial n_j} - c_i \sum_k^n \frac{c_k}{n_k} \frac{\partial n_k}{\partial n_j} \\ &= \frac{N}{\rho} \frac{\rho_i}{N_i} \delta_{ij} - c_i \sum_k^{n-1} \left(\frac{c_k}{n_k} - \frac{c_n}{n_n} \right) \frac{\partial n_k}{\partial n_j} \\ &= \frac{N}{\rho} M_i \delta_{ij} - c_i \frac{N}{\rho} \sum_k^{n-1} \left(\frac{\rho_k}{N_k} - \frac{\rho_n}{N_n} \right) \frac{\partial n_k}{\partial n_j} \end{aligned}$$

$$= \frac{N}{\rho} \sum_{j}^{n-1} [M_j \delta_{ij} + c_i(M_n - M_j)]$$

$$= \frac{N}{\rho} \sum_{j}^{n-1} \left[\delta_{ij} + \frac{c_i(M_n - M_j)}{M_j} \right] M_j$$

$$= \frac{N}{\rho} B_{ij}^{cn} M_j. \tag{16}$$

Equations (14—16) in the form of $\partial y / \partial z$ are useful in changing partial differential coefficients of chemical potentials from one concentration unit y to another concentration unit z,

$$\frac{\partial \mu}{\partial y} \frac{\partial y}{\partial z} = \frac{\partial \mu}{\partial z}. \tag{17}$$

Equation (17) can be abbreviated as

$$\mu^y \left(\frac{\partial y}{\partial z} \right) = \mu^z. \tag{18}$$

Also

$$\text{grad}\, y = \left(\frac{\partial y}{\partial z} \right) \text{grad}\, z. \tag{19}$$

Thus we can change the partial coefficients of chemical potential and concentration gradient independently,

$$\text{grad}\, \mu = \mu^y \,\text{grad}\, y = \mu^z \,\text{grad}\, z. \tag{20}$$

Conversion between Atomic Basis and Mass Basis

So far the diffusion flux has been given in terms of partial mass density,

$$\boldsymbol{J}_i^a = \rho_i(\boldsymbol{v}_i - \boldsymbol{v}^a). \tag{21}$$

We may also define

$$\boldsymbol{J}_i^{\bar{a}} = N_i(\boldsymbol{v}_i - \boldsymbol{v}^a), \tag{22}$$

where the bar refers to an atomic basis. Comparing eqs. (21) and (22),

$$J_i^{\bar{a}} = \frac{J_i^a}{M_i}. \tag{23}$$

Instead of eq. (4,28), we have the restraint

$$\sum_0^n \frac{a_i}{n_i} J_i^{\bar{a}} = 0. \tag{24}$$

Specifically [comparing eqs. (4, 41—4, 44)],

$$\sum_k^n M_k J_k^{\bar{c}} = 0, \tag{25}$$

$$\sum_k^n J_k^{\bar{n}} = 0, \tag{26}$$

$$\sum_k^n \bar{V}_k J_k^{\bar{0}} = \sum_k^n V_k M_k J_k^{\bar{0}} = 0, \tag{27}$$

$$J_n^{\bar{r}} = 0. \tag{28}$$

From eq. (23), it is clear that the corresponding relation between forces on an atomic basis and those on a mass basis is

$$X^{\bar{a}} = MX^a. \tag{29}$$

Similar relations exist in chemical potential and volume,

$$\bar{\mu} = M\mu, \quad \frac{1}{N} = \bar{V} = MV = \frac{M}{\rho}. \tag{30}$$

For example,

$$\bar{\mu}^N = M\mu^N = M\mu^\rho \frac{\partial \rho}{\partial N} = M\mu^\rho M. \tag{31}$$

Of the matrix B in eq. (7),

$$B_{ik}^{\overline{ab}} = \delta_{ik} + \frac{n_i}{n_k} \sum_k^{n-1} \left(\frac{b_k}{b_n} a_n - a_k \right), \quad i, k = 1, 2, \cdots, (n-1). \tag{32}$$

Or

$$B^{\overline{ab}} = M^{-1} B^{ab} M. \tag{33}$$

Similarly of matrix A in eq. (11),

$$A^{\bar{a}} = MA^{a}M^{-1}. \tag{34}$$

Because the product of matrices A and μ is used so frequently, we abbreviate it as

$$G = A\mu. \tag{35}$$

For example,

$$G^{\overline{aN}} = A^{\bar{a}}\bar{\mu}^{N} = MA^{a}M^{-1}M\mu^{\rho}M = MG^{a\rho}M, \tag{36}$$

eqs. (31) and (34) having been used.

Diffusion Coefficients and the Rules for Transformations

From the last equality of eq. (10), in the region of linearity,

$$\begin{aligned} \boldsymbol{J}^{a} &= -T^{-1}L^{a}\boldsymbol{X}^{a} = -T^{-1}L^{a}A^{a}\operatorname{grad}\mu = -T^{-1}L^{a}A^{a}\mu^{y}\operatorname{grad}y \\ &= -T^{-1}L^{a}G^{ay}\operatorname{grad}y = -D^{ay}\operatorname{grad}y, \end{aligned} \tag{37}$$

where diffusion coefficient D is defined as

$$D^{ay} = T^{-1}L^{a}A^{a}\mu^{y} = T^{-1}L^{a}G^{ay}. \tag{38}$$

If $\boldsymbol{J}$ is transformed in a certain way such that

$$\boldsymbol{J}^{a} = B^{ab}\boldsymbol{J}^{b}, \tag{39}$$

$\boldsymbol{X}$ is transformed contragrediently with respect to $\boldsymbol{J}$,

$$\boldsymbol{X}^{a} = [(B^{ab})^{-1}]^{\mathrm{T}}\boldsymbol{X}^{b} = (B^{ba})^{\mathrm{T}}\boldsymbol{X}^{b},$$

so that the rate of entropy-production is invariant. It is seen that A is also transformed contragrediently with respect to $\boldsymbol{J}$,

$$A^{a} = (B^{ba})^{\mathrm{T}}A^{b}. \tag{40}$$

The first suffix of D is transformed cogrediently with respect to $\boldsymbol{J}$,

$$D^{ay} = B^{ab}D^{by}. \tag{41}$$

The second suffix of D varies according to

$$D^{ay} = D^{az}\left(\frac{\partial z}{\partial y}\right). \tag{42}$$

Because

$$T^2\sigma = [L^a \boldsymbol{X}^a | \boldsymbol{X}^a] = [L^a (B^{ba})^{\mathrm{T}} \boldsymbol{X}^b | (B^{ba})^{\mathrm{T}} \boldsymbol{X}^b]$$
$$= [B^{ba} L^a (B^{ba})^{\mathrm{T}} \boldsymbol{X}^b | \boldsymbol{X}^b], \tag{43}$$

and

$$T^2\sigma = [L^b \boldsymbol{X}^b | \boldsymbol{X}^b]. \tag{44}$$

Thus

$$L^b = B^{ba} L^a (B^{ba})^{\mathrm{T}}. \tag{45}$$

Or L is transformed congruently.

Binary Diffusion

In binary diffusion, eq. (7) is simplified to

$$B_{11}^{ab} = \frac{a_2}{b_2}, \tag{46}$$

eq. (11) is simplified to

$$A_{11}^{a} = \frac{1}{a_2}, \tag{47}$$

eq. (14) is simplified to

$$\frac{\partial \rho_1}{\partial N_1} = M_1, \tag{48}$$

eq. (15) is simplified to

$$\frac{\partial \rho_1}{\partial c_1} = \rho B_{11}^{0c} = \rho\,\frac{\rho_2 V_2}{c_2} = \rho^2 V_2, \tag{49}$$

eq. (16) is simplified to

$$\frac{\partial c_1}{\partial n_1} = \frac{N}{\rho} M_1 B_{11}^{cn} = \frac{N}{\rho} M_1 \frac{c_2}{n_2} = \frac{c_1 c_2}{n_1 n_2}. \tag{50}$$

No Onsager reciprocal relationship exists in binary diffusion, because there is only one diffusion coefficient, D_{11} or D_{22}. Only in $D^{0\rho}$, the two diffusion coefficients are equal, as is seen from

$$V_1\boldsymbol{J}_1^0 + V_2\boldsymbol{J}_2^0 = -V_1D_{11}^{0\rho}\,\text{grad}\,\rho_1 - V_2D_{22}^{0\rho}\,\text{grad}\,\rho_2$$
$$= -D_{11}^{0\rho}V_1\,\text{grad}\,\rho_1 - D_{22}^{0\rho}V_2(-V_1\,\text{grad}\,\rho_1)/V_2 = 0. \quad (51)$$

This is one of the reasons why $D^{0\rho}$ is called the standard diffusion coefficient (The experimentally determined velocities $\boldsymbol{v}_1$ and $\boldsymbol{v}_2$ are independent, i.e., convection-free. The restraint on $\boldsymbol{v}_1$ and $\boldsymbol{v}_2$ in eq. (4,35) is supposed to eliminate this independence and $\sum_i^2 V_i\boldsymbol{J}_i^0$ is supposed to approach the volume flow relative to a fixed point of the apparatus[21]). D without any suffix is understood to mean $D^{0\rho}$. For other diffusion coefficients D_{11} is simply related to D_{22}. For example,

$$D_{22}^{c\rho} = \frac{V_2}{V_1}D_{11}^{c\rho}, \quad (52)$$

$$D_{22}^{n\rho} = \frac{M_2V_2}{M_1V_1}D_{11}^{n\rho}. \quad (53)$$

Because

$$\mu_1 = \frac{\bar{\mu}_1}{M_1} = \frac{RT}{M_1}\ln n_1\gamma_1 + \text{constant}, \quad (54)$$

where γ_1 is the activity coefficient,

$$\mu_{11}^n = \frac{RT}{M_1}\left(\frac{\partial \ln n_1}{\partial n_1} + \frac{\partial \ln \gamma_1}{\partial n_1}\right)$$
$$= \frac{RT}{n_1M_1}\left(1 + \frac{\partial \ln \gamma_1}{\partial \ln n_1}\right). \quad (55)$$

Consequently

$$D_{11}^{0\rho} = B_{11}^{0a}D_{11}^{a\rho} = \frac{\rho_2V_2}{a_2}T^{-1}L_{11}^aA_{11}^a\mu_{11}^\rho$$
$$= \frac{\rho_2V_2}{a_2}T^{-1}L_{11}^aA_{11}^a\mu_{11}^n\frac{\partial n_1}{\partial \rho_1}$$
$$= \frac{n_1n_2}{a_2\rho_1}T^{-1}L_{11}^aA_{11}^a\mu_{11}^n, \quad (56)$$

eqs. (18), (38), (41), (46), (49) and (50) having been used. Substituting eq. (55) in eq. (56),

$$D_{11}^{0\rho} = \frac{Rn_2}{a_2^2 M_1 \rho_1} L_{11}^{a} \left(1 + \frac{\partial \ln \gamma_1}{\partial \ln n_1}\right). \tag{57}$$

Thus we know the standard coefficient if the phenomenological coefficient with respect to any reference velocity and the activity coefficient are known.

Multi-component Diffusion

Because the reference velocity a or b may be $0, c, n,$ or r and the unit of concentration y or z may be $c, n, \rho,$ or N, there are 16 equations like

$$\boldsymbol{J}^a = -D^{ay} \operatorname{grad} y, \tag{58}$$

64 equations like

$$\boldsymbol{J}^a = -D^{ay} \frac{\partial y}{\partial z} \operatorname{grad} z, \tag{59}$$

or 256 equations like

$$\boldsymbol{J}^a = -B^{ab} D^{by} \frac{\partial y}{\partial z} \operatorname{grad} z. \tag{60}$$

Other than binary diffusion, our choice of diffusion coefficients is no longer unique. Although eq. (58) is very simple, any conversion involves the complicated triple product of matrices in eq. (60), the use of which is limited. The convenient choice is dictated by having $\partial y / \partial z$ in eq. (60) in a form as simple as possible. We also broaden our choice by admitting the matrix of coefficients having the same characteristic roots as that of standard coefficients, i.e., by similarity transformation. We distinguish these diffusion coefficients from the standard coefficients by writing the former as $\mathscr{D}$.

The equation

$$\boldsymbol{J}^0 = -D^{0\rho} \operatorname{grad} \rho \tag{61}$$

is simple. Written in full,

$$\boldsymbol{J}_i^0 = -\sum_j^{n-1} D_{ij}^{0\rho} \operatorname{grad} \rho_j. \tag{62}$$

Dividing eq. (62) by M_i,

$$\frac{\boldsymbol{J}_i^0}{M_i} = -\frac{1}{M_i}\sum_{jk}^{n-1} D_{ij}^{0\rho} \frac{\partial \rho_j}{\partial N_k} \operatorname{grad} N_k$$

$$= -\frac{1}{M_i}\sum_j^{n-1} D_{ij}^{0\rho} M_j \operatorname{grad} N_j. \tag{63}$$

From eq. (23),

$$\boldsymbol{J}_i^{\bar{0}} = -\mathscr{D}_{ij}^{\overline{0N}} \operatorname{grad} N_j, \tag{64}$$

where

$$\mathscr{D}^{\overline{0N}} = M^{-1} D^{0\rho} M. \tag{65}$$

The bar on "0" is consistent with $\boldsymbol{J}^{\bar{0}}$ and the bar on "N" indicates atomic chemical potential.

The equation

$$\boldsymbol{J}^c = -D^{cc} \operatorname{grad} c \tag{66}$$

is simple. Referring to $D^{0\rho}$,

$$\boldsymbol{J}^c = -\left(B^{c0} D^{0\rho} \frac{\partial \rho}{\partial c}\right) \operatorname{grad} c = -\rho (B^{c0} D^{0\rho} B^{0c}) \operatorname{grad} c$$

$$= -\rho \mathscr{D}^{cc} \operatorname{grad} c, \tag{67}$$

where

$$\mathscr{D}^{cc} = B^{c0} D^{0\rho} (B^{c0})^{-1}, \tag{68}$$

and eq. (15) having been used.

The equation

$$\boldsymbol{J}^n = -D^{nn} \operatorname{grad} n \tag{69}$$

is simple. Referring to $D^{0\rho}$,

$$J^n = -B^{n0}D^{0\rho}\frac{\partial\rho}{\partial n}\,\text{grad}\,n = -N(B^{n0}D^{0\rho}B^{0n})M\,\text{grad}\,n$$

$$= -N\mathscr{D}^{nn}M\,\text{grad}\,n, \tag{70}$$

where

$$\mathscr{D}^{nn} = B^{n0}D^{0\rho}(B^{n0})^{-1}, \tag{71}$$

and eqs. (15) and (16) having used. Equation (70) can be simplified as

$$J^{\bar{n}} = -N\mathscr{D}^{\overline{nn}}\,\text{grad}\,n, \tag{72}$$

where

$$\mathscr{D}^{\overline{nn}} = M^{-1}\mathscr{D}^{nn}M = M^{-1}B^{n0}D^{0\rho}(B^{n0})^{-1}M. \tag{73}$$

The bar on the first "n" is consistent with $J^{\bar{n}}$ and the bar on the second "n" indicates the atomic chemical potential.

The equation

$$J^r = -D^{r\rho}\,\text{grad}\,\rho \tag{74}$$

is simple. Referring to $D^{0\rho}$,

$$J^r = -B^{r0}D^{0\rho}(B^{0r}B^{r0})\,\text{grad}\,\rho = -\mathscr{D}^{r\rho}B^{r0}\,\text{grad}\,\rho, \tag{75}$$

where

$$\mathscr{D}^{r\rho} = B^{r0}D^{0\rho}(B^{r0})^{-1}. \tag{76}$$

Written in full,

$$J^r_i = -\sum_k^{n-1}\mathscr{D}^{r\rho}_{ik}\left(\delta_{kj} + \sum_j^{n-1}\frac{\rho_k V_j}{\rho_n V_n}\right)\text{grad}\,\rho_j$$

$$= -\sum_k^{n-1}\mathscr{D}^{r\rho}_{ik}\left[\text{grad}\,\rho_k - \frac{\rho_k(\text{grad}\,\rho_n)}{\rho_n}\right]$$

$$= -\sum_k^{n-1}\mathscr{D}^{r\rho}_{ik}\left[\text{grad}\,\rho_k - \frac{c_k(\text{grad}\,\rho_n)}{c_n}\right], \tag{77}$$

where eq. (7) and the Gibbs-Duhem relation at constant temperature and pressure have been used.

To summarize, $D^{0\rho}$ in eq. (61), $\mathscr{D}^{0\overline{N}}$ in eq. (64), $\mathscr{D}^{cc}$ in eq. (67) and $\mathscr{D}^{\overline{nn}}$ in eq. (72) are good choices in simplifying the diffusion equations. In spite of the fact that A^r is the unit matrix, the diffusion equation (77) is not simple.

By definition,

$$D = T^{-1}LA\mu = T^{-1}LG. \tag{78}$$

From the Onsager law, L is symmetric. Or

$$DG^{-1} = (G^{-1})^{\mathrm{T}}D^{\mathrm{T}}. \tag{79}$$

Or

$$G^{\mathrm{T}}D = D^{\mathrm{T}}G. \tag{80}$$

Written in full,

$$\sum_{j}^{n-1} G_{ji}D_{jk} = \sum_{j}^{n-1} D_{ji}G_{jk}. \tag{81}$$

Conversely, if eq. (80) is true, the Onsager law is verified. We see that, besides diffusion coefficients, the detailed thermodynamic information is required.

In general, neither D nor G is symmetric; there are some choices of G that it is symmetric. What are these choices?

The Gibbs equation is

$$dG = -SdT + Vdp + \sum_{i}^{n-1}(\mu_i - \mu_n)dc_i. \tag{82}$$

Using Maxwell's relation,

$$\frac{\partial(\mu_i - \mu_n)}{\partial c_j} = \frac{\partial(\mu_j - \mu_n)}{\partial c_i}. \tag{83}$$

Applying the Gibbs-Duhem relation,

$$\mu_{ij}^c + \frac{1}{c_n}\sum_{k}^{n-1} c_k\mu_{kj}^c = \mu_{ji}^c + \frac{1}{c_n}\sum_{k}^{n-1} c_k\mu_{ki}^c. \tag{84}$$

Using Kronecker's delta,

$$\sum_{k}^{n-1}\left(\delta_{ik}+\frac{c_k}{c_n}\right)\mu_{kj}^c=\sum_{k}^{n-1}\left(\delta_{jk}+\frac{c_k}{c_n}\right)\mu_{ki}^c. \tag{85}$$

Using the A matrix,

$$A_{ik}^c\mu_{kj}^c=A_{jk}^c\mu_{ki}^c. \tag{86}$$

In condensed form,

$$A^c\mu^c=(\mu^c)^{\mathrm{T}}(A^c)^{\mathrm{T}}. \tag{87}$$

Thus G^{cc} is symmetric. From eqs. (18), (40), and (87),

$$(B^{0c})^{\mathrm{T}}A^0\mu^\rho\frac{\partial\rho}{\partial c}=\left(\frac{\partial\rho}{\partial c}\right)^{\mathrm{T}}(\mu^\rho)^{\mathrm{T}}(A^0)^{\mathrm{T}}B^{0c}. \tag{88}$$

From eq. (15),

$$\frac{1}{\rho}\left(\frac{\partial\rho}{\partial c}\right)^{\mathrm{T}}A^0\mu^\rho\frac{\partial\rho}{\partial c}=\frac{1}{\rho}\left(\frac{\partial\rho}{\partial c}\right)^{\mathrm{T}}(\mu^\rho)^{\mathrm{T}}(A^0)^{\mathrm{T}}\frac{\partial\rho}{\partial c}. \tag{89}$$

Or

$$A^0\mu^\rho=(\mu^\rho)^{\mathrm{T}}(A^0)^{\mathrm{T}}. \tag{90}$$

Thus $G^{0\rho}$ is symmetric. From eqs. (18), (40) and (87),

$$(B^{nc})^{\mathrm{T}}A^n\mu^n\frac{\partial n}{\partial c}=\left(\frac{\partial n}{\partial c}\right)^{\mathrm{T}}(\mu^n)^{\mathrm{T}}(A^n)^{\mathrm{T}}B^{nc}. \tag{91}$$

From eq. (16),

$$\frac{N}{\rho}\left(\frac{\partial n}{\partial c}\right)^{\mathrm{T}}M^{\mathrm{T}}A^n\mu^n\frac{\partial n}{\partial c}=\frac{N}{\rho}\left(\frac{\partial n}{\partial c}\right)^{\mathrm{T}}(\mu^n)^{\mathrm{T}}(A^n)^{\mathrm{T}}M\frac{\partial n}{\partial c}. \tag{92}$$

Or $MA^n\mu^n$ is symmetric. Or $MA^n\,M^{-1}M\,\mu^n$ is symmetric.

Or $G^{nn}=A^{\bar{n}}\bar{\mu}^n$ is symmetric from eqs. (30) and (34). Similarly we can prove that $G^{\overline{0N}}$ is also symmetric. To summarize, our choice of $D^{0\rho}$, $\mathscr{D}^{cc}$, $\mathscr{D}^{\overline{0N}}$, and $\mathscr{D}^{\overline{nn}}$ has the further advantage that $G^{0\rho}$, G^{cc}, $G^{\overline{0N}}$, and $G^{\overline{nn}}$ are symmetric.

Ternary Diffusion

The Onsager law, eq. (81), is simplified in ternary diffusion to

$$G_{11}D_{12} + G_{21}D_{22} = D_{11}G_{12} + D_{21}G_{22}. \qquad (93)$$

Before inserting the values of diffusion coefficients into eq. (93), a bit of information is helpful in checking thermodynamical data, namely, $G^{\alpha\beta}$ is symmetric if $D^{\alpha\beta}$ are used as diffusion coefficients, where $\alpha\beta$ are limited to 0ρ, cc, $\overline{0N}$ and $\overline{nn}$. Because

$$A^c = \begin{bmatrix} 1+\dfrac{c_1}{c_3} & \dfrac{c_2}{c_3} \\ \dfrac{c_1}{c_3} & 1+\dfrac{c_2}{c_3} \end{bmatrix}, \quad \mu^c = \begin{bmatrix} \mu_{11}^c & \mu_{12}^c \\ \mu_{21}^c & \mu_{22}^c \end{bmatrix}, \qquad (94)$$

and

$$A^0 = \begin{bmatrix} 1+\dfrac{\rho_1 V_1}{\rho_3 V_3} & \dfrac{\rho_2 V_1}{\rho_3 V_3} \\ \dfrac{\rho_1 V_2}{\rho_3 V_3} & 1+\dfrac{\rho_2 V_2}{\rho_3 V_3} \end{bmatrix}, \quad \mu^\rho = \begin{bmatrix} \mu_{11}^\rho & \mu_{12}^\rho \\ \mu_{21}^\rho & \mu_{22}^\rho \end{bmatrix}, \qquad (95)$$

the statement $G_{12}^{cc} = G_{21}^{cc}$ is the same as

$$(1-c_2)\mu_{12}^c + c_2\mu_{22}^c = c_1\mu_{11}^c + (1-c_1)\mu_{21}^c, \qquad (96)$$

by multiplication of the two matrices in eq. (94). The statement $G_{12}^{0\rho} = G_{21}^{0\rho}$ is the same as

$$(1-\rho_2 V_2)\mu_{12}^\rho + \rho_2 V_1\mu_{22}^\rho = \rho_1 V_2\mu_{11}^\rho + (1-\rho_1 V_1)\mu_{21}^\rho, \quad (97)$$

by multiplication of the two matrices in eq. (95). The statement $G_{12}^{\overline{nn}} = G_{21}^{\overline{nn}}$ is the same as eq. (96) except that c is replaced by n and μ is barred. The statement $G_{12}^{\overline{0N}} = G_{21}^{\overline{0N}}$ is the same as eq. (97) except that ρ is replaced by N and μ and V are barred. So far the Onsager law has been tested rigorously in only a few systems[3–5].

A survey of ternary systems[6] with four diffusion coefficients determined shows that

$$D_{11} > D_{12}, \tag{98}$$

$$D_{22} > D_{21}, \tag{99}$$

$$\frac{D_{22} - D_{21}}{D_{11} - D_{12}} > -\frac{D_{21}}{D_{11}}, \tag{100}$$

$$\frac{D_{11} - D_{12}}{D_{22} - D_{21}} > -\frac{D_{12}}{D_{22}}. \tag{101}$$

The systems Na_2SO_4–H_2SO_4–H_2O[7] and polystyrene-cyclohexane-toluene[8] are interesting. For certain concentrations, both cross coefficients are negative and therefore smaller than the main coefficients. In one concentration of the latter system, the absolute value of D_{12} is even greater than that of D_{11}. Table 1 illustrates that inequalities (100) and (101) hold in these two systems.

Fick's Law

Fick's first law was first formulated in terms of concentration gradient. Furthermore, it was confined to binary systems and the question of interaction between different components was not raised. In recognizing the fact that "uphill" diffusion of a multi-component system is possible, the law is amended to the present form as expressed in eq. (37), which admits the interaction between different components. In case of interaction, we still maintain the convention that the diffusion of any component is "downhill" according to its own chemical potential gradient. With this convention the main diffusion coefficients are always positive. It must be emphasized that this is not a proof.

From eqs. (4, 25) and (4, 45) we have Fick's second law

$$\frac{\partial \rho_i}{\partial t} = -\operatorname{div} \rho_i \boldsymbol{v}_i \approx -\operatorname{div} \boldsymbol{J}_i^a = \operatorname{div} (D_{ij}^{ay} \operatorname{grad} y_j)$$

$$\approx D_{ij}^{ay} \operatorname{div} \operatorname{grad} y_j, \tag{102}$$

provided that velocity $\boldsymbol{v}^a$ is small and D_{ij}^{ay} is uniform.

Table 1

Check of Inequalities of Diffusion Coefficients in Na_2SO_4-H_2SO_4-H_2O and Polystyrene-cyclohexane-toluene Systems

N_1 or c_1	N_2 or c_2	$\frac{D_{22}-D_{21}}{D_{11}-D_{12}}$	$\frac{-D_{21}}{D_{11}}$	$\frac{D_{11}-D_{12}}{D_{22}-D_{21}}$	$\frac{-D_{12}}{D_{22}}$
0.5[a]	0.5	2.04	0.57	0.490	0.20
1.0[a]	0.5	1.97	0.37	0.509	0.24
0.5[a]	1.0	2.19	0.49	0.456	0.13
1.0[a]	1.0	2.12	0.44	0.471	0.20
0.05[b]	0.05	20.2	1.0	0.0495	0.0079
0.05[c]	0.05	21.2	0.5	0.047	0.037

a 1 is Na_2SO_4, 2 is H_2SO_4, 3 is H_2O; units of N in moles per liter.

b 1 is polystyrene, 2 is cyclohexane, 3 is toluene; units of c_1 and c_2 in mass fractions.

c 1 is polystyrene, 2 is toluene, 3 is cyclohexane; units of c_1 and c_2 in mass fractions.

References

[1] G. J. Hooyman, H. Holtan, Jr., P. Mazur and S. R. De Groot: *Physica*, **19**, 1095 (1953).

[2] J. N. Agar: *Trans. Faraday Soc.*, **56**, 776 (1960).

[3] P. J. Dunlop: *J. Phys. Chem.*, **63**, 612 (1959).

[4] L. A. Woolf, D. G. Miller and L. J. Gosting: *J. Am. Chem. Soc.*, **84**, 317 (1962).

[5] D. G. Miller: *J. Phys. Chem.*, **69**, 3374 (1965).

[6] Y. L. Yao: *J. Chem. Phys.*, **45**, 110 (1966).

[7] R. P. Wendt: *J. Phys. Chem.*, **66**, 1279 (1962).

[8] E. L. Cussler, Jr., and E. N. Lightfoot: *J. Phys. Chem.*, **69**, 1135 (1965).

§39 *CONDUCTION OF ELECTRICITY*

In this section we introduce electrical charge, current density (charge flux), electrical field (force for conduction of electricity) and conductance (related to phenomenological coefficients). In the conduction of electricity, the electroneutrality overshadows the mechanical equilibrium. A certain association of electrical charges and phenomenological coefficients is the so-called transference number, which is meaningful because of the Onsager law. While the definition of mobility is artificial, a certain association of electrical charges and mobilities, called electrical mobility, is meaningful in electrolytes, because there is no interaction between ions in infinite dilution.

Let us consider an electric conductor, be it a metallic conductor, a molten electrolyte or an electrolytic solution. In the presence of a static electrical field, the total charge density is zero,

$$\sum_{k}^{n} Z_k \rho_k = \sum_{k}^{n} \bar{z}_k N_k = 0, \tag{1}$$

where n refers to all charged species, Z_k and $\bar{z}_k$ are the specific and the molar charge in faradays of the component k respectively, and ρ_k and N_k are the partial mass and the molar density respectively. The sign of $\bar{z}$ or Z is positive for cations but negative for electrons or anions. Furthermore, $\bar{z}$ or Z is proportional to the valency of the ions,

$$\bar{z}_k = M_k Z_k = \phi_k \mathscr{F}, \tag{2}$$

where M_k is the mole of the component k, ϕ_k its valency, and $\mathscr{F}$ (faraday) the charge in coulombs on one mole of a univalent ion. If we consider dissociation of an electrolyte as an electrochemical reaction,

$$\sum_{k}^{n} \bar{z}_k \bar{\Lambda}_k = \sum_{k}^{n} \phi_k \bar{\Lambda}_k \mathscr{F} = 0, \text{ or } \sum_{k}^{n} \phi_k \bar{\Lambda}_k = 0, \tag{3}$$

where $\bar{\Lambda}_k$ is the atomic stoichiometric coefficient. Thus the electroneutrality stems from the neutrality of valences.

We may extend the electroneutrality to dynamic conditions

when an electric field $\boldsymbol{E}$ is changing,

$$\boldsymbol{I} = \sum_{k}^{n} \boldsymbol{J}_k^c \bar{z}_k = \sum_{k}^{n} \boldsymbol{J}_k^c Z_k = 0, \tag{4}$$

where $\boldsymbol{I}$ is the total current density (charge flux). This is equivalent to saying that no space charge is accumulated or withdrawn. Due to strong coulombian forces between charged species, except for very poor conductors, the space charge is negligible. Furthermore, it takes little time to reach the steady state. Thus in the steady state, eq. (4) is usually valid. Even if there is space charge, the ground level of space charge is arbitrary and can be shifted accordingly.

Equation (4) can be written as

$$\sum_{k}^{n} \rho_k Z_k \boldsymbol{v}_k = \sum_{k}^{n} N_k \bar{z}_k \boldsymbol{v}_k = 0. \tag{5}$$

Thus we have the interesting conclusion that, if electroneutrality is valid, it is immaterial whichever reference velocity we choose, even when mechanical equilibrium is not established. A corollary of this is that we can always reduce the number of charged species in an electrolytic solution to one less (see § 40). To avoid confusion we may add that electrons, except in a semiconductor, play a predominant or an insignificant role. In metallic conductors, the charged species are electrons and positive ions, but in electrolytic solutions, the charged species are anions and cations. Thus the number of the effective charged species is usually $n-1$.

The mechanical force in a static electric field is

$$Z_k \boldsymbol{E} = -Z_k \operatorname{grad} \zeta, \tag{6}$$

where ζ is the potential.

$$\boldsymbol{J}_i^a = \frac{1}{T} \sum_{k}^{n} L_{ik}^a Z_k \operatorname{grad} \zeta, \tag{7}$$

$$\boldsymbol{I}_j = Z_j \boldsymbol{J}_j^a = \frac{1}{T} Z_j \sum_k^n L_{jk}^a Z_k \operatorname{grad} \zeta . \tag{8}$$

We define the transference number as

$$t_j = \frac{\boldsymbol{I}_j}{\sum_j^n \boldsymbol{I}_j} = \frac{Z_j \sum_k^n L_{jk}^a Z_k}{\sum_j^n \sum_k^n Z_j L_{jk}^a Z_k} . \tag{9}$$

The total current is

$$\boldsymbol{I} = \sum_j^n I_j = \frac{1}{T} \sum_j^n \sum_k^n Z_j L_{jk}^a Z_k \operatorname{grad} \zeta = k \operatorname{grad} \zeta = -k\boldsymbol{E}, \tag{10}$$

where k is the specific conductivity. This is Ohm's law in vector form. Its truth depends on the fact that L is independent of the electric field (not true in extremely high fields), although it can be a function of temperature, pressure, and concentration, etc. Because of the Onsager law,

$$t_j = \frac{Z_k \sum_j^n Z_j L_{kj}^a}{\sum_j^n \sum_k^n Z_k L_{kj}^a Z_j} . \tag{11}$$

In eq. (9) Z_k may be considered as a column vector, but in eq. (11) it is a row vector. Without the Onsager law, we would have to distinguish between two kinds of transference number. The transference number defined by eq. (9) or (11) is called the Hittorf transference number.

We also define mobility in absolute quantities as velocity per unit force acting on a mole,

$$\boldsymbol{v}_j - \boldsymbol{v}^a = M_j \sum_k^n W_{kk}^a \boldsymbol{F}_k = M_j \sum_k^n W_{kk}^a Z_k \operatorname{grad} \zeta, \tag{12}$$

then

$$\boldsymbol{J}_j^a = \rho_j M_j \sum_k^n W_{kk}^a Z_k \,\mathrm{grad}\,\zeta, \tag{13}$$

$$\boldsymbol{I} = \sum_j^n \rho_j M_j Z_j \sum_k^n W_{kk}^a Z_k \,\mathrm{grad}\,\zeta. \tag{14}$$

We also define electrical mobility as

$$U_{jj}^a = M_j \sum_k^n W_{kk}^a Z_k. \tag{15}$$

Thus eq. (14) is simplified to

$$\boldsymbol{I} = \sum_j^n \rho_j Z_j U_{jj}^a \,\mathrm{grad}\,\zeta. \tag{16}$$

The definition of electrical mobility is artificial because we simply distribute the interactions among the direct responses. However, by determining the equivalent conductance of a large number of electrolytes with a common ion at infinite dilution, the equivalent conductance (or the specific conductance) of an electrolyte at infinite dilution is equal to the sum of conductances of the ions. This could only mean that at infinite dilution, besides complete dissociation, there is no interaction between the movement of the different ions. Or

$$k = \frac{\boldsymbol{I}}{\mathrm{grad}\,\zeta} = \rho_+ Z_+ U'_+ + \rho_- Z_- U'_- = N_+ \bar{z}_+ U'_+ + N_- \bar{z}_- U'_-. \tag{17}$$

In electrolytic solutions $\boldsymbol{v}_n$ is referred to the solvent. Because the ions are solvated, it is desirable to refer to the velocity of "free" solvent, $\boldsymbol{v}_n^*$. The total current remains unchanged. Let s_k, the solvent number, be the mass of solvent per unit mass of the species k. The momentum of the solvent is conserved,

$$\rho_n \boldsymbol{v}_n = \left(\rho_n - \sum_k^n s_k \rho_k\right) \boldsymbol{v}_n^* + \sum_k^n s_k \rho_k \boldsymbol{v}_k. \tag{18}$$

Or

$$v_n = v_n^* + \sum_k^n \frac{s_k \rho_k}{\rho_n}(v_k - v_n^*). \qquad (19)$$

We name t^* as the Washburn transference number in contrast to t as the Hittorf transference number,

$$t - t^* = \frac{1}{I}[Z_k \rho_k (v_k - v_n)] - \frac{1}{I}[Z_k \rho_k (v_k - v_n^*)]$$

$$= -\frac{Z_k \rho_k}{\rho_n} \sum_j^n \frac{1}{I}[s_j \rho_j (v_j - v_n^*)]$$

$$= -\frac{Z_k \rho_k}{\rho_n} \sum_j^n \frac{s_j t_j^*}{Z_j}. \qquad (20)$$

Unfortunately, there is no method free of controversy to determine the Washburn transference number. In infinite dilution, of course, there is no difference between the Hittorf and the Washburn transference number.

§40 *DIFFUSION IN ELECTROLYTIC SOLUTIONS*

In spite of the fact that the active components are ions, diffusion in a mixture of ions and non-electrolytes always can be treated as diffusion in an equivalent mixture of neutral components.

Because diffusion is a slow process, the postulate of local equilibrium, §2, with respect to the dissociation of electrolyte, is valid. Because of electroneutrality, § 39, the system of independent ions in a neutral mixture of non-electrolytes can be reduced to an equivalent system with one less of the number of ions. This, added to the number of non-electrolytes will give the total number of neutral components. For example, a mixture of H_2O, raffinose, and urea is a three-component system. A mixture of H_2O, Na^+, K^+, Li^+, and Cl^- can be replaced by the system H_2O–NaCl–LiCl–KCl. The electroneutrality relation is $[Na^+] + [K^+] + [Li^+] = [Cl^-]$, where brackets represent molal concentrations, eq. (41, 37). A mixture of H_2O, Na^+, K^+, Cl^-, and Br^- can be replaced by the system H_2O–NaCl–KBr–NaBr or H_2O–NaCl–KBr–KCl. The electroneutrality relation is $[Na^+] + [K^+] = [Cl^-] + [Br^-]$.

A mixture of H_2O, Na^+, K^+, Cl^-, Br^-, KCl, and NaBr is still a four-component system. The additional restraints are

$$[K] + [K^+] = [Cl] + [Cl^-]$$

and

$$[Na] + [Na^+] = [Br] + [Br^-].$$

Thus the presence of undissociated electrolytes does not affect the number of neutral components.

A mixture of H_2O, H^+, and $H_2PO_4^-$ is a two-component system. A mixture of H_2O, $H_2PO_4^-$, H^+, $H_2PO_4^{2-}$, PO_4^{3-} is still a two-component system. The additional restraints are

$$[H_2PO_4^-] = [H^+] \times [HPO_4^{2-}] \times \text{constant}$$

and

$$[HPO_4^{2-}] = [H^+] \times [PO_4^{3-}] \times \text{constant}.$$

Thus whether an electrolyte is strong or weak does not affect the number of neutral components.

In conclusion, to account for the number of independent neutral components, it is possible to omit the undissociated electrolyte or the ions of a weak electrolyte beyond primary ionization. In a sense, the number of independent neutral components is like the degrees of freedom in the phase rule. On the other hand, if we consider ions as components, the number of diffusion coefficients is one more than the number of diffusion coefficients of neutral components, but they are not independent.

§41 DIFFUSION COEFFICIENTS OF ELECTROLYTES

Let us consider a binary electrolyte $A_{\bar{\lambda}_+} B_{\bar{\lambda}_-}$. According to § 40, a mixture of H_2O-anion-cation may be replaced by that of H_2O-electrolyte. There is only one diffusion coefficient, say

$$\boldsymbol{J}_1^0 = -D \operatorname{grad} \rho_1, \tag{1}$$

where the subscript 1 refers to the electrolyte. Because the active diffusion species are ions, we would like to derive a formula for D in terms of the electrical mobility of the ions.

In general, the mobilities of ions are different. A charge separation will be induced as soon as diffusion starts. After a short time, the powerful coulombian effects speed up the slow ones and slow down the fast ones so that no electrical current flows in the system. We can set up the diffusion fluxes as

$$\boldsymbol{J}_+^r = \rho_+(\boldsymbol{v}_+ - \boldsymbol{v}_2) \tag{2}$$

and

$$\boldsymbol{J}_-^r = \rho_-(\boldsymbol{v}_- - \boldsymbol{v}_2), \tag{3}$$

nstead of

$$\boldsymbol{J}_1^r = \rho_1(\boldsymbol{v}_1 - \boldsymbol{v}_2), \tag{4}$$

where the subscript 2 refers to water. Of course $\boldsymbol{J}_+^r$ and $\boldsymbol{J}_-^r$ are not independent. Although the velocities of the ions are the same,

$$\boldsymbol{v}_+ = \boldsymbol{v}_- = \boldsymbol{v}_1, \tag{5}$$

the partial densities, the mass stoichiometric coefficients, and electrical charges are different:

$$\frac{\rho_+}{\rho_-} = \frac{\Lambda_+}{\Lambda_-} = -\frac{Z_-}{Z_+}, \tag{6}$$

therefore,

$$\frac{Z_+}{\Lambda_-} = -\frac{Z_-}{\Lambda_+} = Z_+ - Z_-, \tag{7}$$

$$\frac{\rho_+}{\Lambda_+} = \frac{\rho_-}{\Lambda_-} = \rho_+ + \rho_- = \rho_1. \tag{8}$$

Also the specific chemical potentials are not the same,

$$\mu_1 = \Lambda_+\mu_+ + \Lambda_-\mu_- = \frac{\Lambda_+}{Z_-}(Z_-\mu_+ - Z_+\mu_-)$$

$$= \frac{Z_-\mu_+ - Z_+\mu_-}{Z_- - Z_+}, \tag{9}$$

eq. (7) having been used. The equations in molar basis corresponding to eqs. (5)—(8) are

$$\frac{N_+}{N_-} = \frac{\bar{\Lambda}_+}{\bar{\Lambda}_-} = -\frac{\bar{z}_-}{\bar{z}_+}, \tag{10}$$

$$\frac{\bar{z}_+}{\bar{\Lambda}_-} = -\frac{\bar{z}_-}{\bar{\Lambda}_+} = \frac{\bar{z}_+ - \bar{z}_-}{\bar{\Lambda}}, \tag{11}$$

$$\frac{N_+}{\bar{\Lambda}_+} = \frac{N_-}{\bar{\Lambda}_-} = \frac{N_+ + N_-}{\bar{\Lambda}} = N_1, \tag{12}$$

and

$$\bar{\mu}_1 = \frac{\bar{\Lambda}_+}{\bar{z}_-}(\bar{z}_-\bar{\mu}_+ - \bar{z}_+\bar{\mu}_-)$$

$$= \frac{\bar{\Lambda}(\bar{z}_-\bar{\mu}_+ - \bar{z}_+\bar{\mu}_-)}{\bar{z}_- - \bar{z}_+}, \tag{13}$$

where

$$\bar{\Lambda} = \bar{\Lambda}_+ + \bar{\Lambda}_-.$$

We set up the linear relations between the fluxes and the forces,

$$J_i^r = -\sum_j^n L_{ij}\,(\operatorname{grad}\mu_j + Z_j \operatorname{grad}\zeta),\ i, j = +, -. \tag{14}$$

Because of electroneutrality,

$$\sum_{i}^{n} Z_i J_i^r = 0. \tag{15}$$

Because of the Gibbs-Duhem relation at constant temperature and pressure,

$$\sum_{j}^{n} \rho_j \operatorname{grad} \mu_j = 0. \tag{16}$$

Because of eq. (6),

$$\sum_{j}^{n} \rho_j Z_j = 0. \tag{17}$$

Thus both fluxes and forces are not independent. According to Theorem 4 of § 18, the Onsager law is not necessarily valid. Instead of eliminating one dependent flux and one dependent force, we use the theorem to construct four new phenomenological coefficients and specify

$$L_{+-} = L_{-+}. \tag{18}$$

However, L_{--} is arbitrary. From eqs. (39, 10) and (39, 16) (let $L'/T = L$),

$$\frac{1}{\rho_i} \sum_{k}^{n} L_{ik}^r Z_k = U_i^r, \tag{19}$$

$$\sum_{i}^{n} \sum_{k}^{n} Z_i Z_k L_{ik}^r = k;\ i,\ k = +, -. \tag{20}$$

Written in matrix form,

$$\begin{bmatrix} \frac{Z_+}{\rho_+} & \frac{Z_-}{\rho_+} & 0 \\ 0 & \frac{Z_+}{\rho_-} & \frac{Z_-}{\rho_-} \\ Z_+^2 & 2Z_+Z_- & Z_-^2 \end{bmatrix} \begin{bmatrix} L_{++} \\ L_{+-} \\ L_{--} \end{bmatrix} = \begin{bmatrix} U_+ \\ U_- \\ k \end{bmatrix}. \tag{21}$$

The 3 × 3 matrix in eq. (21) is of the rank 2. Using the first two rows of the matrix, we can express L_{++} and L_{+-} in terms of L_{--},

$$L_{++} = \frac{\rho_+}{Z_+}\left(U_+ + U_- - \frac{Z_- L_{--}}{\rho_-}\right), \tag{22}$$

$$L_{+-} = \frac{\rho_-}{Z_+}\left(U_- - \frac{Z_- L_{--}}{\rho_-}\right). \tag{23}$$

The third row yields

$$\rho_+ Z_+ U_+ + \rho_- Z_- U_- = k, \tag{24}$$

after eq. (6) has been used. Equation (24) gives no new information other than the definition of electroneutrality, eq. (39,17). The arbitrariness of L_{--} is settled by letting

$$L_{--} = \frac{\rho_- U_-}{Z_-}. \tag{25}$$

Then

$$L_{+-} = L_{-+} = 0, \tag{26}$$

and

$$L_{++} = \frac{\rho_+ U_+}{Z_+}. \tag{27}$$

Substituting eq. (15) in eq. (14), we obtain the expression for the diffusion potential,

$$\operatorname{grad} \zeta = -\frac{\sum_i^n Z_i \sum_j^n L_{ij} \operatorname{grad} \mu_j}{\sum_i^n \sum_j^n Z_i L_{ij} Z_j}$$

$$= -\frac{\sum_j^n Z_j \sum_i^n L_{ij} \operatorname{grad} \mu_i}{\sum_i^n \sum_j^n Z_i L_{ij} Z_j}; \quad i, j = +, -. \tag{28}$$

Just as for the transference number, eqs. (39, 9) and (39, 11), we emphasize that the second equality in eq. (28) is true because of the Onsager law. Otherwise we have to distinguish between two kinds of diffusion potential. Written in full,

$$\operatorname{grad} \zeta = -\frac{L_{++}Z_+ \operatorname{grad} \mu_+ + L_{--}Z_- \operatorname{grad} \mu_-}{L_{++}Z_+^2 + L_{--}Z_-^2}. \tag{29}$$

Substituting eq. (29) in eq. (14) and using eq. (26),

$$J_+^r = -LZ_-(Z_- \operatorname{grad} \mu_+ - Z_+ \operatorname{grad} \mu_-), \tag{30}$$

$$J_-^r = LZ_+(Z_- \operatorname{grad} \mu_+ - Z_+ \operatorname{grad} \mu_-), \tag{31}$$

where

$$L = \frac{L_{++}L_{--}}{L_{++}Z_+^2 + L_{--}Z_-^2}. \tag{32}$$

Therefore,

$$J_1^r = J_+^r + J_-^r = -L(Z_- - Z_+)^2 \operatorname{grad} \mu_1, \tag{33}$$

where eq. (9) has been used. Or

$$D^{r\rho} = L(Z_- - Z_+)^2 \mu_{11}^{\rho}. \tag{34}$$

For a strong electrolyte in aqueous solution, the measured activity of the solute is related to the activity of the ions by

$$a_1 = (a_+^{\bar{\lambda}_+})(a_-^{\bar{\lambda}_-}). \tag{35}$$

The standard state is so chosen that the equilibrium constant of the dissociation is unity. Because no commonly accepted procedure has been devised for measuring the activities of the individual ions, the mean activity is defined as the geometric mean of the activities of the ions,

$$a_\pm = [(a_+^{\bar{\lambda}_+})(a_-^{\bar{\lambda}_-})]^{1/\bar{\lambda}} = a_1^{1/\bar{\lambda}}. \tag{36}$$

The molality $\mathscr{M}$ is the number of moles of the solute dissolved in 1 kg of the solvent,

$$n_1 = \frac{\mathscr{M}}{M + \dfrac{1000}{M_2}}, \tag{37}$$

where M_2 is the mole of the solvent. The mean molality $\mathscr{M}_{\pm}$ is defined as

$$\mathscr{M}_{\pm} = [(M_{+}^{\bar{\Lambda}+})(\mathscr{M}_{-}^{\bar{\Lambda}-})]^{1/\bar{\Lambda}} = \mathscr{M}\bar{\Lambda}_{\pm}, \tag{38}$$

where

$$\bar{\Lambda}_{\pm} = [(\Lambda_{+}^{\bar{\Lambda}+})(\bar{\Lambda}_{-}{}^{\bar{\Lambda}-})]^{1/\bar{\Lambda}}.$$

Both the mean activity and the mean molality are defined so that the mean activity coefficient $\gamma_{\pm}$ is given by

$$\gamma_{\pm} = [(\gamma_{+}^{\bar{\Lambda}+})(\gamma_{-}^{\bar{\Lambda}-})]^{1/\bar{\Lambda}} = a_{\pm}/\mathscr{M}_{\pm}. \tag{39}$$

Thus

$$\begin{aligned}\bar{\mu}_1 &= \text{constant} + RT\ln a_1 = \text{constant} + \bar{\Lambda}RT\ln(\mathscr{M}_{\pm}\gamma_{\pm}) \\ &= \text{constant} + \bar{\Lambda}RT\ln(\mathscr{M}\bar{\Lambda}_{\pm}\gamma_{\pm}),\end{aligned} \tag{40}$$

eqs. (36), (38) and (39) having been used. $\gamma_{\pm}$ is the value of the activity coefficient recorded in the literature. Then

$$\begin{aligned}\bar{\mu}_{11}^{N} &= \frac{\partial\bar{\mu}_1}{\partial\mathscr{M}}\frac{\partial\mathscr{M}}{\partial n_1}\frac{\partial n_1}{\partial N_1} \\ &= \frac{\bar{\Lambda}RT}{\mathscr{M}}\left(1 + \frac{\partial\ln\gamma_{\pm}}{\partial\ln\mathscr{M}}\right)\frac{\mathscr{M}}{n_1 n_2}\frac{n_1 n_2}{c_1 c_2}\frac{M_1}{\rho^2 V_2} \\ &= \frac{\bar{\Lambda}RT}{N_1\rho_2 V_2}\left(1 + \frac{\partial\ln\gamma_{\pm}}{\partial\ln\mathscr{M}}\right),\end{aligned} \tag{41}$$

eqs. (38, 48—50) having been used. Accordingly,

$$\mu_{11}^{\rho} = \frac{\partial\mu_1}{\partial N_1}\frac{\partial N_1}{\partial\rho_1} = \frac{\bar{\Lambda}RT}{M_1^2 N_1\rho_2 V_2}\left(1 + \frac{\partial\ln\gamma_{\pm}}{\partial\ln\mathscr{M}}\right). \tag{42}$$

Substituting eqs. (25) and (27) in eq. (32) and using eqs. (6—8),

$$L = \frac{\rho_1}{(Z_{+} - Z_{-})Z_{+}Z_{-}}\frac{U_{+}U_{-}}{U_{+} - U_{-}}. \tag{43}$$

Substituting eqs. (42) and (43) in eq. (34),

$$D^{r\rho} = -\frac{\bar{\Lambda}}{M_1}\frac{Z_{-} - Z_{+}}{\rho_2 V_2 Z_{+}Z_{-}}\frac{U_{+}U_{-}}{U_{+} - U_{-}}RT\left(1 + \frac{\partial\ln\gamma_{\pm}}{\partial\ln\mathscr{M}}\right). \tag{44}$$

From eqs. (12) and (6—8),

$$\frac{\bar{\Lambda}}{M_1}=\frac{N_+ + N_-}{N_1 M_1}=\frac{1}{\rho_1}\left(\frac{\rho_+}{M_+}+\frac{\rho_-}{M_-}\right)$$

$$=\frac{\bar{z}_+ - \bar{z}_-}{Z_+ - Z_-}M_+ M_-. \tag{45}$$

From eqs. (38, 41) and (38, 46),

$$D^{0\rho}=B^{0r}D^{r\rho}=\rho_2 V_2 D^{r\rho}. \tag{46}$$

Substituting eqs. (44) and (45) in eq. (46),

$$D=\frac{\bar{z}_+ - \bar{z}_-}{\bar{z}_+ \bar{z}_-}\frac{U_+ U_-}{U_+ - U_-}RT\left(1+\frac{\partial \ln \gamma_\pm}{\partial \ln \mathscr{M}}\right). \tag{47}$$

For a uni-uni electrolyte at infinite dilution,

$$\bar{z}_+=\mathscr{F},\quad \bar{z}_-=-\mathscr{F},\quad \gamma_\pm=1,$$

$$D=\frac{2RT}{\mathscr{F}}\frac{U_+\|U_-\|}{U_+ + \|U_-\|}. \tag{48}$$

This is Nernst's limiting law of diffusion coefficients in an electrolytic solution.

§42 *ISOTHERMAL OXIDATION OF METALS*

The best established mechanism for the isothermal oxidation of metals is scale thickening controlled by migration due to the electrochemical potential gradient. According to the postulate of local equilibrium, §2, local equilibrium prevails at every thin slice of the oxide. However, the whole oxide is not at equilibrium because of the non-uniformity of the electrochemical potential. The irreversible equalization of the potential of ions and electrons, not all independent, is responsible for the growth of the oxide. In this section we derive the formula for the parabolic law of oxidation[1] by assuming either the absence or the presence of intreaction. It turns out that the absence of interaction is not equivalent to the disposal of the arbitrariness of L_{nn}. Incidentally, in contrast to the preceding section, the molar basis is used in this section because a chemical reaction is involved. Because the electrolytic dissociation may be considered as a chemical reaction too, the reader can see that the treatment in the preceding section would be simpler if the molar basis were used. On the other hand, the mass basis will be advantageous if mechanical forces are involved. One should be familiar with either basis.

We start with the equation

$$\boldsymbol{J}_i^{\bar{c}} = -\sum_k^n L_{ik}^{\bar{c}}\left(\operatorname{grad}\bar{\mu}_k + \bar{z}_k \operatorname{grad}\zeta\right). \tag{1}$$

From eq. (38, 25),

$$\sum_i^n M_i \boldsymbol{J}_i^{\bar{c}} = 0. \tag{2}$$

Because of the Gibbs-Duhem relation at constant temperature and pressure,

$$\sum_k^n N_k \operatorname{grad}\bar{\mu}_k = 0. \tag{3}$$

Because of electroneutrality,

$$\sum_k^n N_k \bar{z}_k = 0. \tag{4}$$

From eqs. (2—4), both fluxes and forces are not independent. Instead of eliminating the dependency, we construct n new phenomenological coefficients and specify

$$L_{ik}^{\bar{c}} = L_{ki}^{\bar{c}}; \quad i, k = 1, 2, \cdots, n. \tag{5}$$

$L_{nn}^{\bar{c}}$ is arbitrary except not to make the determinant of the matrix $L^{\bar{c}}$ negative.

Before there is any separation of the ionic species, i. e.,

before any concentration gradient of the ionic species is established (grad $\bar{\mu}_i = 0$), the individual electric current density may be expressed as

$$\bar{z}_i \boldsymbol{J}_i^{\bar{c}} = -t^{(i)} k \operatorname{grad} \zeta, \tag{6}$$

where k is the specific conductivity and t the transference number. From eq. (39, 9),

$$t^{(i)} = \frac{\bar{z}_i \sum_k^n L_{ik}^{\bar{c}} \bar{z}_k}{\sum_i^n \sum_k^n \bar{z}_i L_{ik}^{\bar{c}} \bar{z}_k}. \tag{7}$$

Evidently,

$$\sum_k^n t^{(i)} = 1. \tag{8}$$

From eqs. (1) and (6),

$$\sum_k^n L_{ik}^{\bar{c}} \bar{z}_k = \frac{t^{(i)} k}{\bar{z}_i}. \tag{9}$$

Equation (1) becomes

$$\boldsymbol{J}_i^{\bar{c}} = -\sum_i^n \frac{t^{(i)} k}{\bar{z}_i^2} (\operatorname{grad} \bar{\mu}_i + \bar{z}_i \operatorname{grad} \zeta) \tag{10}$$

if we assume $L_{ik}^{\bar{c}} = 0, i \neq k$. Equation (10) is applicable after there is separation but just before the steady state of no current exists.

It takes little time to reach the steady state of no current, i. e.,

$$\sum_i^n \bar{z}_i \boldsymbol{J}_i^{\bar{c}} = 0. \tag{11}$$

Substituting eq. (11) in eq. (1), the diffusion potential is

$$\operatorname{grad}\zeta = -\frac{\sum_i^n \bar{z}_i \sum_k^n L_{ik}^{\bar{c}} \operatorname{grad}\bar{\mu}_k}{\sum_i^n \sum_k^n \bar{z}_i L_{ik}^{\bar{c}} \bar{z}_k}$$

$$= -\frac{\sum_i^n \sum_k^n \operatorname{grad}\bar{\mu}_i L_{ik}^{\bar{c}} \bar{z}_k}{\sum_i^n \sum_k^n \bar{z}_i L_{ik}^{\bar{c}} \bar{z}_k} = -\sum_i^n \frac{t^{(i)}}{\bar{z}_i} \operatorname{grad}\bar{\mu}_i. \quad (12)$$

In the second equality, we swapped the indices i and k, and in the third equality we used eq. (7).

Substituting eq. (12) in eq. (10),

$$J_i^{\bar{c}} = -\frac{t^{(i)}k}{\bar{z}_i^2}\left[\operatorname{grad}\bar{\mu}_i - \bar{z}_i \sum_j^n \frac{t^{(j)}}{\bar{z}_j} \operatorname{grad}\bar{\mu}_j\right]. \quad (13)$$

In the preceding section, dependency exists between positive ions (+) and negative ions (−) of a binary electrolyte. In this section dependency exists among metallic ions (+), oxygen ions (−) and electrons (⊖) of a semi-conductor. Because we are dealing with oxidation, it is convenient to use neutral oxygen atoms as one parameter. The formula of the oxide is $Me_{\bar{\Lambda}_+} O_{\bar{\Lambda}_-}$, small deviations from stoichiometry being neglected. We assume that instantaneous equilibrium is established at the metal-oxide and the oxide-oxygen interfaces.

$$\bar{\mu}_{\text{oxide}} = \bar{\Lambda}_+ \bar{\mu}_{\text{me}} + \bar{\Lambda}_- \bar{\mu}_{\text{ox}}, \quad (14)$$

$$\bar{\mu}_{\text{me}} = \bar{\mu}_+ - \left(\frac{\bar{z}_+}{\bar{z}_\ominus}\right) \bar{\mu}_\ominus, \quad (15)$$

$$\bar{\mu}_{\text{ox}} = \bar{\mu}_- - \left(\frac{\bar{z}_-}{\bar{z}_\ominus}\right) \bar{\mu}_\ominus. \quad (16)$$

Combining eqs. (14—16) and using eq. (41, 11),

$$\begin{bmatrix} \operatorname{grad}\bar{\mu}_+ \\ \operatorname{grad}\bar{\mu}_- \end{bmatrix} = \begin{bmatrix} \bar{z}_+/\bar{z}_- & \bar{z}_+/\bar{z}_\ominus \\ 1 & \bar{z}_-/\bar{z}_\ominus \end{bmatrix} \begin{bmatrix} \operatorname{grad}\bar{\mu}_{\text{ox}} \\ \operatorname{grad}\bar{\mu}_\ominus \end{bmatrix}. \quad (17)$$

Substituting eq. (17) in eq. (13) for $J_+^{\bar{c}}$ and $J_-^{\bar{c}}$,

$$\frac{\bar{z}_+}{\bar{z}_\ominus} J_+^{\bar{c}} + \frac{\bar{z}_-}{\bar{z}_\ominus} J_-^{\bar{c}} = -\left[\frac{(t_+ + t_-)t_\ominus k}{\bar{z}_- \bar{z}_\ominus}\right] \operatorname{grad} \bar{\mu}_{ox}. \tag{18}$$

Similarly,

$$J_\ominus^{\bar{c}} = \left[\frac{(t_+ + t_-)t_\ominus k}{\bar{z}_- \bar{z}_\ominus}\right] \operatorname{grad} \bar{\mu}_{ox}. \tag{19}$$

We see that the two opposite streams are equal in absolute quantity, as they should be. The left side of eq. (18) is equal to the sum of equivalents of the outgoing metallic ions and the incoming oxygen ions, which may be expressed as

$$\frac{dq}{dt} = \frac{1}{X}(t_+ + t_-)\frac{t_\ominus k}{\bar{z}_-}(\bar{\mu}_{ox}^{\mathrm{II}} - \bar{\mu}_{ox}^{\mathrm{I}}), \tag{20}$$

where X is the instantaneous thickness of the scale, I is the metal-oxide boundary, and II is the oxygen-oxide boundary. Or

$$\frac{dX}{dt} = \frac{1}{X}\frac{(t_+ + t_-)t_\ominus k}{2\bar{\Lambda}_- \bar{z}_- A N_{\mathrm{oxide}}}(\bar{\mu}_{ox}^{\mathrm{II}} - \bar{\mu}_{ox}^{\mathrm{I}}), \tag{21}$$

where N_{oxide} is the density of the oxide in moles per unit volume, A is the area of the cross section perpendicular to the thickness, and $2\bar{\Lambda}_-$ is the number of equivalents per mole of the oxide. Equation (21) is of the form

$$dX/dt = b/X, \tag{22}$$

where b is a constant if the composition of the oxide does not change. On integration,

$$X^2 = 2bt. \tag{23}$$

In a plot of thickness against time, we expect a parabola and we speak of the parabolic law, which was derived by Tammann[2] and by Pilling and Bedworth[3] independently.

In the alternative treatment we do not assume the absence of interaction. From eq. (9) we obtain three equations containing six elements of the $L^{\bar{c}}$ matrix. From eqs. (2) and (6),

$$\sum_{i=1}^{3} \frac{t^{(i)} M_i}{\bar{z}_i} = 0. \tag{24}$$

From eqs. (9) and (24), we obtain three more equations. These six equations are written in matrix form as follows:

$$\begin{bmatrix} \bar{z}_+ & 0 & \bar{z}_- & \bar{z}_\ominus & 0 & 0 \\ 0 & \bar{z}_- & \bar{z}_+ & 0 & \bar{z}_\ominus & 0 \\ 0 & 0 & 0 & \bar{z}_+ & \bar{z}_- & \bar{z}_\ominus \\ M_+ & 0 & M_- & M_\ominus & 0 & 0 \\ 0 & M_- & M_+ & 0 & M_\ominus & 0 \\ 0 & 0 & 0 & M_+ & M_- & M_\ominus \end{bmatrix} \begin{bmatrix} L^{\bar{c}}_{++} \\ L^{\bar{c}}_{--} \\ L^{\bar{c}}_{+-} \\ L^{\bar{c}}_{+\ominus} \\ L^{\bar{c}}_{-\ominus} \\ L^{\bar{c}}_{\ominus\ominus} \end{bmatrix} = \begin{bmatrix} kt_+/\bar{z}_+ \\ kt_-/\bar{z}_- \\ kt_\ominus/\bar{z}_\ominus \\ 0 \\ 0 \\ 0 \end{bmatrix}. \tag{25}$$

It can be shown that the matrix of coefficients and the augmented matrix in eq. (25) are of rank 5. Let us say that $L^{\bar{c}}_{\ominus\ominus}$ is arbitrary.

Substituting eq. (17) in eq. (1) for $\boldsymbol{J}^{\bar{c}}_+$ and $\boldsymbol{J}^{\bar{c}}_-$ and using eqs. (12) and (25),

$$\frac{\bar{z}_+ \boldsymbol{J}^{\bar{c}}_+}{\bar{z}_\ominus} + \frac{\bar{z}_- \boldsymbol{J}^{\bar{c}}_-}{\bar{z}_\ominus}$$

$$= [-kt_\ominus(t_+ + t_-) + kt_\ominus - L^{\bar{c}}_{\ominus\ominus}\bar{z}^2_\ominus] \operatorname{grad} \frac{\bar{\mu}_{ox}}{\bar{z}_- \bar{z}_\ominus}$$

$$= (kt^2_\ominus - L^{\bar{c}}_{\ominus\ominus}\bar{z}^2_\ominus) \operatorname{grad} \frac{\bar{\mu}_{ox}}{\bar{z}_- \bar{z}_\ominus}. \tag{26}$$

Similarly,

$$\boldsymbol{J}^{\bar{c}}_\ominus = (-kt^2_\ominus + L^{\bar{c}}_{\ominus\ominus}\bar{z}^2_\ominus) \operatorname{grad} \frac{\bar{\mu}_{ox}}{\bar{z}_- \bar{z}_\ominus}. \tag{27}$$

Thus $\sum_{i=1}^{3} \bar{z}_i \boldsymbol{J}^{\bar{c}}_i = 0$ is satisfied irrespective of the value of $L^{\bar{c}}_{\ominus\ominus}$.

We have not used eqs. (10) and (13), i. e., not used the assumption $L^{\bar{c}}_{ik} = 0$, $i \neq k$.

One natural way of disposing the arbitrariness of $L^{\bar{c}}_{\ominus\ominus}$ is to set

$$L^{\bar{c}}_{\ominus\ominus} = \frac{kt_{\ominus}}{z^{\bar{c}}_{\ominus}}. \tag{28}$$

Then eqs. (19) and (27) are identical. In the preceding section, we dealt with a 3×3 matrix of rank 2. That $L^{\bar{c}}_{--}$ has a certain definite value is equivalent to $L^{\bar{c}}_{+-} = L^{\bar{c}}_{-+} = 0$. In the present section we have dealt with a 6×6 matrix of rank 5. That $L^{\bar{c}}_{\ominus\ominus}$ has a certain definite value is equivalent to certain definite values of $L^{\bar{c}}_{+-}$, $L^{\bar{c}}_{+\ominus}$ and $L^{\bar{c}}_{-\ominus}$.

REFERENCES

[1] Y. L., Yao: *J. Chem. Phys.*, **43**, 3050 (1965).
[2] G. Tammann: *Z. anorg. Chem.*, **111**, 178 (1920).
[3] N. B. Pilling and R. E. Bedworth: *J. Inst. Metals*, **29**, 529 (1922).

§ 43 CONCENTRATION CELLS WITH TRANSFERENCE

Diffusion potential is not directly measurable. In § 41 we employ electrical mobility of ions to obtain the diffusion coefficient and in § 42 we employ specific conductivity and transference numbers to obtain the rate constant of parabolic oxidation. In this section we employ activity coefficients and transference numbers to obtain the EMF of a concentration cell with transference.

Let us consider a galvanic concentration cell, which has two identical electrodes immersed in two solutions of the same electrolyte with different concentrations and having one ion in common with the electrode. The scheme of a concentration cell may be represented as follows:

(I) $\mathrm{A} | \mathrm{A}_{\bar{\lambda}_+}\mathrm{B}_{\bar{\lambda}_-}(\mu_1)_\alpha |_\beta \mathrm{A}_{\bar{\lambda}_+}\mathrm{B}_{\bar{\lambda}_-}(\mu_2) | A$

(II) $\mathrm{A} | \mathrm{AB(solid)}, \mathrm{C}_{\bar{\lambda}_+}\mathrm{B}_{\bar{\lambda}_-}(\mu_1)_\alpha |_\beta \mathrm{C}_{\bar{\lambda}_+}\mathrm{B}_{\bar{\lambda}_-}(\mu_2), \mathrm{AB(solid)} | \mathrm{A},$

where $\mathscr{M}$ is molality, and α and β are two boundaries in which the composition of the solution varies in a continuous but arbitrary way from $\mathscr{M}_1$ to $\mathscr{M}_2$. In the first scheme the electrodes are reversible to cations, but in the second scheme they are reversible to anions.

Suppose that an infinitesimal amount of positive current passes from left to right. This can be considered as a forward stream of positive ions or a backward stream of electrons. In determining the EMF of the concentration cell we have to use copper leads from the electrodes to a potentiometer. In considering the stream of electrons, the contact potential of Cu|A on the left is exactly balanced by A|Cu on the right because of the symmetry of the arrangement of the electrodes. In considering the stream of ions, the electrode potential on the left is not cancelled by that on the right because the concentrations of the two solutions are not equal,

$$\left(\frac{\bar{\mu}_i^s - \bar{\mu}_i^e}{\bar{z}_i}\right)^{\text{right}} - \left(\frac{\bar{\mu}_i^s - \bar{\mu}_i^e}{\bar{z}_i}\right)^{\text{left}} = \int_\alpha^\beta \frac{d\bar{\mu}_i^s}{\bar{z}_i}, \tag{1}$$

where the suffix s refers to the solution, the suffix e refers to the electrode and i refers to the ions. The equality of eq. (1) is because the concentration of the ions in the electrode, and therefore the chemical potential, is negligible.

In considering the stream of ions, the diffusion potential in the region between the two solutions, from eq. (42, 12), is

$$-\sum_k^n \frac{t^{(k)}}{\bar{z}_k} d\bar{\mu}_k .$$

Hence the EMF of the concentration cell is

$$E = \int_\alpha^\beta \frac{d\bar{\mu}_i}{\bar{z}_i} - \int_\alpha^\beta \sum_k^n \frac{t^{(k)}}{\bar{z}_k} d\bar{\mu}_k . \tag{2}$$

Suppose that the concentration cell is reversible to cations. From eqs. (41, 13) and (41, 40),

$$\frac{d\bar{\mu}_+}{\bar{z}_+} - \frac{t_+ d\bar{\mu}_+}{\bar{z}_+} - \frac{t_- d\bar{\mu}_-}{\bar{z}_-} = t_-\left(\frac{d\bar{\mu}_+}{\bar{z}_+} - \frac{d\bar{\mu}_-}{\bar{z}_-}\right)$$

$$= \frac{t_-}{\bar{z}_+ \bar{\Lambda}_+} d\bar{\mu} = \bar{\Lambda} RT \frac{t_-}{\bar{z}_+ \bar{\Lambda}_+} d \ln \mathscr{M}_\pm \gamma_\pm . \tag{3}$$

From eqs. (2), (3) and (41, 11),

$$E = RT\left(\frac{1}{\bar{z}_+} - \frac{1}{\bar{z}_-}\right)\int_\alpha^\beta t_- d \ln \mathscr{M}_\pm \gamma_\pm . \tag{4}$$

Similarly for the concentration cell reversible to anions,

$$E = -RT\left(\frac{1}{\bar{z}_+} - \frac{1}{\bar{z}_-}\right)\int_\alpha^\beta t_+ d \ln \mathscr{M}_\pm \gamma_\pm . \tag{5}$$

§ 44 *SEDIMENTATION*

In this section we derive relations among the sedimentation coefficient (defined on p. 285), diffusion coefficient and mobility.

According to eq. (8, 11), for diffusion alone with mechanical forces,

$$\sigma = -\frac{1}{T}\sum_k^n [\boldsymbol{J}_k | \operatorname{grad}_T \mu_k - \boldsymbol{F}_k], \tag{1}$$

because

$$\operatorname{grad}_T \mu_k = \operatorname{grad}_{Tp} \mu_k + \frac{\partial \mu_k}{\partial p} \operatorname{grad} p$$

$$= \operatorname{grad}_{Tp} \mu_k + V_k \operatorname{grad} p . \tag{2}$$

At mechanical equilibrium,

$$\sum_k^n \rho_k \boldsymbol{F}_k = \operatorname{grad} p . \tag{3}$$

Substituting eq. (3) in eq. (2),

$$\operatorname{grad}_T \mu_k = \operatorname{grad}_{Tp} \mu_k + V_k \sum_k^n \rho_k \boldsymbol{F}_k . \tag{4}$$

Substituting eq. (4) in eq. (1),

$$\sigma = \frac{1}{T} \sum_{k}^{n} \left[\boldsymbol{J}_k | \boldsymbol{F}_k - V_k \sum_{k}^{n} \rho_k \boldsymbol{F}_k - \mathrm{grad}_{Tp}\, \mu_k \right]. \tag{5}$$

In gravitational field modified by rotation, there are centrifugal and Coriolis forces, eq. (7, 38),

$$\boldsymbol{F}_k = \omega^2 \boldsymbol{r} + 2\boldsymbol{v}_k \times \boldsymbol{\omega}. \tag{6}$$

Substituting eq. (6) in eq. (5),

$$\begin{aligned} \sigma &= \frac{1}{T} \sum_{k}^{n} [\boldsymbol{J}_k | (\omega^2 \boldsymbol{r} + 2\boldsymbol{v}_k \times \boldsymbol{\omega}) \\ &\quad - \rho V_k (\omega^2 \boldsymbol{r} + 2\boldsymbol{v} \times \boldsymbol{\omega}) - \mathrm{grad}_{Tp}\, \mu_k] \\ &= \frac{1}{T} \sum_{k}^{n} [\boldsymbol{J}_k | (1 - \rho V_k)(\omega^2 \boldsymbol{r} + 2\boldsymbol{v} \times \omega) - \mathrm{grad}_{Tp}\, \mu_k], \end{aligned} \tag{7}$$

where we have used eqs. (4, 49), (4, 50) and

$$\sum_{k}^{n} [\boldsymbol{J}_k | \boldsymbol{v}_k \times \boldsymbol{\omega}] = \sum_{k}^{n} [\boldsymbol{J}_k | \boldsymbol{v} \times \boldsymbol{\omega}]. \tag{8}$$

Although $\sum_{k}^{n} [\boldsymbol{J}_k | \omega^2 \boldsymbol{r} + 2\boldsymbol{v} \times \boldsymbol{\omega}]$ vanishes, we refrain from striking out this expression in eq. (7) because it is convenient to retain this for reducing dimensions from n to $(n-1)$. As a matter of fact, the n fluxes and forces in eq. (7) are dependent:

$$\sum_{k}^{n} \boldsymbol{J}_k = 0, \tag{9}$$

$$\sum_{k}^{n} c_k (1 - \rho V_k) = 0, \tag{10}$$

$$\sum_{k}^{n} c_k\, \mathrm{grad}_{Tp}\, \mu_k = 0. \tag{11}$$

Thus the dependency of fluxes and forces in diffusion with

mechanical forces is the same as that of fluxes and forces in diffusion without mechanical forces. We could use the "A" matrix of §38 to reduce dimensions from n to $(n-1)$ and apply the Prigogine theorem of § 10 to choose any reference velocity at mechanical equilibrium,

$$\boldsymbol{J}_i^a = -\frac{1}{T}\sum_k^{n-1}\sum_l^{n-1} L_{ik}^a A_k^a \Big[(1-\rho V_l)(\omega^2\boldsymbol{r} + 2\boldsymbol{v}\times\boldsymbol{\omega}) - \sum_m^{n-1} \mu_{lm}^y \operatorname{grad} y_m\Big]. \tag{12}$$

At static equilibrium,

$$\boldsymbol{X}_k^a = 0. \tag{13}$$

For centrifuge or supercentrifuge experiments,

$$\boldsymbol{v}\times\boldsymbol{\omega} \ll \omega^2\boldsymbol{r}. \tag{14}$$

From eqs. (12—14),

$$(1-\rho V_l)\omega^2\boldsymbol{r} = \sum_m^{n-1} \mu_{lm}^y \operatorname{grad} y_m. \tag{15}$$

One important use of eq. (15) is to determine the molecular weight of high polymers if the partial volume and variation of chemical potential with concentration are known. In the binary system, eq. (15) is simple if molar concentration is used,

$$(1-\rho V_1)\omega^2\boldsymbol{r} = \frac{\partial\mu_1}{\partial n_1}\operatorname{grad} n_1. \tag{16}$$

Because

$$\bar{\mu}_1 = RT\ln n_1\gamma_1, \tag{17}$$

eq. (16) becomes

$$(1-\rho V_1)\omega^2\boldsymbol{r} = \frac{RT}{M_1}\left(1+\frac{\partial\ln\gamma_1}{\partial\ln n_1}\right)\frac{\operatorname{grad} n_1}{n_1}. \tag{18}$$

For other units of concentration, eq. (15) is not so simple. If mass concentration is used, from eqs. (38, 19) and (38, 50),

$$\frac{\text{grad}\, n_1}{n_1} = \frac{\text{grad}\, c_1}{c_1\left[1 - c_1\left(1 - \dfrac{M_2}{M_1}\right)\right]}. \tag{19}$$

If partial density is used, from eqs. (38, 19), (38, 49) and (38, 50),

$$\frac{\text{grad}\, n_1}{n_1} = \frac{\text{grad}\, \rho_1}{\rho_1} \frac{1 - n_1}{1 - \rho_1 V_1}. \tag{20}$$

If partial molar density is used, from eqs. (38, 19) and (38, 48—50),

$$\frac{\text{grad}\, n_1}{n_1} = \frac{\text{grad}\, N_1}{N_1} \frac{1 - n_1}{1 - N_1 \overline{V}_1}. \tag{21}$$

In the multi-component system, we assume that the system is ideal,

$$\gamma_i = 1.$$

Equation (15) is simple if the molar concentration is used,

$$\sum_k^{n-1} \frac{\text{grad}\, n_k}{M_k n_k} = (1 - \rho V_k) \frac{\omega^2 \mathbf{r}}{RT}. \tag{22}$$

Equation (15) is not so simple if other concentration units are used. If the mass concentration, for example, is used. From eq. (38, 16),

$$\text{grad}\, c_k = \frac{\partial c_k}{\partial n_k} \text{grad}\, n_k = \frac{N}{\rho} M_k B_{kk}^{cn} \,\text{grad}\, n_k$$

$$= \frac{N}{\rho} \left[M_k - \sum_j^{n-1} c_j (M_j - M_n)\right] \text{grad}\, n_k. \tag{23}$$

Or

$$\frac{\text{grad}\, n_k}{n_k} = \frac{M_k \text{grad}\, c_k}{c_k \left[M_k - \sum_j^{n-1} c_j (M_j - M_n)\right]}. \tag{24}$$

If the system under centrifugal force is not at equilibrium, both fluxes and forces are not zero. For the fluxes, from eq. (39, 13),

$$\boldsymbol{J}_i^a = \sum_j^{n-1} \rho_i M_i W_{jj}^a \boldsymbol{F}_j. \tag{25}$$

For centrifugal force, $\boldsymbol{F}_j$ is independent of j. We call $M_i W_{jj}^a$ in absolute quantities the sedimentation coefficient S_{ii}^a. From eq. (38, 38), in matrix form,

$$D = T^{-1} L A \mu. \tag{26}$$

Substituting eqs. (25) and (26) in eq. (12),

$$\rho_i S_{ii}^a \omega^2 \boldsymbol{r} = \sum_k^{n-1} D_{ik}^{ay} \left[\sum_m^{n-1} \frac{(1-\rho V_m)\omega^2 \boldsymbol{r}}{\mu_{km}^y} - \text{grad } y_k \right]. \tag{27}$$

Thus in a strong gravitational field, i. e., a centrifuge with a high speed of revolution, sedimentation will occur. Because the force exerted on a particle is proportional to the molecular weight of the particle, the heavier ones sediment faster than the lighter ones until sedimentation equilibrium is reached. The mixture is non-uniform with various species distributed along the distance from the center according to molecular weights. Because sedimentation is a very slow process, except at the extreme end for the rotational axis, there is a large region of homogeneous composition. Setting grad $y_k = 0$ in eq. (27) and using D as a standard diffusion coefficient of a binary system,

$$\frac{D_{11}}{\|S_{11}^0\|} = \frac{\rho_1 \mu_{11}^\rho}{1 - \rho V_1}. \tag{28}$$

This is the Svedberg equation[1]. From eq. (38, 41) (It is clear that D, S, W or U in §39 transform cogrediently with respect to $\boldsymbol{J}$),

$$S_{11}^0 = B_{11}^{0r} S_{11}^r = \rho_2 V_2 S_{11}^r. \tag{29}$$

Substituting D_{11} from eq. (41, 47), μ_{11}^ρ from eq. (41, 42), and

S_{11}^{0} from eq. (29) in (28) and using eq. (41, 11),

$$S_{11}^{r} = \left(\frac{M_1}{\bar{z}_+\bar{\Lambda}_+}\right)\left(\frac{U_+\|U_-\|}{U_+ - \|U_-\|}\right)(1 - \rho V_1), \tag{30}$$

where $\frac{M_1\mathscr{F}}{\bar{z}_+\bar{\Lambda}_+}$ or $\frac{M_1\mathscr{F}}{\|\bar{z}_-\|\bar{\Lambda}_-}$ is the number of equivalents per mole of solute.

For a dilute solution of homogeneous composition, eq. (27) becomes

$$\sum_{k}^{n-1} \rho_i M_i W_{ii}^{a} = \sum_{j}^{n-1}\sum_{k}^{n-1} \frac{B_{ij}^{ab} D_{ik}^{by}}{\mu_{km}^{y}}. \tag{31}$$

In a binary system, if we choose n for a, 0 for b and ρ for y,

$$\rho_1 M_1 W_{11}^{n} = \frac{B_{11}^{n0} D_{11}^{0\rho}}{\mu_{11}^{\rho}}. \tag{32}$$

If we assume that the solution is ideal, i. e.,

$$\bar{\mu}_1 = RT \ln n_1, \tag{33}$$

and use eqs. (38, 46), (38, 49), and (38, 50),

$$D_{11} = RTW_{11}^{n}. \tag{34}$$

In 1905 Einstein formulated eq. (34) in his theory of Brownian motion.

Reference

[1] T. Svedberg: *Kolloid-Z.*, **36**, 53 (1925).

§ 45 *GRAVITATION CELLS*

Up to § 43, we have dealt with the diffusion potential of an electrolyte without external mechanical forces. For the diffusion potential with gravitational field we have to modify the formula for the diffusion potential. In this section we relate a modified diffusion potential to other thermodynamical quantities to find the EMF of a gravitation cell.

From eq. (42, 12),

$$\operatorname{grad}\zeta = -\sum_i^n \frac{t_i}{Z_i}\operatorname{grad}_{Tp}\mu_i. \tag{1}$$

From eq. (44, 7), we see that the external mechanical force to be inserted besides $\operatorname{grad}_{Tp}\mu_i$ is $(1-\rho V_i)\omega^2\boldsymbol{r}$,

$$\operatorname{grad}\zeta' = \sum_i^n \frac{t_i}{Z_i}[(1-\rho V_i)\omega^2\boldsymbol{r} - \operatorname{grad}_{Tp}\mu_i]. \tag{2}$$

The diffusion potential so modified is called the sedimentation potential. At mechanical equilibrium

$$\operatorname{grad} p = \sum_k^n \rho_k \boldsymbol{F}_k = \rho\omega^2\boldsymbol{r}. \tag{3}$$

Substituting eq. (3) in eq. (2),

$$\operatorname{grad}\zeta' = \sum_i^n \Gamma_i\left[\left(\frac{1}{\rho} - V_i\right)\operatorname{grad} p - \operatorname{grad}_{Tp}\mu_i\right], \tag{4}$$

where Γ_i is the mass transference number ($\Gamma_i = t_i/Z_i$; Γ_i is indeterminate for a neutral component).

A gravity or a centrifuge cell consists of two identical electrodes with an electrolytic solution between them. The electrodes are placed in different heights of a gravitational field or in different distances from the rotational axis of a centrifuge cell. During the experiment the temperature of the cell is maintained constant and uniform. The scheme for a gravitational cell is

$$\text{electrode}\,\underset{\alpha}{|}\,\text{solution}\,\underset{\beta}{|}\,\text{electrode}$$

where α and β are the boundaries of the solution and the electrodes are under different pressures. In this sense both gravity and centrifuge cells are pressure cells.

In contrast to concentration cells, the overall contact potential of the electrodes and the leads of gravitational cells is no longer zero because of the pressure difference. It is

$$\int_{p}^{p+\Delta p}\left(\frac{V_{\ominus}^{e}}{Z_{\ominus}}-\frac{V_{\ominus}^{\mathrm{Cu}}}{Z_{\ominus}}\right)dp+\int_{\alpha}^{\beta}\left(\frac{d_{Tp}\mu_{\ominus}^{e}}{Z_{\ominus}}-\frac{d_{Tp}\mu_{\ominus}^{\mathrm{Cu}}}{Z_{\ominus}}\right), \tag{5}$$

where the suffix e refers to the electrode and the suffix Cu refers to the leads. The first integral represents the pressure change and the second represents the concentration change. It is electrons which conduct the current in the leads; because of the very small mass of electrons, the second integral in eq. (5) is negligible. The over-all electrode potential between the electrode and the solution is controlled by the movement of the ion i reversible to the electrode. It is

$$\int_{p}^{p+\Delta p}\left\{\frac{V_{i}^{s}}{Z_{i}}-\frac{V_{i}^{e}}{Z_{i}}\right\}dp+\int_{\alpha}^{\beta}\left\{\frac{d_{Tp}\mu_{i}^{s}}{Z_{i}}-\frac{d_{Tp}\mu_{i}^{e}}{Z_{i}}\right\}, \tag{6}$$

where the suffix s refers to the solution. Because of the small concentration of the ion in the electrode, the second term in the second integral of eq. (6) is negligible. Combining eqs. (4—6), and using

$$V_{i,\mathrm{neu}}=V_{i}-\frac{Z_{i}}{Z_{\ominus}}V_{\ominus}\ \text{[comparing eqs. (42, 15 and 16)]}, \tag{7}$$

the EMF of the gravitational cell is

$$\int_{p}^{p+\Delta p}\left[\frac{V_{i}^{s}}{Z_{i}}-\frac{V_{i,\mathrm{neu}}^{e}}{Z_{i}}-\frac{V_{\ominus}^{\mathrm{Cu}}}{Z_{\ominus}}+\sum_{k}^{3}\Gamma_{k}\left(\frac{1}{\rho^{s}}-V_{k}^{s}\right)\right]dp$$
$$+\int_{\alpha}^{\beta}\left[\frac{d_{Tp}\mu_{i}^{s}}{Z_{i}}-\sum_{k}^{3}\Gamma_{k}d_{Tp}\mu_{k}^{s}\right]. \tag{8}$$

Suppose that the cation is reversible to the electrode. For the first square bracket in eq. (8),

$$\frac{V_{+}^{s}}{Z_{+}}-\frac{V_{+,\mathrm{neu}}^{e}}{Z_{+}}-\frac{V_{\ominus}^{\mathrm{Cu}}}{Z_{\ominus}}+\frac{1-t_{-}}{Z_{+}}\left(\frac{1}{\rho^{s}}-V_{+}^{s}\right)$$

$$+ \frac{t_-}{Z_-}\left(\frac{1}{\rho^s} - V^s_-\right) + \Gamma_1\left(\frac{1}{\rho^s} - V^s_1\right)$$

$$= -\frac{V^{Cu}_{\ominus}}{Z_{\ominus}} - \frac{V^e_{+,neu}}{Z_+} + \frac{1}{Z_+\rho^s} - \frac{t_-}{\rho^s}\left(\frac{1}{Z_+} - \frac{1}{Z_-}\right)$$

$$+ t_-\left(\frac{V^s_+}{Z_+} - \frac{V^s_-}{Z_-}\right) + \Gamma_1\left(\frac{1}{\rho^s} - V^s_1\right)$$

$$= -\frac{V^{Cu}_{\ominus}}{Z_{\ominus}} - \frac{V^e_{+,neu}}{Z_+} + \frac{1}{Z_+\rho^s}$$

$$-\left(\frac{t_-}{Z_+\Lambda_+} - \Gamma_1\right)\left(\frac{1}{\rho^s} - V^s_1\right), \tag{9}$$

eqs. (41, 7) and (41, 9) having been used. For the second square bracket in eq. (8), with the superscript s and the subscript Tp omitted,

$$\frac{d\mu_+}{Z_+} - \frac{t_+d\mu_+}{Z_+} - \frac{t_-d\mu_-}{Z_-} - \Gamma_1 d\mu_1$$

$$= t_-\left(\frac{Z_-d\mu_+ - Z_+d\mu_-}{Z_+Z_-}\right) - \Gamma_1 d\mu_1$$

$$= \frac{t_-d\mu_1}{Z_+\Lambda_+} - \Gamma_1 d\mu_1 = \left(\frac{t_-}{Z_+\Lambda_+} - \Gamma_1\right) d\mu_1, \tag{10}$$

eq. (41, 9) having been used. If we let

$$\mathscr{D}_- = t_- + Z_-\Lambda_-\Gamma_1, \tag{11}$$

$$\mathscr{D}_+ = t_+ + Z_+\Lambda_+\Gamma_1, \tag{12}$$

evidently

$$\mathscr{D}_+ + \mathscr{D}_- = 1. \tag{13}$$

If i is the ion reversible to the electrode, from eqs. (9—12), the EMF of the gravitational cell is

$$\int_p^{p+\Delta p}\left[-\frac{1}{\rho^{Cu}_{\ominus}Z_{\ominus}} - \frac{1}{\rho^e_{i,neu}Z_i} + \frac{1}{Z_i\rho^s}\right.$$

$$\left.-\left(\frac{1-\mathscr{D}_i}{Z_i\Lambda_i}\right)\left(\frac{1}{\rho^s} - V^s_1\right)\right] dp$$

$$+ \frac{1}{Z_i \Lambda_i} \int_{\alpha}^{\beta} (1 - \mathscr{D}_i) d_{Tp} \mu_i^s . \tag{14}$$

If $dp = 0$, the EMF of a gravitational cell is reduced to that of a concentration cell, eq. (43, 3) ($1 - \mathscr{D}_i$ is an abbreviation: Γ_i vanishes). At $t = 0$, the concentration gradient is not set up and the second integral of eq. (14) is dropped. At $t = \infty$, the concentration gradient is fully set up and there is no sedimentation potential. From eq. (4),

$$d_{Tp} \mu_i^s = \left(\frac{1}{\rho^s} - V_1^s \right) dp. \tag{15}$$

The EMF of a gravitational cell is reduced to the first three terms of the first integral of eq. (14). Transference numbers are no longer needed.

§ 46 *CONDUCTION OF HEAT*

In this section we introduce heat current (heat flux), thermal gradient (force for the conduction of heat), and thermal conductivity (related to phenomenological coefficient). In spite of three definitions of heat flux, there is only one definition of thermal conductivity if the irreversible process of heat conduction alone occurs. However, the interaction between the conduction of heat and the transport of matter gives two more definitions of heat conductivity. We also describe the simplest verification of the Onsager law, in heat conduction for certain point groups of hexagonal, tetragonal and trigonal crystals.

From eq. (8, 17) with heat conduction alone and the assumption of linearity,

$$\boldsymbol{J}_q^r = - L_{qq} \frac{\operatorname{grad} T}{T^2} = - \theta \operatorname{grad} T, \tag{1}$$

where $\theta = L_{qq}/T^2$ is the thermal conductivity. As there is no diffusion, $\boldsymbol{J}'_q$ is the same as $\boldsymbol{J}_q$ or $T\boldsymbol{J}_S$. There is only one definition of thermal conductivity.

From eq. (8, 17) with heat conduction and diffusion only and assumption of linearity,

$$\boldsymbol{J}'_q = -L_{qq}\boldsymbol{X}_q - \sum_k^{n-1} L_{qk}\boldsymbol{X}_k, \tag{2}$$

$$\boldsymbol{J}_i = -L_{iq}\boldsymbol{X}_q - \sum_k^{n-1} L_{ik}\boldsymbol{X}_k, \tag{3}$$

where $\boldsymbol{X}_q = \dfrac{\text{grad}\, T}{T^2}$, and $\boldsymbol{X}_k = \sum_j^n A_{kj}\dfrac{\text{grad}\,\mu_j}{T}$. A_{kj} is the matrix A defined in eq. (38, 11). At the beginning of heat conduction, the concentration gradient grad y_m is not set up, i. e.,

$$\text{grad}\,\mu_j = \sum_m^{n-1} \mu^y_m\, \text{grad}\, y_m = 0,$$

$$\begin{aligned}\boldsymbol{J}_q &= \boldsymbol{J}'_q + \sum_k^n H_k\boldsymbol{J}_k = \boldsymbol{J}'_q + \sum_k^{n-1} (H_k - H_n)\boldsymbol{J}_k \\ &= -\left[L_{qq} + \sum_k^{n-1} (H_k - H_n)L_{kq}\right]\boldsymbol{X}_q.\end{aligned} \tag{4}$$

If we define

$$\boldsymbol{J}_q = -\theta'\, \text{grad}\, T\ (\text{grad}\, y_m = 0), \tag{5}$$

$$\theta' = \frac{1}{T^2}\left[L_{qq} + \sum_k^{n-1} (H_k - H_n)L_{kq}\right]. \tag{6}$$

From eq. (3),

$$\boldsymbol{X}_k = -\sum_i^{n-1} (L^{-1})_{ik}(\boldsymbol{J}_i + L_{iq}\boldsymbol{X}_q). \tag{7}$$

Substituting eq. (7) in eq. (2),

$$\boldsymbol{J}'_q = -\left[L_{qq} - \sum_k^{n-1}\sum_i^{n-1} L_{qk}(L^{-1})_{ik}L_{iq}\right]\boldsymbol{X}_q$$

$$+\sum_{k}^{n-1}\sum_{i}^{n-1} L_{qk}(L^{-1})_{ik}\boldsymbol{J}_i. \tag{8}$$

If $\boldsymbol{J}_i = 0$,

$$\boldsymbol{J}_q = \boldsymbol{J}_q^r = -\left[L_{qq} - \sum_{k}^{n-1}\sum_{i}^{n-1} L_{qk}(L^{-1})_{ik}\, L_{iq}\right]\boldsymbol{X}_q. \tag{9}$$

If we define

$$\boldsymbol{J}_q = \boldsymbol{J}_q^r = -\,\theta''\mathrm{grad}\, T(\boldsymbol{J}_i = 0), \tag{10}$$

$$\theta'' = \frac{1}{T^2}\left[L_{qq} - \sum_{k}^{n-1}\sum_{i}^{n-1} L_{qk}(L^{-1})_{ik}L_{iq}\right]. \tag{11}$$

So far the Onsager law has not been used. With the application of the Onsager law, a new concept of the heat of transport can be developed, which is to be discussed in the next section.

By lining up the crystallographic axes with the coordinate axes as far as possible, the matrix of the general thermal conductivity tensor

$$\begin{bmatrix} \theta_{11} & \theta_{12} & \theta_{13} \\ \theta_{21} & \theta_{22} & \theta_{23} \\ \theta_{31} & \theta_{32} & \theta_{33} \end{bmatrix} \tag{12}$$

can be simplified. The more symmetric is the crystal, the greater is the simplification. In all 5 point groups of the cubic system, it can be simplified to

$$\begin{bmatrix} \theta_{11} & 0 & 0 \\ 0 & \theta_{11} & 0 \\ 0 & 0 & \theta_{11} \end{bmatrix}. \tag{13}$$

In four groups of the hexagonal system (622, 6mm, $\bar{3}$, 6/mmm) and four groups of the tetragonal system (422, 4mm, $\bar{4}2m$, 4/ mmm), it can be simplified to

$$\begin{bmatrix} \theta_{11} & 0 & 0 \\ 0 & \theta_{11} & 0 \\ 0 & 0 & \theta_{33} \end{bmatrix}. \tag{14}$$

In all three groups of the rhombic system and three groups of the trigonal system (32, $3m$, $\bar{3}m$), it can be simplified to

$$\begin{bmatrix} \theta_{11} & 0 & 0 \\ 0 & \theta_{22} & 0 \\ 0 & 0 & \theta_{33} \end{bmatrix}. \tag{15}$$

In the above nineteen groups, the consideration of geometrical symmetry alone has resulted in a symmetric matrix quite independently of the Onsager law. Thus a test for the Onsager law could come only from the remaining thirteen groups. In all two groups of the triclinic system it admits no simplification. In all three groups of the monoclinic system it can be simplified to

$$\begin{bmatrix} \theta_{11} & \theta_{12} & 0 \\ \theta_{21} & \theta_{22} & 0 \\ 0 & 0 & \theta_{33} \end{bmatrix}. \tag{16}$$

In three groups of the hexagonal system (6, $\bar{6}$, $6/m$), three groups of the tetragonal system (4, $\bar{4}$, $4/m$), and two groups of the trigonal system (3, $\bar{3}$), it can be simplified to

$$\begin{bmatrix} \theta_{11} & \theta_{12} & 0 \\ -\theta_{12} & \theta_{11} & 0 \\ 0 & 0 & \theta_{33} \end{bmatrix}. \tag{17}$$

Hence the simplest cases are those of the eight point groups of the hexagonal, tetragonal, and trigonal systems which have the same form of matrix, matrix (17). In § 15 we have shown how the Onsager law is used to justify the symmetric matrix determined experimentally. For the remainder of this section, we describe Voigt's method[1], suggested also by Curie[2], for the direct verification of the Onsager law.

Suppose that a fixed temperature difference is applied to the ends of a long, narrow and thin plate of a crystal, of which the matrix of the thermal conductivity tensor is the same as matrix (17). Let the axis x_1 be along the length, x_2 along the width, and the principal axis x_3 perpendicular to the plate. We have

$$\begin{bmatrix} (J_q)_1 \\ (J_q)_2 \\ (J_q)_3 \end{bmatrix} = \begin{bmatrix} \theta_{11} & \theta_{12} & 0 \\ -\theta_{12} & \theta_{11} & 0 \\ 0 & 0 & \theta_{33} \end{bmatrix} \begin{bmatrix} \partial T/\partial x_1 \\ \partial T/\partial x_2 \\ \partial T/\partial x_3 \end{bmatrix}. \tag{18}$$

Because there is no heat applied in the x_2 or x_3 direction, $(J_q)_2$ and $(J_q)_3$ vanish (Clearly we are dealing with isotropy or anisotropy in the plane. What happens in the x_3 direction is irrelevant). From eqs. (18),

$$\frac{\theta_{12}}{\theta_{11}} = \frac{\dfrac{\partial T}{\partial x_2}}{\dfrac{\partial T}{\partial x_1}} = \tan \alpha, \tag{19}$$

where α is the angle which the isothermal straight line (If the plate were heated at the center only, the isotherms would be circles, ellipses or spirals depending on the degree of isotropy) makes with the normal of heat flow, i. e., the x_2 direction. If this line is inclined away from the normal, the matrix (17) is not symmetric.

Suppose that the plate has been covered with a thin layer of wax. When a point in the plate is heated, the wax will melt in the region where the temperature is higher than its freezing point. The boundary line between the melted and solid wax is an isotherm. The trace of an isotherm becomes visible as a raised edge.

Owing to the possible heat loss from the edges, it is more accurate to use Voigt's "twin plate" method. Suppose that the plate is sawed in half along the x_1 axis and one piece is rotated 180° about the x_2 axis. Instead of measuring α,

$$\beta = \pi - 2\alpha$$

is measured. Then

$$\frac{\theta_{12}}{\theta_{11}} = \tan\frac{\pi - \beta}{2}. \tag{20}$$

Voigt found that for suitable crystals of apatite and dolomite (both hexagonal), with a heated solid copper slab on one end and another slab at room temperature on the other end, the isothermal lines were straight and parallel to x_2. More precisely, β was 180° with an error of not more than 4 minutes. Thus α is less than 2 minutes. This implies that $\theta_{12} = 0$ to less than 0.05%.

References

[1] W. Voigt: Lehrbuch der Kristallphysik, Teubner, Leipzig (1910).
[2] P. Curie: *Arch. Sci.* (Geneva) **29**, 342 (1893).

§ 47 *HEAT OF TRANSPORT*

In this section we introduce the concept of heat of transport, as the result of the interaction between conduction of heat and transport of matter. Because we could define heat flux in three different ways for n or $n-1$ dimensions, there are six kinds of heat of transport. The heat of transport derived from the reduced flux of heat is again preferred, because it is invariant in both n and $n-1$ dimensions.

If we define

$$\sum_{k}^{n-1} L_{qk}(L^{-1})_{ki} = Q_i^{r*}, \tag{1}$$

or, because of the Onsager law,

$$\sum_{k}^{n-1} (L^{-1})_{ik}L_{kq} = Q_i^{r*}, \tag{2}$$

eq. (46, 8) becomes

$$\boldsymbol{J}_q^r = -\left(L_{qq} - \sum_i^{n-1} Q_i^{r*} L_{iq}\right) \boldsymbol{X}_q + \sum_i^{n-1} Q_i^{r*} \boldsymbol{J}_i. \tag{3}$$

At uniform temperature,

$$\boldsymbol{X}_q = 0. \tag{4}$$

Then

$$\boldsymbol{J}_q^r = \sum_i^{n-1} Q_i^{r*} \boldsymbol{J}_i. \tag{5}$$

Physically Q_i^{r*} is the heat of transport of the component i associated with isothermal diffusion. Mathematically it is the matrix of phenomenological coefficients of thermal diffusion divided by the matrix of coefficients of ordinary diffusion. Because of the Onsager law, we need not differentiate between the two kinds of matrix division in eqs. (1) and (2).

In § 8 we have given the definitions of $\boldsymbol{J}_q^r$, $\boldsymbol{J}_q$, the $\boldsymbol{J}_S$. It is not surprising that we could define three kinds of heat of transport. From one viewpoint, the flux of heat is indistinguishable from the flux of heat of transport. In the expression of $\boldsymbol{J}_q^r + \sum_k^n H_k \boldsymbol{J}_k$, we single out the enthalpy to be responsible for the heat of transport. In the expression

$$\frac{\boldsymbol{J}_q^r + \sum_k^n S_k \boldsymbol{J}_k}{T},$$

we single out the entropy to be responsible for the heat of transport. Physically there is only one entity. There are relations among the different definitions of heat flux. In § 8 we define

$$\boldsymbol{J}_q - \sum_i^n H_i \boldsymbol{J}_i = \boldsymbol{J}_q^r, \quad \text{and } \boldsymbol{J}_q = T\boldsymbol{J}_S + \sum_i^n \mu_i \boldsymbol{J}_i. \tag{6}$$

Therefore

$$J_q^r = TJ_S - T\sum_i^n S_iJ_i. \tag{7}$$

The other kinds of heat of transport similar to eq. (5) are

$$J_q = \sum_i^{n-1} Q_i^* J_i \tag{8}$$

and

$$TJ_S = T\sum_i^{n-1} S_i^* J_i. \tag{9}$$

From eqs. (6) and (7) the relations between different kinds of heat of transport or between heat of transport and entropy of transport [comparing eqs. (22, 25—26), where $Q^* = Q^{r*}$, and $H^* = Q^*$] are

$$Q_i^* = Q_i^{r*} + H_i - H_n, \tag{10}$$

$$S_i^* = \frac{1}{T}(Q_i^* - \mu_i + \mu_n) = \frac{Q_i^{r*}}{T} + S_i - S_n. \tag{11}$$

So far the different definitions of heat of transport are given in $(n-1)$ dimensions, because n members of J are dependent. The dependency can be removed by defining J in absolute sense,

$$J_i^{\text{abs}} = \rho_i v_i = J_i + \rho_i v. \tag{12}$$

Similarly we define

$$J_q^{\text{abs}} = J_q + \sum_i^n \rho_i H_i v = J_q + \rho H v, \tag{13}$$

$$J_S^{\text{abs}} = J_S + \sum_i^n \rho_i S_i v = J_S + \rho S v. \tag{14}$$

For the moment we leave $J_q^{r,\text{abs}}$ undefined. Combining eqs. (6), (13) and (14),

$$J_q^r = J_q^{\text{abs}} - \sum_i^n H_i J_i^{\text{abs}} = TJ_S^{\text{abs}} - T\sum_i^n S_i J_i^{\text{abs}}. \tag{15}$$

Thus J_q^r is invariant in n and $(n-1)$ dimensions.

Corresponding to eqs. (5), (8), and (9), we define

$$\boldsymbol{J}_q^r = \sum_i^n Q_i^{r,\mathrm{abs}*} \boldsymbol{J}_i^{\mathrm{abs}}, \tag{16}$$

$$\boldsymbol{J}_q^{\mathrm{abs}} = \sum_i^n Q_i^{\mathrm{abs}*} \boldsymbol{J}_i^{\mathrm{abs}}, \tag{17}$$

$$\boldsymbol{J}_S^{\mathrm{abs}} = \sum_i^n S_i^{\mathrm{abs}*} \boldsymbol{J}_i^{\mathrm{abs}}. \tag{18}$$

Because

$$\begin{aligned} \boldsymbol{J}_q^r &= \sum_i^n Q_i^{r,\mathrm{abs}*}(\boldsymbol{J}_i + \rho_i \boldsymbol{v}) \\ &= \sum_i^{n-1} Q_i^{r,\mathrm{abs}*} \boldsymbol{J}_i + Q_n^{r,\mathrm{abs}*} \boldsymbol{J}_n + \sum_i^n Q_i^{r,\mathrm{abs}*} \rho_i \boldsymbol{v} \\ &= \sum_i^{n-1} (Q_i^{r,\mathrm{abs}*} - Q_n^{r,\mathrm{abs}*}) \boldsymbol{J}_i + \sum_i^n Q_i^{r,\mathrm{abs}*} \rho_i \boldsymbol{v}, \end{aligned} \tag{19}$$

comparing eq. (19) to eq. (5) shows that the equations for heat of transport for reduced heat flux are

$$Q_i^{r,\mathrm{abs}*} - Q_n^{r,\mathrm{abs}*} = Q_i^{r*}, \tag{20}$$

$$\sum_i^n Q_i^{r,\mathrm{abs}*} \rho_i = 0. \tag{21}$$

Similarly the equations for heat of transport for heat flux are

$$Q_i^{\mathrm{abs}*} - Q_n^{\mathrm{abs}*} = Q_i^*, \tag{22}$$

$$\sum_i^n Q_i^{\mathrm{abs}*} \rho_i = \rho H. \tag{23}$$

Also the equations of entropy of transport for entropy flux are

$$S_i^{\mathrm{abs}*} - S_n^{\mathrm{abs}*} = S_i^*, \tag{24}$$

$$\sum_i^n S_i^{\mathrm{abs}*} \rho_i = \rho S. \tag{25}$$

§ 48 THE SORET EFFECT AND THE DUFOUR EFFECT

In this section we deal with the simultaneous flow of heat and matter. The Soret effect and the Dufour effect are the cross effects.

When an initially uniform liquid solution is placed in a nonuniform temperature field, which consists of two parallel plates, with the warm plate above the cold one to prevent convection, the temperature gradient will be built up before there is any appreciably separation. As a matter of fact, the time required to set up the temperature gradient is much less than the characteristic time for diffusion, which depends on the distance between the plates. When the temperature gradient is set up, the thermal diffusion will start, forming gradually a concentration gradient. Then the ordinary diffusion will start, which tends to diminish the separation. After a while, a stationary state exists for which thermal diffusion exactly balances ordinary diffusion. That a concentration gradient is established as a result of the temperature gradient is called the Soret effect. The ratio of thermal diffusion to ordinary diffusion is called the Soret coefficient. Because the thermal diffusion can be in either of the two opposite directions, the sign of the Soret coefficient is indeterminate even with an arbitrary direction agreed as positive. Furthermore, because there are many definitions of the ordinary diffusion coefficients, there are also many definitions of the Soret coefficients.

The reciprocal phenomenon of the Soret effect is called the Dufour effect, the transport of heat by concentration gradients.

The simplest system in which the Soret effect can be determined is a binary one,

$$J_1 = -L_{1q}\frac{\text{grad } T}{T^2} - \frac{L_{11}^c A_{11}^c \mu_{11}^c}{T}\text{grad } c_1. \tag{1}$$

If $\text{grad } T = 0$, $J_1 = -\rho D \text{ grad } c_1$ and

$$L_{11}^c = Tc_2(\mu_{11}^c)^{-1}\rho D. \tag{2}$$

Let the thermal diffusion coefficient be D^{th},

$$\frac{D^{\text{th}}}{D} = -\frac{\rho \text{ grad } c_1}{\rho c_1 c_2 \text{ grad } T}. \tag{3}$$

Equation (1) becomes

$$J_1 = -\rho c_1 c_2 D^{\text{th}} \text{grad } T - \rho D \text{ grad } c_1,$$

and

$$L_{1q} = \rho c_1 c_2 T^2 D^{\text{th}}. \tag{4}$$

At steady state,

$$\frac{D^{\text{th}}}{D} = -\frac{1}{c_1 c_2}\frac{\text{grad } c_1}{\text{grad } T} = -\frac{1}{n_1 n_2}\frac{\text{grad } n_1}{\text{grad } T}. \tag{5}$$

From eqs. (38, 41), (38, 42), (38, 46) and (38, 49),

$$D^{cc} = \rho D, \tag{6}$$

$$\frac{D^{\text{th}}}{D^{cc}} = -\frac{1}{\rho c_1 c_2}\frac{\text{grad } c_1}{\text{grad } T}. \tag{7}$$

From eqs. (38, 41), (38, 46) and (38, 49),

$$D^{r\rho} = \frac{1}{\rho_2 V_2} D, \tag{8}$$

$$\frac{D^{\text{th}}}{D^{r\rho}} = -\frac{\rho_2 V_2}{c_1 c_2}\frac{\text{grad } c_1}{\text{grad } T} = -\frac{1}{\rho_1}\frac{\text{grad } \rho_1}{\text{grad } T}. \tag{9}$$

If we reserve the Soret coefficient for

$$\mathscr{S} = \frac{D^{\text{th}}}{D} = -\frac{1}{c_1 c_2}\frac{\text{grad } c_1}{\text{grad } T}, \tag{10}$$

we call

$$\mathscr{S}T = \alpha \tag{11}$$

the thermal diffusion factor and

$$\mathscr{S}Tc_1c_2 = r \tag{12}$$

the thermal diffusion ratio. For the binary mixture,

$$\boldsymbol{J}_q^r = -L_{qq}\frac{\text{grad}\,T}{T^2} - \frac{L_{q1}\mu_{11}^c}{c_2T}\text{grad}\,c_1, \tag{13}$$

and

$$\frac{L_{qq}}{T^2} = \theta, \quad L_{1q} = L_{q1}. \tag{14}$$

Because

$$L_{qq}L_{11}^c - L_{1q}L_{q1} > 0, \tag{15}$$

$$D_{\text{th}}^2 < \frac{\theta D}{T\rho c_1^2 c_2 \mu_{11}^c}. \tag{16}$$

Because

$$Q_{11}^* = \frac{L_{1q}}{L_{11}^c} = -\frac{\mu_{11}^c T\,\text{grad}\,c_1}{c_2\,\text{grad}\,T} = c_1\mu_{11}^c T\mathscr{S}, \tag{17}$$

the heat of transport can be found indirectly from the Soret coefficient.

§ 49 *THERMOELECTRICITY*

In this and the next section we deal with the simultaneous flow of heat and electricity. There are a number of thermoelectric effects depending on the degree of homogeneity and isotropy of the system. In this section, we limit ourselves to isotropic thermocouples without a magnetic field.

For a metal or an alloy, the components are metallic ions and electrons. If we use the velocity of the metallic ions as the reference velocity and assign it zero value, we are dealing only with the electrons. No subscripts are necessary for the electrical charge Z, the chemical potential μ, the velocity $\boldsymbol{v}$ or the diffusin flux $\boldsymbol{J}$. Furthermore, there is no difference between the relative current density $\boldsymbol{i}$ and the absolute current density $\boldsymbol{I}$.

Equation (10, 10) is reduced to

$$T\sigma = -[\boldsymbol{J}_{S,\mathrm{tot}}|\operatorname{grad} T] + [\rho\boldsymbol{v}|\boldsymbol{F} - \operatorname{grad}\mu]. \tag{1}$$

With

$$Z\rho\boldsymbol{v} = Z\boldsymbol{J} = \boldsymbol{I}, \tag{2}$$

$$\frac{\boldsymbol{F} - \operatorname{grad}\mu}{Z} = \boldsymbol{E} - \frac{\operatorname{grad}\mu}{Z} = -\left[\operatorname{grad}\zeta + \frac{\operatorname{grad}\mu}{Z}\right], \tag{3}$$

where $\operatorname{grad}\zeta + (\operatorname{grad}\mu)/Z$ is the electrochemical or thermo-diffusion potential. If the hypothesis of linearity is valid,

$$\boldsymbol{J}_{S,\mathrm{tot}} = -\frac{L_{11}\operatorname{grad} T}{T} - \frac{L_{12}}{T}\left(\frac{\operatorname{grad}\mu}{Z} + \operatorname{grad}\zeta\right), \tag{4}$$

$$\boldsymbol{I} = -\frac{L_{21}\operatorname{grad} T}{T} - \frac{L_{22}}{T}\left(\frac{\operatorname{grad}\mu}{Z} + \operatorname{grad}\zeta\right). \tag{5}$$

Our first task is to relate the phenomenological coefficients to experimentally determinable quantities. If the temperature is uniform, or $\operatorname{grad} T = 0$, from eq. (5),

$$\boldsymbol{I} = -\frac{L_{22}}{T}\left(\frac{\operatorname{grad}\mu}{Z} + \operatorname{grad}\zeta\right). \tag{6}$$

Or

$$L_{22} = \frac{T}{R}, \tag{7}$$

where R is the specific resistance. If the electrons have not separated, there is no electrical potential and there is no concentration gradient of electrons, $(\operatorname{grad}\mu)/Z + \operatorname{grad}\zeta = 0$; from eq. (4),

$$\boldsymbol{J}_{S,\mathrm{tot}} = -\frac{L_{11}\operatorname{grad} T}{T}. \tag{8}$$

From eq. (46, 5),

$$\boldsymbol{J}_{S,\mathrm{tot}} = -\frac{\theta'\operatorname{grad} T}{T}. \tag{9}$$

Hence

$$L_{11} = \theta', \tag{10}$$

where θ' is the thermal conductivity if the concentration gradient vanishes. Eliminating $(\operatorname{grad}\mu)/Z + \operatorname{grad}\zeta$ from eqs. (4) and (5),

$$\boldsymbol{J}_{S,\mathrm{tot}} = \frac{L_{12}\boldsymbol{I}}{L_{22}} + \left(\frac{L_{12}^2}{L_{22}} - L_{11}\right)\frac{\operatorname{grad} T}{T}, \tag{11}$$

the Onsager law

$$L_{12} = L_{21} \tag{12}$$

having been used. From eqs. (46, 8) and (47, 1),

$$\boldsymbol{J}_{S,\mathrm{tot}} = S^*\boldsymbol{J} - \frac{\theta'' \operatorname{grad} T}{T} = \frac{S^*\boldsymbol{I}}{Z} - \frac{\theta'' \operatorname{grad} T}{T}, \tag{13}$$

where θ'' is the thermal conductivity if there is no flow of current and S^* is the entropy of transport for the electrons. Comparing eqs. (11) and (13),

$$L_{12} = S^*T/ZR, \tag{14}$$

$$\frac{\theta'}{T} - \frac{\theta''}{T} = \frac{(S^*)^2}{Z^2R}. \tag{15}$$

Because the matrix of phenomenological coefficients is positive definite,

$$L_{11} > 0, \quad L_{22} > 0, \quad L_{11}L_{22} - L_{12}^2 > 0. \tag{16}$$

In terms of physical quantities,

$$\theta' > 0, \quad R > 0, \quad (S^*)^2 < \frac{\theta' Z^2 R}{T}. \tag{17}$$

While the first two parts of eq. (17) merely confirm the positiveness of thermal conductivity and specific resistance, the third part sets the upper limit of the magnitude of entropy of transport for electrons. Using eqs. (7), (10), (14) and (15), we rewrite eqs. (4) and (5) as

$$\boldsymbol{I} = -\frac{S^* \operatorname{grad} T}{ZR} - \frac{1}{R}\left(\frac{\operatorname{grad} \mu}{Z} + \operatorname{grad} \zeta\right), \tag{18}$$

$$\boldsymbol{J}_{S,\mathrm{tot}} = -\frac{\theta' \operatorname{grad} T}{T} - \frac{S^*}{ZR}\left(\frac{\operatorname{grad} \mu}{Z} + \operatorname{grad} \zeta\right)$$

$$= \frac{S^* \boldsymbol{I}}{Z} - \frac{\theta'' \operatorname{grad} T}{T}. \tag{19}$$

Except for the absolute value of S^*, all other physical quantities can be measured.

From eqs. (8, 9) and (8, 21),

$$\rho \frac{dS}{dt} = -\operatorname{div} \boldsymbol{J}_S - \frac{1}{T}(\boldsymbol{J}_{S,\mathrm{tot}} \cdot \operatorname{grad} T)$$

$$- \boldsymbol{I} \cdot \frac{1}{T}\left(\frac{\operatorname{grad} \mu}{Z} + \operatorname{grad} \zeta\right). \tag{20}$$

Using eqs. (4, 11) and (10, 11),

$$\frac{\partial \rho S}{\partial t} = -\operatorname{div} \boldsymbol{J}_{S,\mathrm{tot}} - \frac{1}{T}(\boldsymbol{J}_{S,\mathrm{tot}} \cdot \operatorname{grad} T)$$

$$- \boldsymbol{I} \cdot \frac{1}{T}\left(\frac{\operatorname{grad} \mu}{Z} + \operatorname{grad} \zeta\right). \tag{21}$$

Substituting eqs. (18) and (19) in eq. (21) and using $\operatorname{div} \boldsymbol{I} = 0$,

$$\frac{\partial \rho S}{\partial t} = \frac{1}{T} \operatorname{div}(\theta'' \operatorname{grad} T) + \frac{I^2 R}{T}$$

$$+ \frac{\boldsymbol{I}}{ZT} \cdot (S^* \operatorname{grad} T - \operatorname{grad} TS^*). \tag{22}$$

Or

$$T \frac{\partial \rho S}{\partial t} = \operatorname{div}(\theta'' \operatorname{grad} T) + I^2 R - \frac{\boldsymbol{I}}{Z} \cdot (T \operatorname{grad} S^*)$$

$$= \operatorname{div}(\theta'' \operatorname{grad} T) + I^2 R$$

$$- \frac{\boldsymbol{I}}{Z}\left[(T \operatorname{grad} S^*_{T=\mathrm{constant}}) + T(\operatorname{grad} T)\frac{\partial S^*}{\partial T}\right]. \tag{23}$$

The left side of eq. (23) is dissipated energy per unit volume per unit time. The first term on the right side is due to conduction of heat. The second term is the joulean heat. The third term must be related to some heat quantity. We shall return to this later.

Consider a thermocouple, the components of which are materials A and B. One junction is at temperature T_1 and the other at $T_2, T_2 > T_1$. The scheme of the thermocouple is

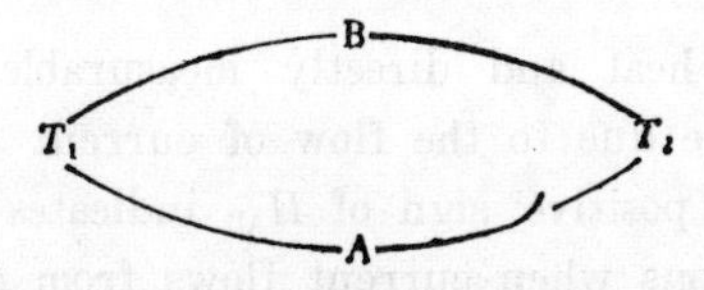

From eq. (18),

$$\frac{\dfrac{\operatorname{grad}\mu}{Z} + \operatorname{grad}\zeta}{\operatorname{grad} T_{I=0}} = -\frac{S^*}{Z}. \qquad (24)$$

Because $(\operatorname{grad}\mu)/Z + \operatorname{grad}\zeta$ is continuous at junctions,

$$\frac{\Delta\mu}{Z} + \Delta\zeta = (S_B^* - S_A^*)\frac{\Delta T}{Z} \text{ for } I = 0. \qquad (25)$$

But at junctions $\Delta\mu = 0$,

$$\left(\frac{\Delta\zeta}{\Delta T}\right)_{I=0} = \frac{S_B^* - S_A^*}{Z}. \qquad (26)$$

Consider the thermocouple as a whole,

$$\left(\frac{d\zeta}{dT}\right)_{I=0} = -\frac{S_A^* - S_B^*}{Z} = \pi_{AB}, \qquad (27)$$

π is the thermoelectric power of the Seebeck effect and directly measurable. It is defined as the potential difference per unit temperature difference under the condition of no flow of current. The cold junction is on the left and A is the lower member of the couple. The positive sign of π_{AB} means the flow of current is in the clockwise direction. If the position of A is switched with that of B, the sign of π is reversed.

From eq. (19)

$$\left(\frac{T\boldsymbol{J}_{S,\mathrm{tot}}}{\boldsymbol{I}}\right)_{\mathrm{grad}T=0} = TS^*/Z\,. \tag{28}$$

But TS^*/Z is continuous at junctions,

$$T\boldsymbol{J}_{S,\mathrm{tot}} = T(S^*_{\mathrm{A}} - S^*_{\mathrm{B}})\,\frac{\boldsymbol{I}}{Z} = \Pi_{\mathrm{AB}}\boldsymbol{I} \quad \text{if} \quad \mathrm{grad}\;T = 0. \tag{29}$$

Π is the Peltier heat and directly measurable. It is defined as the heat change due to the flow of current under isothermal conditions. The positive sign of Π_{AB} indicates that the heat is absorbed at junctions when current flows from A to B. Another measurable quantity is the Thomson heat. $\Sigma_{\mathrm{A}}\Delta T$ is the heat change when unit current traverses a temperature gradient. The positive sign of Σ_{A} means that heat is absorbed when unit current flows from the cold to the hot junction.

Because of the Onsager law, only three phenomenological coefficients are independent, or that three independent physical quantities can be measured. This means only one independent physical quantity besides thermal conductivity and specific resistance. Therefore, there are two relations amongst the thermoelectric power, the Peltier effect and the Thomson heat. Comparing eqs. (27) and (29),

$$\pi = \frac{d\zeta}{dT} = -\frac{\Pi}{T}\,. \tag{30}$$

Equation (30) depicts the relation between thermoelectric power and Peltier effect. It is the same as Thomson's second relation[1], admittedly an unjustifiable derivation. It is the consequence of the Onsager law; otherwise the meaning of S^* is not unique. In returning to the last term in eq. (23), we notice that the first part is flow of current under isothermal conditions, or Peltier heat; and that the second part is flow of current in a temperature gradient, or Thomson heat. If current is not allowed to flow, it appears as a potential. Thus

$$-d\Pi-(\Sigma_A-\Sigma_B)dT=d\zeta. \tag{31}$$

From eq. (30),

$$T\frac{d^2\zeta}{dT^2}=T\frac{d\left(\frac{-\Pi}{T}\right)}{dT}=-\frac{d\Pi}{dT}+\frac{\Pi}{T}$$

$$=-\frac{d\Pi}{dT}-\frac{d\zeta}{dT}. \tag{32}$$

Combining eqs. (31) and (32),

$$T\frac{d^2\zeta}{dT^2}=\Sigma_A-\Sigma_B. \tag{33}$$

Equation (33) depicts the relation between the thermoelectric power and the Thomson effect. It is the same as Thomson's first relation. It is the consequence of the law of conservation of energy.

It is easily seen that the relative entropy of transport of two components of a thermocouple can be determined from thermoelectric power at any temperature. The zero point can be fixed by setting the relative entropy of transport zero at zero degree Kelvin.

Reference

[1] W. Thomson: *Proc. Royal Soc.* (Edinburgh) **3**, 225 (1854).

§ 50 *GALVANOMAGNETISM AND THERMOMAGNETISM*

In this section, we are dealing with thermoelectrical effects of an isotropic material in presence of a magnetic field.

As is shown in § 46, for certain crystals of the hexagonal, tetragonal or trigonal system, the matrix of the thermal conductivity tensor can be simplified to

$$\begin{bmatrix} L_{11} & L_{12} & 0 \\ -L_{12} & L_{11} & 0 \\ 0 & 0 & L_{33} \end{bmatrix}, \tag{1}$$

if the principal axis is aligned with the x_3-axis. This holds true also for the electrical resistivity tensor, the (absolute) Peltier coefficient tensor and the (absolute) thermoelectrical power tensor. For an isotropic crystal, which has the property of invariance in physical properties for rotations at any angle with respect to any direction, the matrix of phenomenological coefficients with the x_3-axis fixed, will have symmetry at least in the form

$$\begin{bmatrix} L_{11} & L_{12} & 0 & L_{14} & L_{15} & 0 \\ -L_{12} & L_{11} & 0 & -L_{15} & L_{14} & 0 \\ 0 & 0 & L_{33} & 0 & 0 & L_{36} \\ L_{41} & L_{42} & 0 & L_{44} & L_{45} & 0 \\ -L_{42} & L_{41} & 0 & -L_{45} & L_{44} & 0 \\ 0 & 0 & L_{63} & 0 & 0 & L_{66} \end{bmatrix}, \tag{2}$$

the upper left corner being for thermal conductivity, the lower left corner being for thermoelectrical power, the upper right corner being for Peltier coefficient, and the lower right corner being for electrical resistivity. To summarize, due to geometrical symmetry, the 36 components have been reduced to 12 independent ones.

When a magnetic field $\boldsymbol{B}$ is present, an isotropic crystal is no longer so because the direction of the magnetic field is a special one. However, the plane perpendicular to the magnetic field is still isotropic. In other words, if $\boldsymbol{B}$ is applied in a direction parallel to the x_3-axis, the x_3-axis is a two-fold axis of rotation.

If we transform

$$x_1 \to -x_1, \quad x_2 \to x_2, \quad x_3 \to -x_3, \quad \text{and } \boldsymbol{B} \to -\boldsymbol{B},$$

the physical properties are invariant. Comparing these with the entries of the matrix (1),

$$L_{11}(\boldsymbol{B}) = L_{11}(-\boldsymbol{B}), \quad L_{12}(\boldsymbol{B}) = -L_{12}(-\boldsymbol{B}),$$

and

$$L_{33}(\boldsymbol{B}) = L_{33}(-\boldsymbol{B}).$$

Thus in matrix (2) L_{11}, L_{33}, L_{14}, L_{36}, L_{41}, L_{63}, L_{44}, and L_{66} are even functions of $\boldsymbol{B}$, but L_{12}, L_{15}, L_{42}, and L_{45} are odd functions of $\boldsymbol{B}$. To summarize, due to geometrical symmetry in presence of a magnetic field, the single appearance of 1, 3, 4, or 6 gives an odd function of $\boldsymbol{B}$, but otherwise we have an even function.

From § 18 the Onsager-Casimir law assumes the form $L_{ij}(\boldsymbol{B}) = \varepsilon_i \varepsilon_j L_{ji}(-\boldsymbol{B})$, in the presence of a magnetic field. Experimentally it is more convenient to use grad T and $\boldsymbol{I}$ as forces. Thus ε_1, ε_2, or ε_3 for grad T is $+1$, but ε_4, ε_5, or ε_6 for $\boldsymbol{I}$ is -1. Combining geometrical symmetry and the Onsager-Casimir law,

$$L_{12}(\boldsymbol{B}) = L_{21}(-\boldsymbol{B}) = -L_{12}(-\boldsymbol{B}), \quad \text{or } L_{21} = -L_{12}; \tag{3}$$

$$L_{14}(\boldsymbol{B}) = -L_{41}(-\boldsymbol{B}) = L_{14}(-\boldsymbol{B}), \quad \text{or } L_{41} = -L_{14}; \tag{4}$$

$$L_{15}(\boldsymbol{B}) = -L_{51}(-\boldsymbol{B}) = -L_{15}(-\boldsymbol{B}), \quad \text{or } L_{51} = L_{15}; \tag{5}$$

$$L_{36}(\boldsymbol{B}) = -L_{63}(-\boldsymbol{B}) = L_{36}(-\boldsymbol{B}), \quad \text{or } L_{63} = -L_{36}; \tag{6}$$

$$L_{42}(\boldsymbol{B}) = -L_{24}(-\boldsymbol{B}) = -L_{42}(-\boldsymbol{B}), \quad \text{or } L_{42} = L_{24}; \tag{7}$$

$$L_{45}(\boldsymbol{B}) = L_{54}(-\boldsymbol{B}) = -L_{45}(-\boldsymbol{B}), \quad \text{or } L_{54} = -L_{45}. \tag{8}$$

While eqs. (3) and (8) give no new information to the matrix (2), the new information provided by eqs. (4)—(6) allows us to change matrix (2) into matrix (9) [Equation (7) is superfluous to eq. (5). We know already from the matrix (2) that $L_{24} = -L_{15}$ and $L_{51} = -L_{42}$. That $L_{51} = L_{15}$ implies that $L_{42} = L_{24}$, which is incorporated already in matrix (9)],

$$\begin{bmatrix} L_{11} & L_{12} & 0 & L_{14} & L_{15} & 0 \\ -L_{12} & L_{11} & 0 & -L_{15} & L_{14} & 0 \\ 0 & 0 & L_{33} & 0 & 0 & L_{36} \\ -L_{14} & -L_{15} & 0 & L_{44} & L_{45} & 0 \\ L_{15} & -L_{14} & 0 & -L_{45} & L_{44} & 0 \\ 0 & 0 & -L_{36} & 0 & 0 & L_{66} \end{bmatrix}. \tag{9}$$

From eqs. (49, 1)—(49, 3),

$$T\sigma = -[\boldsymbol{J}_{S,\text{tot}} | \operatorname{grad} T] - \left[\operatorname{grad} \zeta + \frac{\operatorname{grad} \mu}{Z} \middle| \boldsymbol{I}\right]. \tag{10}$$

If we use $\boldsymbol{I}$ instead of $\operatorname{grad} \zeta + (\operatorname{grad} \mu)/Z = -\boldsymbol{E} + (\operatorname{grad} \mu)/Z$ as force,

$$\boldsymbol{J}_{S,\text{tot}} = L_{11} \frac{(-\operatorname{grad} T)}{T} - \frac{L_{12}}{T}(\boldsymbol{I}), \tag{11}$$

$$\boldsymbol{E} - \frac{\operatorname{grad} \mu}{Z} = -L_{21} \frac{(-\operatorname{grad} T)}{T} + \frac{L_{22}}{T}(\boldsymbol{I}). \tag{12}$$

The phenomenological coefficients (9) assume a different form,

$$\begin{bmatrix} (J_{S,\text{tot}})_x \\ (J_{S,\text{tot}})_y \\ (J_{S,\text{tot}})_z \\ E_x - \dfrac{\operatorname{grad} \mu}{Z} \\ E_y - \dfrac{\operatorname{grad} \mu}{Z} \\ E_z - \dfrac{\operatorname{grad} \mu}{Z} \end{bmatrix} = \begin{bmatrix} L_{11} & L_{12} & 0 & \dfrac{L_{14}}{T} & \dfrac{L_{15}}{T} & 0 \\ -L_{12} & L_{11} & 0 & -\dfrac{L_{15}}{T} & \dfrac{L_{14}}{T} & 0 \\ 0 & 0 & L_{33} & 0 & 0 & -\dfrac{L_{36}}{T} \\ L_{14} & L_{15} & 0 & \dfrac{L_{44}}{T} & \dfrac{L_{45}}{T} & 0 \\ -L_{15} & L_{14} & 0 & -\dfrac{L_{45}}{T} & \dfrac{L_{44}}{T} & 0 \\ 0 & 0 & L_{36} & 0 & 0 & \dfrac{L_{66}}{T} \end{bmatrix}$$

$$\begin{bmatrix} -\dfrac{(\text{grad } T)_x}{T} \\ -\dfrac{(\text{grad } T)_y}{T} \\ -\dfrac{(\text{grad } T)_z}{T} \\ I_x \\ I_y \\ I_z \end{bmatrix}. \tag{13}$$

From eqs. (49, 18), (49, 19), (49, 27) and (49, 30),

$$\boldsymbol{J}_{S,\text{tot}} = -\frac{\theta'' \text{ grad } T}{T} + \frac{\Pi \boldsymbol{I}}{T} = \theta'' \frac{(-\text{grad } T)}{T} - \pi(\boldsymbol{I}), \tag{14}$$

$$E - \frac{\text{grad}\,\mu}{Z} = \pi T \frac{(-\text{grad } T)}{T} + R(\boldsymbol{I}). \tag{15}$$

The phenomenological equations (13) in terms of physical quantities (with double primes omitted from θ'') are

$$\begin{bmatrix} (J_{S,\text{tot}})_x \\ (J_{S,\text{tot}})_y \\ (J_{S,\text{tot}})_z \\ E_x - \dfrac{\text{grad } \mu}{Z} \\ E_y - \dfrac{\text{grad } \mu}{Z} \\ E_z - \dfrac{\text{grad } \mu}{Z} \end{bmatrix} \begin{bmatrix} \theta_{xx} & \theta_{xy} & 0 & -\pi_{xx} & -\pi_{xy} & 0 \\ -\theta_{xy} & \theta_{xx} & 0 & \pi_{xy} & -\pi_{xx} & 0 \\ 0 & 0 & \theta_{zz} & 0 & 0 & -\pi_{zz} \\ \pi_{xx}T & \pi_{xy}T & 0 & R_{xx} & R_{xy} & 0 \\ -\pi_{xy}T & \pi_{xx}T & 0 & -R_{xy} & R_{xx} & 0 \\ 0 & 0 & \pi_{zz}T & 0 & 0 & R_{zz} \end{bmatrix}$$

$$\begin{bmatrix} -\frac{(\mathrm{grad}\,T)_x}{T} \\ -\frac{(\mathrm{grad}\,T)_y}{T} \\ -\frac{(\mathrm{grad}\,T)_z}{T} \\ I_x \\ I_y \\ I_z \end{bmatrix}. \tag{16}$$

We see that there are only 9 independent physical quantities in matrix (16). If 12 physical quantities can be determined, clearly there are 3 independent relations among them.

If these quantities are caused by electrical currents, they are called galvanomagnetic effects. If these are caused by temperature gradients, they are called thermomagnetic effects. If the direction of the cause is the same as that of the effect, the effect is longitudinal. If the directions are perpendicular to each other, the effect is transverse. If the temperature gradient perpendicular to the cause is zero, the effect is isothermal. If the heat flux perpendicular to the cause is zero, the effect is adiabatic.

From the phenomenological equations (16), 12 physical quantities can be determined:

(I) Transverse Magnetic Field

(*A*) *Longitudinal*

(a) Thermomagnetic

(1) $$\theta_1 = \left[-\frac{T(J_{S,\mathrm{tot}})_x}{(\mathrm{grad}\,T)_x} \right]_{(\mathrm{grad}T)_y = I = 0} = \theta_{xx}.$$

This is the isothermal conductivity.

$$\pi_1 = \left[\frac{E_x - \dfrac{\text{grad}\,\mu}{Z}}{(\text{grad}\,T)_x}\right]_{(\text{grad}T)_y = I = 0} = -\pi_{xx}. \tag{2}$$

This is the isothermal Ettinghausen-Nernst effect.

(b) Galvanomagnetic

$$\eta_1 = \left[\frac{(\text{grad}\,T)_x}{I_x}\right]_{I_y = (\text{grad}T)_y = (J_{S,\text{tot}})_x = 0} = -\frac{\pi_{xx} T}{\theta_{xx}}. \tag{3}$$

This is Peltier effect for unit thermal conductivity.

$$R_1 = \left[\frac{E_x - \dfrac{\text{grad}\,\mu}{Z}}{I_x}\right]_{\text{grad}T = I_y = 0} = R_{xx}. \tag{4}$$

This is isothermal electrical resistivity.

(*B*) *Transverse*

(a) Thermomagnetic

$$(RL) = \left[\frac{(\text{grad}\,T)_y}{(\text{grad}\,T)_x}\right]_{(J_{S,\text{tot}})_y = I = 0} = \frac{\theta_{xy}}{\theta_{xx}}. \tag{5}$$

This is the Righi-Leduc effect.

$$N = \left[\frac{E_y - \dfrac{\text{grad}\,\mu}{Z}}{(\text{grad}\,T)_x}\right]_{(\text{grad}T)_y = I = 0} = \pi_{xy}. \tag{6}$$

This is the isothermal Nernst effect.

(b) Galvanomagnetic

$$(ET) = \left[\frac{(\text{grad}\,T)_y}{I_x}\right]_{I_y = (J_{S,\text{tot}})_y = (\text{grad}\,T)_x = 0} = \frac{\pi_{xy} T}{\theta_{xx}}. \tag{7}$$

This is the Ettinghausen effect.

$$H = \left[\frac{E_y - \dfrac{\text{grad}\,\mu}{Z}}{I_x}\right]_{I_y = \text{grad}\,T = 0} = -R_{xy}. \tag{8}$$

This is the isothermal Hall effect.

(II) Longitudinal Magnetic Field

(a) Thermomagnetic

(9) $$\theta_3 = \left[\frac{-T(J_{S,\text{tot}})_z}{(\text{grad } T)_z}\right]_{I_z=0} = \theta_{zz}.$$

This is the thermal conductivity.

(10) $$\pi_3 = \left[\frac{E_z - \dfrac{\text{grad } \mu}{Z}}{(\text{grad } T)_z}\right]_{I_z=0} = -\pi_{zz}.$$

This is the thermoelectrical power.

(b) Galvanomagnetic

(11) $$\eta_3 = \left[\frac{(\text{grad } T)_z}{I_z}\right]_{(\boldsymbol{J}_{S,\text{tot}})_z=0} = -\frac{\pi_{zz} T}{\theta_{zz}}.$$

This is the Peltier effect for unit thermal conductivity.

(12) $$R_3 = \left[\frac{E_z - \dfrac{\text{grad } \mu}{Z}}{I_z}\right]_{(\text{grad} T)_z=0} = R_{zz}.$$

This is the electrical resistivity.

The three independent relations among these effects are

$$\theta_1 = (ET) = NT, \tag{17}$$

$$\theta_1 \eta_1 = \pi_1 T, \tag{18}$$

$$\theta_3 \eta_3 = \pi_3 T. \tag{19}$$

Equation (17) is known as the Bridgman equation[1].

If we confine ourselves to the plane perpendicular to the magnetic field, five additional adiabatic effects are determinable. The adiabatic effects are

adi θ_1 = thermal conductivity

$$= \left[\frac{-T(\boldsymbol{J}_{S,\text{tot}})_x}{(\text{grad } T)_x}\right]_{(\boldsymbol{J}_{S,\text{tot}})_y = \boldsymbol{I} = 0} = \theta_{xx} + \frac{\theta_{xy}^2}{\theta_{xx}},$$

adi π_1 = Ettinghausen-Nernst effect

$$= \left[\frac{E_x - \dfrac{\text{grad}\, \mu}{Z}}{(\text{grad}\, T)_x} \right]_{(J_{S,\text{tot}})_y = I = 0} = -\pi_{xx} - \pi_{xy} \frac{\theta_{xy}}{\theta_{xx}},$$

adi R_1 = electrical resistivity

$$= \left[\frac{E_x - \dfrac{\text{grad}\, \mu}{Z}}{I_x} \right]_{I_y = (J_{S,\text{tot}})_y = (\text{grad} T)_x = 0} = R_{xx} - \frac{\pi_{xy}^2 T}{\theta_{xx}},$$

adi N = Nernst effect

$$= \left[\frac{E_y - \dfrac{\text{grad}\, \mu}{Z}}{(\text{grad}\, T)_x} \right]_{(J_{S,\text{tot}})_y = I = 0} = \pi_{xy} - \frac{\theta_{xy} \pi_{xx}}{\theta_{xx}},$$

adi H = Hall effect

$$= \left[\frac{E_y - \dfrac{\text{grad}\, \mu}{Z}}{I_x} \right]_{I_y = (\text{grad}\, T)_x = (J_{S,\text{tot}})_y = 0}$$

$$= -R_{xy} - \frac{\pi_{xy} \pi_{xx} T}{\theta_{xx}}.$$

Also five additional relations exist. They are

$$\text{adi}\ \theta_1 = \theta_1 + (RL)^2 \theta_1, \tag{20}$$

$$\text{adi}\ \pi_1 = \pi_1 - N(RL), \tag{21}$$

$$\text{adi}\ R_1 = R_1 - N(ET), \tag{22}$$

$$\text{adi}\ N = N + \pi_1(RL), \tag{23}$$

$$\text{adi}\ H = H + \pi_1(ET). \tag{24}$$

Different authors use different definitions and the definitions may be the same except for sign or a factor of $\boldsymbol{B}$. For example, the Peltier effect for unit thermal conductivity may be called the Nernst effect, the Nernst effect may be called the Ettinghausen-Nernst effect and the Ettinghauser-Nernst effect may be called the thermoelectrical power. Our convention for the Peltier effect for unit thermal conductivity is that the magnetic field is downward, the positive current flows from left to right and the heat is absorbed from top to bottom.

Reference

[1] P. Bridgman, Thermodynamics of Electrical Phenomena in Metals, MacMillan, New York (1934).

§ 51 *HEAT CONDUCTION, DIFFUSION AND CHEMICAL REACTIONS*

In this section we introduce the Planck potential to simplify the interactions between conduction of heat and diffusion for an isotropic system. Moreover, the Planck potentials instead of chemical affinities are used as forces in chemical reactions.

If there is no viscosity and if no external forces are present, from eq. (8, 11),

σ for heat conduction, diffusion and chemical reactions

$$= \boldsymbol{J}_q \cdot \operatorname{grad}\left(\frac{1}{T}\right) - \sum_i^n \boldsymbol{J}_i \cdot \operatorname{grad}\left(\frac{\mu_i}{T}\right) - \frac{1}{T}\sum_k^r j_r A_k. \tag{1}$$

So far we have reduced the dependency of forces of diffusion by using the Gibbs-Duhem relation at constant temperature and pressure to reduce dimensions from n to $n-1$. We can reduce the dependency of forces of thermal conduction and diffusion by using the Gibbs-Duhem relation at constant pressure to reduce dimensions from $n+1$ to n,

$$SdT + \sum_i^n c_i d\mu_i = 0. \tag{2}$$

By definition,

$$S = \frac{1}{T}\left(H - \sum_i^n c_i\mu_i\right). \tag{3}$$

Substituting eq. (3) in eq. (2),

$$H \operatorname{grad}\left(\frac{1}{T}\right) - \sum_{i}^{n} c_i \operatorname{grad}\left(\frac{\mu_i}{T}\right) = 0. \tag{4}$$

Substituting eq. (4) in eq. (1) without chemical reactions,

$$\sigma \text{ for heat conduction and diffusion}$$
$$= \sum_{i}^{n} \left(\frac{c_i}{H}\right) \boldsymbol{J}_q \cdot \operatorname{grad}\left(\frac{\mu_i}{T}\right) - \sum_{i}^{n} \boldsymbol{J}_i \cdot \operatorname{grad}\left(\frac{\mu_i}{T}\right)$$
$$= - \sum_{i}^{n} \boldsymbol{K}_i \cdot \operatorname{grad}\left(\frac{\mu_i}{T}\right), \tag{5}$$

where

$$\boldsymbol{K}_i = \boldsymbol{J}_i - \frac{c_i}{H} \boldsymbol{J}_q. \tag{6}$$

If we let

$$\boldsymbol{J}_i = \boldsymbol{K}_i - c_i \sum_{j}^{n} \boldsymbol{K}_j, \tag{7}$$

from eqs. (6) and (7),

$$\boldsymbol{J}_q = - H \sum_{j}^{n} \boldsymbol{K}_j, \tag{8}$$

from eqs. (7) and (8),

$$\boldsymbol{J}_q^r = - \sum_{i}^{n} H_i \boldsymbol{K}_i. \tag{9}$$

From eqs. (7)—(9), we see that $\boldsymbol{K}_i$ and $\boldsymbol{J}_q$ or $\boldsymbol{J}_q^r$ are independent in n dimensions.

Now we turn our attention to the last term of eq. (1). Because the rank of fluxes of chemical reactions involving n substances never exceeds $n - 1$, let us say that the rank is r,

$$\sigma \text{ for chemical reactions} = - \frac{1}{T} \sum_{k}^{r} j_k A_k$$

$$= \frac{1}{T^2} \sum_k^r \sum_l^r L_{kl} A_l A_k \text{ [from eq. (25, 12)]}$$

$$= \sum_k^r \sum_l^r \sum_i^r L_{kl} \left(- \frac{\Lambda_{lk}\mu_k}{T} \right) \left(- \frac{\Lambda_{ik}\mu_i}{T} \right)$$

$$\text{[from eq. (25, 10)].} \tag{10}$$

Combining eqs. (5) and (10),

$$\sigma = \sum_i^n \boldsymbol{K}_i \cdot \text{grad}\,\phi_i + \sum_i^r \sum_k^r b_{ik}\phi_i\phi_k, \tag{11}$$

where ϕ is the Planck potential,

$$\phi_i = - \frac{\mu_i}{T}; \tag{12}$$

and

$$b_{ik} = \sum_l^r \Lambda_{ik} L_{kl} \Lambda_{lk}. \tag{13}$$

If we let

$$\boldsymbol{K}_i = \sum_k^n a_{ik} \text{grad}\,\phi_k, \tag{14}$$

eq. (11) becomes

$$\sigma = \sum_i^n a_{ik} \text{grad}\,\phi_i \cdot \text{grad}\,\phi_k + \sum_i^r \sum_k^r b_{ik}\phi_i\phi_k. \tag{15}$$

The first term on the right of eq. (15) is the usual quadratic form for the rate of entropy-production of heat conduction and diffusion and thus a is symmetric. Also b is symmetric because L is symmetric in eq. (13). Written in the double contracted form,

$$\sigma = a\colon \text{grad}\,\phi\,\text{grad}\,\phi + b\colon \phi\phi \geqq 0. \tag{16}$$

If we ignore for the moment the equality sign in eq. (16), matrix a is positive definite. Furthermore, both matrix a and

b are symmetric. It is possible to find a non-singular matrix Q to diagonalize matrix a to the unit matrix and to diagonalize b to f, where f are the characteristic roots of the matrix $a^{-1}b$. Or

$$Q^T a Q = 1, \tag{17}$$

and

$$Q^T b Q = f. \tag{18}$$

If we let

$$n_i = \sum_{l}^{r} b_{il}\phi_l = -\sum_{j}^{r}\sum_{k}^{r}\sum_{l}^{r} \Lambda_{ij}L_{jk}\Lambda_{kl}\frac{\mu_l}{T}$$

$$= -\sum_{j}^{r}\sum_{k}^{r} \Lambda_{ij}L_{jk}\frac{A_k}{T} = \sum_{j}^{r} \Lambda_{ij}j_j, \tag{19}$$

eq. (11) can be written in the condensed form,

$$\sigma = \boldsymbol{K}\cdot \operatorname{grad}\phi + n\phi. \tag{20}$$

In order that σ be invariant, if $\boldsymbol{K}$ is transformed according to

$$\boldsymbol{K}^* = Q^T\boldsymbol{K}, \tag{21}$$

from eq. (20), n is transformed cogradiently,

$$n^* = Q^T n; \tag{22}$$

but ϕ is transformed contragradiently,

$$\phi^* = Q^{-1}\phi, \tag{23}$$

and

$$\operatorname{grad}\phi^* = Q^{-1}\operatorname{grad}\phi. \tag{24}$$

The condensed form of eq. (14) is

$$K = a\operatorname{grad}\phi. \tag{25}$$

Pre-multiplying eq. (25) by Q^T,

$$Q^T K = Q^T a Q Q^{-1}\operatorname{grad}\phi. \tag{26}$$

From eqs. (21), (26), (17) and (24),

$$\boldsymbol{K}^* = \operatorname{grad}\phi^*. \tag{27}$$

The condensed form of eq. (19) is

$$n = b\psi. \tag{28}$$

Pre-multiplying eq. (28) by Q^T,

$$Q^T n = Q^T b Q Q^{-1} \psi. \tag{29}$$

From eqs. (22), (29), (18) and (23),

$$n^* = f\psi^*. \tag{30}$$

The transformed fluxes and forces in eq. (11) can be written as

$$\sigma = \sum_i^n (\operatorname{grad} \psi_i^*)^2 + \sum_i^r f_i(\psi_i^*)^2 \geqq 0. \tag{31}$$

We first consider the equality sign in eq. (31), which is valid for static equilibrium. Equation (31) can be satisfied by

$$(\operatorname{grad} \psi_i^*)^2 = 0, \quad i = r + 1, \cdots, n, \tag{32}$$

and

$$(\operatorname{grad} \psi_i^*)^2 + f_i(\psi_i^*)^2 = 0, \quad i = 1, 2, \cdots, r. \tag{33}$$

Now we consider the inequality sign in eq. (31), which is valid for stationary equilibrium. From eq. (6, 20),

$$\rho \frac{dU}{dt} = -\operatorname{div} \boldsymbol{J}_q = 0. \tag{34}$$

From eq. (4, 52),

$$\rho \frac{dc_i}{dt} = -\operatorname{div} \boldsymbol{J}_i + \sum_j^r \Lambda_{ij} j_j = 0. \tag{35}$$

Substituting eq. (19) in eq. (35),

$$\operatorname{div} \boldsymbol{J}_i = n_i. \tag{36}$$

In transformed form,

$$\operatorname{div} \boldsymbol{J}_i^* = n_i^*. \tag{37}$$

From eqs. (34), (37) and (6),

$$\operatorname{div} \boldsymbol{K}^* = n_i^*. \tag{38}$$

From eqs. (38), (27) and (30),

$$\operatorname{div}\operatorname{grad}\psi_i^* = f_i\psi_i^*. \tag{39}$$

In one dimension,

$$\frac{\partial^2\psi_i^*}{\partial x^2} = f_i\psi_i^*. \tag{40}$$

The solution is

$$\psi_i^* = A\exp\sqrt{f_i}\,x + B\exp -\sqrt{f_i}\,x, \tag{41}$$

where A and B are integration constants. If we choose the origin at $x = 0$ and the walls at $x = \pm d/2$, the boundary conditions are $\boldsymbol{J}_i = 0$ at $x = \pm d/2$. From eq. (6), the boundary conditions are

$$\boldsymbol{K}_i + \frac{c_i}{H}\boldsymbol{J}_q = 0 \text{ at } x = \pm\frac{d}{2}. \tag{42}$$

In terms of the transformed coordinates,

$$\frac{\partial\psi_i^*}{\partial x} + \frac{c_i^*}{H}\boldsymbol{J}_q = 0 \text{ at } x = \pm\frac{d}{2}\,(c^* = Q^Tc). \tag{43}$$

Substituting eq. (43) in eq. (41),

$$A = -B = \frac{-c_i^*J_q}{H\sqrt{f_i}\,(e^{\sqrt{f_i}\,d/2} + e^{-\sqrt{f_i}\,d/2})}. \tag{44}$$

Or

$$\psi_i^* = \frac{-\left(\dfrac{c_i^*J_q}{H\sqrt{f_i}}\right)(e^{\sqrt{f_i}\,x} - e^{-\sqrt{f_i}\,x})}{(e^{\sqrt{f_i}\,d/2} + e^{-\sqrt{f_i}\,d/2})}$$

$$= -\left(\frac{c_i^*J_q}{H\sqrt{f_i}}\right)\left(\frac{\sinh\sqrt{f_i}\,x}{\cosh\sqrt{f_i}\,d/2}\right). \tag{45}$$

CHAPTER IX

TENSORIAL PROCESSES

§ 52 *VISCOUS FLUIDS IN THE PRESENCE OF A FIELD*

From eq. (8, 11) with viscous flow alone,

$$\sigma = -(\Pi - p\overset{\leftrightarrow}{1}) : \frac{\text{Grad}\,\boldsymbol{v}}{T}. \tag{1}$$

In the region of linearity,

$$P_{\alpha\beta} = -\,L_{\alpha\beta\gamma\delta}\,\frac{(\text{Grad}\,v)_{\gamma\delta}}{T}, \quad \alpha,\ \beta,\ \gamma,\ \delta = 1, 2, 3. \tag{2}$$

There are 81 terms. If we assume that both P and Grad $\boldsymbol{v}$ are symmetric,

$$P^{sy}_{\alpha\beta} = -\,L_{\alpha\beta\gamma\delta}\,\frac{(\text{Grad}^{sy}\,v)_{\gamma\delta}}{T}. \tag{3}$$

There are 21 terms. Because the fluid is isotropic, for unrestricted P and Grad $\boldsymbol{v}$ there are only 3 independent phenomenological coefficients (See §13.), which are related to the volume viscosity coefficient η_V, the shear viscosity coefficient η, and the rotational viscosity coefficient η_r. For symmetric $\boldsymbol{P}$ and grad $\boldsymbol{v}$, η_r is dropped out. In this section we deal with a viscous fluid in the presence of a magnetic field, a rotational velocity field, or an electric field.

We can simplify eq. (3) if we adopt the double notation:

$$11 \rightleftarrows 1,\ 22 \rightleftarrows 2,\ 33 \rightleftarrows 3,\ 23 \rightleftarrows 4,\ 31 \rightleftarrows 5,\ 12 \rightleftarrows 6.$$

For simplicity we omit the superscript sy from P^{sy} and $\text{Grad}^{sy} v$.

$$P_\alpha = \frac{-L_{\alpha\alpha}(\text{Grad } v)_\alpha}{T} - \frac{L_{\alpha\beta} 2(\text{Grad } v)_\beta}{T},$$

$$\alpha, \beta = 1, 2, 3, 4, 5, 6. \tag{4}$$

That terms with mixed coefficients, $\alpha \neq \beta$, appear twice, is easily seen from writing eq. (1) in quadratic form, splitting it into 6 terms and recasting the phenomenological equations,

$$\sigma = (\text{Grad } \boldsymbol{v} : L \text{ Grad } \boldsymbol{v}). \tag{5}$$

If a magnetic field is present, the fluid is no longer isotropic. However, the plane perpendicular to the magnetic field, parallel to the x_3 axis, is still isotropic, i.e., plane rotation

$$\begin{bmatrix} \cos\alpha & -\sin\alpha & 0 \\ \sin\alpha & \cos\alpha & 0 \\ 0 & 0 & 1 \end{bmatrix} \tag{6}$$

is still permitted. Because the angle of plane rotation α is arbitrary, we can assign α to an infinitesimal small angle. The matrix (6) is simplified to

$$\begin{bmatrix} 1 & -\alpha & 0 \\ \alpha & 1 & 0 \\ 0 & 0 & 1 \end{bmatrix}. \tag{7}$$

By the procedure similar to obtaining eq. (15, 7), with powers of α higher than 1 neglected, the phenomenological equations are

$$\begin{bmatrix} P_{11} \\ P_{22} \\ P_{33} \\ P_{23} \\ P_{31} \\ P_{12} \end{bmatrix} = \begin{bmatrix} L_{11} & L_{12} & L_{13} & 0 & 0 & L_{16} \\ L_{12} & L_{11} & L_{13} & 0 & 0 & -L_{16} \\ L_{31} & L_{31} & L_{33} & 0 & 0 & 0 \\ 0 & 0 & 0 & L_{44} & L_{45} & 0 \\ 0 & 0 & 0 & -L_{45} & L_{44} & 0 \\ -L_{16} & L_{16} & 0 & 0 & 0 & \dfrac{L_{11} - L_{12}}{2} \end{bmatrix}$$

$$\begin{bmatrix} -\dfrac{(\mathrm{Grad}\,\boldsymbol{v})_{11}}{T} \\ -\dfrac{(\mathrm{Grad}\,\boldsymbol{v})_{22}}{T} \\ -\dfrac{(\mathrm{Grad}\,\boldsymbol{v})_{33}}{T} \\ -\dfrac{2\,(\mathrm{Grad}\,\boldsymbol{v})_{23}}{T} \\ -\dfrac{2\,(\mathrm{Grad}\,\boldsymbol{v})_{31}}{T} \\ -\dfrac{2\,(\mathrm{Grad}\,\boldsymbol{v})_{12}}{T} \end{bmatrix}. \qquad (8)$$

There are 8 independent phenomenological coefficients.

From the reasoning similar to § 50, due to geometric symmetry, the single appearance of 1 or 3 for L in full notation gives an odd function of $\boldsymbol{B}$. Thus L_{11}, L_{12}, L_{13}, L_{31}, L_{33}, and L_{44} are even functions of $\boldsymbol{B}$ but L_{16} and L_{45} are odd functions of $\boldsymbol{B}$. From the Onsager-Casimir law, with

$$\varepsilon_\alpha,\ \varepsilon_\beta(\alpha, \beta = 1, \cdots, 6) = -1,$$

$$L_{31}(\boldsymbol{B}) = L_{13}(-\boldsymbol{B}) = L_{31}(-\boldsymbol{B}) \text{ or } L_{31} = L_{13}, \qquad (9)$$

$$L_{16}(\boldsymbol{B}) = L_{61}(-\boldsymbol{B}) = -L_{16}(-\boldsymbol{B}) \text{ or } L_{61} = -L_{16}, \qquad (10)$$

$$L_{45}(\boldsymbol{B}) = L_{54}(-\boldsymbol{B}) = -L_{45}(-\boldsymbol{B}) \text{ or } L_{54} = -L_{45}. \qquad (11)$$

While eqs. (10) and (11) give no new information on eqs. (8), eq. (9) reduces the matrix in eqs. (8) to 7 independent phenomenological coefficients.

The equations (8) will have a simple physical interpretation if the 6×6 matrix is expanded to a 7×7 matrix. This can be done if we split P into its trace and traceless components,

$$\begin{aligned} &P_{11} + P_{22} + P_{33} + P_{23} + P_{31} + P_{12} \\ &= P^0_{11} + P^0_{22} + P^0_{33} + P^0_{23} + P^0_{31} + P^0_{12} + \mathrm{tr}P. \end{aligned} \qquad (12)$$

Thus

$$P_{ii} = P^0_{ii} + \frac{1}{3}\,\mathrm{tr}\,P \quad (i = 1, 2, 3;\ i \text{ not summed}), \tag{13}$$

$$P_{ij} = P^0_{ij} \quad (i \neq j), \tag{14}$$

$$P^0_{11} + P^0_{22} + P^0_{33} = 0. \tag{15}$$

Similarly,

$$G_{ii} = G^0_{ii} + \frac{1}{3}\,\mathrm{div}\,\boldsymbol{v} \quad (i \text{ not summed}), \tag{16}$$

$$G_{ij} = G^0_{ij} \quad (i \neq j), \tag{17}$$

$$G^0_{11} + G^0_{22} + G^0_{33} = 0. \tag{18}$$

We have abbreviated Grad $\boldsymbol{v}$ as G and $\mathrm{Grad}^0\,\boldsymbol{v}$ as G^0. To preserve the rate of entropy-production, again we resort to the quadratic form of σ. From eqs. (8) and (13—18),

$$\begin{aligned} T\sigma = {} & (L_{11} + L_{33} - 2L_{13})(G^0_{33})^2 + 2L_{44}[(G^0_{23})^2 + (G^0_{31})^2] \\ & + (L_{11}-L_{12})(G^0_{12})^2 + \frac{2}{3}(-L_{11}-L_{12}+L_{13}+L_{33})G^0_{33}\,\mathrm{div}\,\boldsymbol{v} \\ & + 2(L_{12} - L_{11})G^0_{11}G^0_{22} + L_{16}(G^0_{11} - G^0_{22})G^0_{12} \\ & + \frac{1}{9}(2L_{11} + 2L_{12} + 4L_{13} + L_{33})(\mathrm{div}\,\boldsymbol{v})^2. \end{aligned} \tag{19}$$

If we let

$$L_{16} = \eta_3, \tag{20}$$

$$L_{44} = \eta_4, \tag{21}$$

$$L_{45} = \eta_5, \tag{22}$$

$$\frac{1}{9}(2L_{11} + 2L_{12} + 4L_{13} + L_{33}) = \eta_V, \tag{23}$$

$$\frac{1}{9}(-L_{11} - L_{12} + L_{13} + L_{33}) = \varepsilon, \tag{24}$$

$$L_{11} + L_{33} - 2L_{13} = 2\eta_1 + 2\eta_2, \tag{25}$$

$$L_{11} - L_{12} = 4\eta_1 - 2\eta_2, \tag{26}$$

$$\begin{aligned} T\sigma = {} & 2\eta_1[(G^0_{11})^2 + (G^0_{22})^2] + 2\eta_2(G^0_{33})^2 + 2\eta_4[(G^0_{23})^2 \\ & + (G^0_{31})^2] + (4\eta_1 - 2\eta_2)(G^0_{12})^2 + \eta_V\,(\mathrm{div}\,\boldsymbol{v})^2 \end{aligned}$$

$$+ \eta_3(G_{11}^0 - G_{22}^0)G_{12}^0 + 6\varepsilon G_{33}^0 \operatorname{div} \boldsymbol{v}$$
$$+ (4\eta_2 - 4\eta_1)G_{11}^0 G_{22}^0. \tag{27}$$

The new phenomenological equations are

$$\begin{bmatrix} P_{11}^0 \\ P_{22}^0 \\ P_{33}^0 \\ P_{23}^0 \\ P_{31}^0 \\ P_{12}^0 \\ \frac{1}{3}\operatorname{tr}P \end{bmatrix} = \begin{bmatrix} 2\eta_1 & 2\eta_2 - 2\eta_1 & 0 & 0 & 0 & 2\eta_3 & -\varepsilon \\ 2\eta_2 - 2\eta_1 & 2\eta_1 & 0 & 0 & 0 & -2\eta_3 & -\varepsilon \\ 0 & 0 & 2\eta_2 & 0 & 0 & 0 & 2\varepsilon \\ 0 & 0 & 0 & 2\eta_4 & 2\eta_5 & 0 & 0 \\ 0 & 0 & 0 & -2\eta_5 & 2\eta_4 & 0 & 0 \\ -\eta_3 & \eta_3 & 0 & 0 & 0 & 4\eta_1 - 2\eta_2 & 0 \\ -\varepsilon & -\varepsilon & 2\varepsilon & 0 & 0 & 0 & \eta_V \end{bmatrix} \begin{bmatrix} -\frac{G_{11}^0}{T} \\ -\frac{G_{22}^0}{T} \\ -\frac{G_{33}^0}{T} \\ -\frac{G_{23}^0}{T} \\ -\frac{G_{31}^0}{T} \\ -\frac{G_{12}^0}{T} \\ -\frac{\operatorname{div}\boldsymbol{v}}{T} \end{bmatrix}. \tag{28}$$

There are still 7 independent variables. Two of them, η_3 and η_5 are odd functions of $\boldsymbol{B}$. η_1, η_2, η_3, η_4, and η_5 are shear viscosity coefficients, η_V is a volume viscosity coefficient, and ε is a cross-effect viscosity coefficient. If there is no magnetic field, all off-diagonal terms vanish and all diagonal terms in the 6×6 submatrix are equal to each other. We only have η and η_V.

It is pointed out by Hooyman[1] that the results in a magnetic field are in slight disagreement with the results of a kinetic treatment by Chapman and Cowling[2].

Identical results are obtained if the magnetic field $\boldsymbol{B}$ is replaced by a rotational field $\boldsymbol{\omega}$.

If an electrical field $\boldsymbol{E}$ is present, the Onsager law says that

$$L_{31} = L_{13}, \tag{29}$$

$$L_{16} = L_{61}, \tag{30}$$

$$L_{45} = L_{54}. \tag{31}$$

But L_{16} and L_{45} are odd functions of $\boldsymbol{E}$. Thus

$$L_{16} = L_{61} = L_{45} = L_{54} = 0. \tag{32}$$

There are only 5 independent phenomenological coefficients, all being even functions of $\boldsymbol{E}$.

References

[1] G. J. Hooyman, Thesis, University of Leiden (1955).
[2] S. Chapman and T. G. Cowling, The Mathematic Theory of Non-uniform Gases, 2nd Ed., Cambridge University Press (1952).

CHAPTER X

HETEROGENEOUS PROCESSES

§ 53 *DISCONTINUOUS SYSTEMS*

A discontinuous system consists of two subsystems I and II separated by a hole, a capillary, a porous wall or a membrane, collectively called a plug. The opening in the plug must be small compared with the mean free path of all molecules passing through, otherwise the effects of the discontinuous system will be small if not completely lost.

We may illustrate the essential role played by the plug by he following example. Consider an electrolytic cell with a diaphragm to produce chlorine and sodium hydroxide from an aqueous solution of sodium chloride. The chemical reaction, split into two half-reactions, is brought about by electricity and the function of the diaphragm is to minimize the mixing of these two products.

The simplifying assumptions in the discontinuous system are as follows: (1) The two subsystems are homogeneous, by strong stirring for example. There is no concentration gradient, pressure gradient (in the direction perpendicular to the plug), or temperature gradient in either subsystem even if we impose different pressures or temperatures at the left boundary of the subsystem I and at the right boundary of the subsystem II. However, there are differences in properties between these two subsystems. Thus the changes take place discontinuously across the plug. (2) Because the transition occurs in the plug and because the width of the plug is small compared with that of either subsystem, we integrate the forces along the plug as the differences of the forces. In replacing gradients by the differences, we have changed the dimensions of the forces. To maintain the dimensions of σ, we change the $\boldsymbol{j}$'s of the continuous system in quantities per unit area per unit time into $\boldsymbol{j}$'s of the plug in quantities per unit volume per unit time. Because the $\boldsymbol{j}$'s in the plug are independent, $\boldsymbol{j}$'s are expressed in absolute quantities. (3) The differences in forces across the plug are not available experimentally. From the postulate of local equilibrium, §2, the properties on the surfaces of the plug are the same as the corresponding properties in the adjacent solutions. (4) Because the plug is an open system, at steady state the increase in the rate of entropy-production is balanced by the removal of entropy flux in the energy and matter in subsystems I and II. (5) Because the whole system is closed, at steady state the increase in the rate of entropy-production is balanced by the removal of entropy flux in energy from outside of the whole system. (6) The plug is homogeneous on the surfaces and any property will have the same value on all points of the surfaces.

From eqs. (8, 7) (8, 9) and eqs. (47, 12—13),

$$\frac{\partial \rho S}{\partial t} = - \operatorname{div} (\boldsymbol{J}_S + \rho S \boldsymbol{v}) + \sigma$$

$$= - \operatorname{div} \left[\frac{1}{T} \left(\boldsymbol{J}_q - \sum_k^n \mu_k \boldsymbol{J}_k \right) + \rho S \boldsymbol{v} \right] + \sigma$$

$$= - \operatorname{div} \frac{1}{T} \left[\boldsymbol{J}_q + H \rho \boldsymbol{v} - \sum_k^n \mu_k (\boldsymbol{J}_k + \rho_k \boldsymbol{v}) \right] + \sigma$$

$$= - \operatorname{div} \frac{1}{T} \left(\boldsymbol{J}_q^{\text{abs}} - \sum_k^n \mu_k \boldsymbol{J}_k^{\text{abs}} \right) + \sigma. \tag{1}$$

At steady state,

$$\sigma = \boldsymbol{J}_q^{\text{abs}} \cdot \Delta \left(\frac{1}{T} \right) - \sum_k^n \boldsymbol{J}_k^{\text{abs}} \cdot \Delta \left(\frac{\mu_k}{T} \right)$$

$$= \boldsymbol{j}_q \cdot \Delta\left(\frac{1}{T}\right) - \sum_k^n \boldsymbol{j}_k \cdot \Delta\left(\frac{\mu_k}{T}\right). \tag{2}$$

We have carried out the integration by subtracting and replaced $\boldsymbol{J}$'s by $\boldsymbol{j}$'s.

Thermo-molecular Effect

Thermal diffusion in a discontinuous system is collectively called a thermo-molecular effect. The passage of a gas through a porous disk between two vessels I and II is called the thermal effusion. The passage of a gas through a membrane is called the thermoösmosis. The passage of liquid helium–4 through a capillary is called the fountain effect. The mechanism of these effects is certainly different, but the thermodynamical treatment is the same.

From eq. (2),

$$\sigma = \frac{-(j_S \Delta T)}{T} - \sum_k^n \frac{j_k \Delta \mu_k}{T}$$

$$= \frac{-(j_S \Delta T)}{T} - \sum_k^n j_k \frac{-S_k \Delta T + V_k \Delta p}{T}. \tag{3}$$

In steady flow $\boldsymbol{j}$ is stationary, or

$$\frac{\partial \sigma}{\partial \boldsymbol{j}} = 0. \tag{4}$$

By using eq. (22, 22) and limiting to one component,

$$-S^* \Delta T + S \Delta T - V \Delta p = 0. \tag{5}$$

By using eqs. (22, 25—26),

$$\frac{\Delta p}{\Delta T} = -\frac{S^* - S}{V} = -\frac{H^* - H}{VT} = -\frac{Q^*}{VT}. \tag{6}$$

If $\boldsymbol{j} \neq 0$, $p^{\mathrm{I}} \neq p^{\mathrm{II}}$, and $T^{\mathrm{I}} \neq T^{\mathrm{II}}$, the thermo-molecular effect exists.

For ideal gases,

$$\frac{\partial \ln p}{\partial \ln T} = -\frac{T}{p}\left(\frac{S^* - S}{V}\right) = -\frac{M}{R}(S^* - S)$$

$$= -\frac{M}{RT}(H^* - H) = -\frac{MQ^*}{RT}. \tag{7}$$

By plotting ln $(p^{\mathrm{I}}/p^{\mathrm{II}})$ against ln $(T^{\mathrm{I}}/T^{\mathrm{II}})$, the slope is $-MQ^*/RT$. Depending on whether the slope is constant or variable, we can distinguish between linear and nonlinear relationships.

Now we consider one limiting case for a linear relationship. If

$$H^* - H = -\frac{RT}{2M}, \tag{8}$$

$$\frac{\partial \ln p}{\partial \ln T} = \frac{1}{2}. \tag{9}$$

Then

$$\frac{p^{\mathrm{I}}}{p^{\mathrm{II}}} = \sqrt{\frac{T^{\mathrm{I}}}{T^{\mathrm{II}}}}. \tag{10}$$

This was verified by Knudson[1] and others and could be explained by saying that in passing through the barrier, one degree of freedom is lost. The total energy on one mole of gas is $\frac{1}{2}Mv^2$. According to Maxwell's distribution law, the probability density of the velocity v is

$$\phi(\boldsymbol{v}) = \frac{|R'|^{1/2}}{(2\pi)^{n/2}} \exp\left(-\frac{1}{2}\boldsymbol{v}^{\mathrm{T}} \cdot R'\boldsymbol{v}\right), \tag{11}$$

where $R' = M/RT$ is the covariance, n is the number of degrees of freedom and $|R'|^{1/2}/(2\pi)^{n/2}$ is the normalizing factor [Compare eq. (16, 8)]. Hence

$$\left\langle\frac{Mv^2}{2}\right\rangle = \left(\frac{M}{2}\right)\int_{-\infty}^{\infty} \boldsymbol{v}^{\mathrm{T}} \cdot \boldsymbol{v}\phi(\boldsymbol{v})d\boldsymbol{v}$$

$$= \left[\frac{|R'|^{1/2}}{(2\pi)^{n/2}}\right]\left(\frac{M}{2}\right)\int_{-\infty}^{\infty} \boldsymbol{v}^{\mathrm{T}} \cdot \boldsymbol{v} \exp\left(-\frac{1}{2}\boldsymbol{v}^{\mathrm{T}} \cdot R'\boldsymbol{v}\right)d\boldsymbol{v}$$

$$= \left(\frac{M}{2}\right) \operatorname{tr}(R')^{-1} = \frac{RT}{2}. \qquad (12)$$

Or the average energy of one mole of gas per degree of freedom is $RT/2$. Thus $Q^* = H^* - H = -RT/2M$. The enthalpy in the molecular flow is less than that in the bulk by $RT/2M$, which is liberated as heat when the molecules enter the plug, and the same amount of heat is absorbed when they emerge.

Denbigh and Raumann[2] carried out experiments in which a rubber membrane separated two vessels containing CO_2 or H_2. When the surfaces of the membrane are held at different temperatures, CO_2 passes from the cold to the warm side at a pressure difference of several percent for a temperature difference of a few degrees. With H_2, the effect is smaller and in the opposite direction. Both gases are slightly soluble in rubber. The dissolving of the gas at one side of diaphragm will be accompanied by the heat of solution, positive or negative, and there will be a reverse effect when it passes out of the solution at the opposite side.

Ultrafiltration and Reflection Coefficient

With no temperature difference, eq. (2) becomes

$$\sigma = -\sum_{k}^{n} \left[j_k \frac{(\Delta\mu)_{Tp}}{T} + j_k V_k \frac{\Delta p}{T} \right]. \qquad (13)$$

Limiting to 2 components and labeling solute 1 and solvent 2, we have

$$j_1 V_1 + j_2 V_2 = j_V, \qquad (14)$$

and

$$j_1(\Delta\mu_1)_{Tp} + j_2(\Delta\mu_2)_{Tp} = \left(j_2 - \frac{c_2 j_1}{c_1}\right)(\Delta\mu_2)_{Tp} = j'(\Delta\mu_2)_{Tp}, \qquad (15)$$

where

$$j' = j_2 - \frac{c_2}{c_1} j_1 \qquad (16)$$

and the Gibbs-Duhem relation has been used. Thus

$$\sigma = -j' \frac{(\Delta\mu_2)_{Tp}}{T} - j_V \frac{\Delta p}{T}. \tag{17}$$

Assuming linearity,

$$j' = -L_0 \frac{(\Delta\mu_2)_{Tp}}{T} - L_{0f} \frac{\Delta p}{T}, \tag{18}$$

$$j_V = -L_{f0} \frac{(\Delta\mu_2)_{Tp}}{T} - L_f \frac{\Delta p}{T}. \tag{19}$$

Then

$$\frac{L_0}{T} = -\left[\frac{j'}{(\Delta\mu_2)_{Tp}}\right]_{\Delta p=0}$$

is the osmotic flow, or movement of the solvent relative to the solute per unit osmotic pressure π.

$$\frac{L_f}{T} = -\left(\frac{j_V}{\Delta p}\right)_{\pi=0}$$

is the ultrafiltering capacity, or the flow of volume per unit pressure difference.

$$\frac{L_{f0}}{T} = -\left[\frac{j_V}{(\Delta\mu_2)_{Tp}}\right]_{\Delta p=0} = \frac{L_{0f}}{T}$$

is the differential mobility, or the flow of volume per unit osmotic pressure.

Staverman[3] defined

$$\frac{j'}{j_V} = \frac{r}{V_2}, \tag{20}$$

where r is the reflection coefficient. For $r = 1$,

$$\frac{j'}{j_V} = \frac{j_2 - \dfrac{c_2 j_1}{c_1}}{V_1 j_1 + V_2 j_2} = \frac{1}{V_2}. \tag{21}$$

When $j_1 = 0$, the solute is reflected from the membrane and the membrane is impermeable to the solute. For $r = 0$,

$$\frac{j'}{j_V} = 0.$$

When

$$\frac{j_1}{j_2} = \frac{c_1}{c_2},$$

both the solute and the solvent can permeate. There is no osmotic pressure. For $1 > r > 0$ or for $0 > r > -1$, the membrane is leaky and the permeability is not selective. In the former case the transfer of the solvent is more rapid than that of the solute and a positive pressure must be applied the solution to maintain a steady state; but in the latter case the reverse is true. When the application of negative pressure is necessary to maintain the osmosis, the osmosis is anomalous. It is observed in electrolytic solutions with charged membranes. For $r = -1$, the membrane is impermeable to the solvent.

Electrokinetic Phenomena

With no temperature difference and the membrane charged, eq. (2) becomes

$$\sigma = -\frac{i\Delta\zeta}{T} - \frac{j_V \Delta p}{T}, \tag{22}$$

where i is the current. Assuming linearity,

$$\begin{bmatrix} i \\ j_V \end{bmatrix} = \frac{1}{T} \begin{bmatrix} -L_e & -L_{em} \\ -L_{me} & -L_m \end{bmatrix} \begin{bmatrix} \Delta\zeta \\ \Delta p \end{bmatrix}. \tag{23}$$

When the electrodes are short-circuited, i. e., $\Delta\zeta = 0$,

$$\begin{bmatrix} i \\ j_V \end{bmatrix} = \left(\frac{1}{T}\right) \begin{bmatrix} -L_{em} \\ -L_m \end{bmatrix} (\Delta p). \tag{24}$$

$$\frac{i}{\Delta p} = -\frac{L_{em}}{T} \tag{25}$$

is the second streaming current,

$$\frac{j_V}{\Delta p} = -\frac{L_m}{T} \tag{26}$$

is the mechanical conductance, and

$$\frac{i}{j_V} = \frac{L_{em}}{L_m} \tag{27}$$

is the streaming current. When the electrodes are connected by a potentiometer, i. e., $i = 0$,

$$\frac{T}{\Delta}\begin{bmatrix} -L_m & L_{em} \\ L_{me} & -L_e \end{bmatrix}\begin{bmatrix} 0 \\ j_V \end{bmatrix} = \begin{bmatrix} \Delta\zeta \\ \Delta p \end{bmatrix},$$

where Δ is the determinant of the matrix L. Or

$$\begin{bmatrix} \Delta\zeta \\ \Delta p \end{bmatrix} = T\begin{bmatrix} R_{em} \\ R_m \end{bmatrix}(j_V), \tag{28}$$

where R is the inverse of the matrix L.

$$\frac{\Delta\zeta}{j_V} = TR_{em} \tag{29}$$

is the second streaming potential.

$$\frac{\Delta p}{j_V} = TR_m \tag{30}$$

is the reciprocal of mechanical conductance.

$$\frac{\Delta\zeta}{\Delta p} = \frac{R_{em}}{R_m} \tag{31}$$

is the streaming potential. When there is flow of volume under zero pressure difference, i. e., $\Delta p = 0$,

$$\begin{bmatrix} i \\ j_V \end{bmatrix} = \frac{1}{T}\begin{bmatrix} -L_e \\ -L_{me} \end{bmatrix}(\Delta\zeta). \tag{32}$$

$$\frac{i}{\Delta\zeta} = -\frac{L_e}{T} \tag{33}$$

is the electrical conductance,

$$\frac{j_V}{\Delta\zeta} = -\frac{L_{me}}{T} \tag{34}$$

is the second electroösmosis, and

$$\frac{j_V}{i} = \frac{L_{me}}{L_e} \tag{35}$$

is the electroösmosis. When there is no flow of volume but pressure is allowed to develop, i. e., $j_V = 0$,

$$\begin{bmatrix} \Delta\zeta \\ \Delta p \end{bmatrix} = T \begin{bmatrix} R_e \\ R_{me} \end{bmatrix} (i). \tag{36}$$

$$\frac{\Delta\zeta}{i} = TR_e \tag{37}$$

is the reciprocal of electrical conductance,

$$\frac{\Delta p}{i} = TR_{me} \tag{38}$$

is the second electroösmotic pressure, and

$$\frac{\Delta\zeta}{\Delta p} = \frac{R_e}{R_{me}} \tag{39}$$

is the electroösmotic pressure.

From the Onsager law, we have the following Sexen's relations: from eqs. (25) and (34),

$$\left(\frac{i}{\Delta p}\right)_{\Delta\zeta=0} = \left(\frac{j_V}{\Delta\zeta}\right)_{\Delta p=0}, \tag{40}$$

from eqs. (31) and (35),

$$\left(\frac{\Delta\zeta}{\Delta p}\right)_{i=0} = -\left(\frac{j_V}{i}\right)_{\Delta p=0}, \tag{41}$$

from eqs. (29) and (38),

$$\left(\frac{\Delta\zeta}{j_V}\right)_{i=0} = \left(\frac{\Delta p}{i}\right)_{j_V=0}, \tag{42}$$

and from eqs. (27) and (31),

$$\left(\frac{i}{j_V}\right)_{\Delta\zeta=0} = -\left(\frac{\Delta p}{\Delta\zeta}\right)_{i=0}. \tag{43}$$

References

[1] M. Knudson: *Ann. Physik*, **31** (4), 205 (1910).

[2] K. G. Denbigh and G. Raumann: *Proc. Royal Soc.*, series A, **210**, 311, 518 (1951).

[3] A. J. Staverman: *Rec. trav. Chem.*, **70**, 344 (1951).

INDEX